Green Materials for Sustainable Water Remediation and Treatment

RSC Green Chemistry

Series Editors:
James H Clark, *Department of Chemistry, University of York, UK*
George A Kraus, *Department of Chemistry, Iowa State University, Ames, Iowa, USA*
Andrzej Stankiewicz, *Delft University of Technology, The Netherlands*
Peter Siedl, *Federal University of Rio de Janeiro, Brazil*
Yuan Kou, *Peking University, People's Republic of China*

Titles in the Series:

How to obtain future titles on publication:
A standing order plan is available for this series. A standing order will bring delivery of each new volume immediately on publication.

For further information please contact:
Book Sales Department, Royal Society of Chemistry, Thomas Graham House, Science Park, Milton Road, Cambridge, CB4 0WF, UK
Telephone: +44 (0)1223 420066, Fax: +44 (0)1223 420247
Email: booksales@rsc.org
Visit our website at www.rsc.org/books

Green Materials for Sustainable Water Remediation and Treatment

Edited by

Anuradha Mishra
Gautam Buddha University, Greater Noida, India
Email: anuradha_mishra@rediffmail.com

and

James H. Clark
University of York, UK
Email: jhc1@york.ac.uk

RSCPublishing

RSC Green Chemistry No. 23

ISBN: 978-1-84973-621-3
ISSN: 1757-7039

A catalogue record for this book is available from the British Library

Published by The Royal Society of Chemistry,
Thomas Graham House, Science Park, Milton Road,
Cambridge CB4 0WF, UK

Registered Charity Number 207890

For further information see our web site at www.rsc.org

Preface

Life on Earth relies on the gifts of nature. Water is one such gift, without which we and the other creatures on the planet cannot exist. The essential role of water has long been recognized and was worshipped in many countries and civilizations, including ancient India, Egypt, Iran, Greece, Rome, Israel, Syria, Jordan, and Mongolia. Water has been paid high esteem in ancient Indian culture to the extent that it is regarded as "God". Indeed, water is considered sacred in all religions: Christians, Muslims, and Hindus sprinkle holy water on a newborn child. This recognition of the importance of water has led to quotes in their holy books such as "God gives life to a substance by means of water" (Islam: The Holy Quran 21:30), "Whoever believes in me, a stream of living water will pour from within him" (Christianity: John 7:38), and "The life has been created in water" (Hinduism: Atharvaveda, Asthagarideyam).

In his quest to quench a seemingly insatiable hunger, modern man has played havoc with nature and its precious gifts. Out of the total enormous quantity of water available on the Earth, barely a small fraction of it is potable and it is one of the most scarce commodities in some parts of the world. On the other hand, in many other places the gross abuse of water has threatened the very process of obtaining pure potable water from the hydrosphere. The curse of modern industrialization is that while extracting some useful and meaningful substances, significant amounts of pollutants are released into the environment. This is how we have heavily polluted the water, air, and soil. Inadequate access to clean water is one of the most pervasive problems afflicting people throughout the world. Problems with water are expected to grow worse in the coming decades. Global water scarcity is likely to affect even those regions currently considered water-rich. Unless new ways to supply clean water are found, this situation will inevitably eventually lead to wars for water.

The ancient civilizations, knowing water as a vital element for life, were very particular to maintain it pure and free from any kind of pollution. The "Manu

RSC Green Chemistry No. 23
Green Materials for Sustainable Water Remediation and Treatment
Edited by Anuradha Mishra and James H. Clark
© The Royal Society of Chemistry 2013
Published by the Royal Society of Chemistry, www.rsc.org

Smriti", an Indian scripture, stresses at several places the importance to keep water clean. The "Padma Purana" forcefully condemns the person who pollutes water resources. The need for pure water resulted in the development of water purification methods. These methods provided the foundation for the development of modern-day methods of purifying water. Ancient civilizations that developed early water purification methods include those located in Africa, Asia, especially India and the Middle East, and Europe. On the American continent, archeological evidence suggests that the ancient Mayan civilization used an aqueduct technology, similar to that used by the Romans much later, to provide water to urban residents. Methods to assess and maintain water quality and treatment methods for impure water are explained in the Vedas and in Ayurveda, the oldest known health care system. Varahamihira, an ancient Indian scientist, presented methods for obtaining potable water from a contaminated source using plants, metals, and heat. Ayurveda prescribed a water purification method for drinking purposes by using various flowers and fruits. Ayurveda, as per "Sushutra Sutra", also prescribed a few other substances like clearing nuts, Gomedka, lotus bulbs, moss, pearls, thick cloth, *etc.*, with which impurities, including suspended ones, could be removed from water.

> *tatra saptakalusasya prasadhanani santii*
> *tadyatha katakagomedkabhisagranthi-*
> *saivalamula vastrani muktamanisceti*

> – Sushruta Sutra 45.13

Much later, in 2000 BC, the Indians and Greeks started boiling water, sand, and gravel filtration, and straining methods for the purification of water. The main driving force for the earliest water treatment processes was perhaps the taste and turbidity of water. At that time the concept of microorganisms or chemical contaminants was probably unknown. After 1500 BC, the coagulation process, in which a chemical, alum, was used for suspended particle settlement, had been started in ancient Egypt. Pictures of this purification technique were found on the walls of the tombs of Amenophis II and Ramses II. After 500 BC, Hippocrates, the father of modern medicine, discovered the healing powers of water. He invented the practice of sieving water, and created the first bag filter, which was called the "Hippocratic sleeve". The main purpose of the bag was to trap sediments that caused bad tastes or odors in water. Later, in the year 1627, water treatment history continued as Sir Francis Bacon started experimenting with seawater desalination through an unsophisticated form of sand filtration. He did not get the desired success but his work did pave the way for further experimentation by other scientists. In the 1700s the first water filters made of wool, sponge, and charcoal came into existence. In 1804 the first actual municipal water treatment plant based on slow sand filtration was designed by Robert Thom in Scotland. In the 19th century, the effect of disinfectants, such as chlorine, was discovered. In 1854, during the time of a cholera outbreak,

John Snow, a British scientist, applied chlorine to purify water, and this established the route for water disinfection. Along with this, ion exchangers were also developed for water softening. In the late 1890s, America started building large sand filters for water to protect public health. In 1902, calcium hypochlorite and ferric chloride were mixed in a drinking water supply in Belgium, resulting in both coagulation and disinfection. In 1906, ozone was used as a disinfectant for the first time in France. In the 1970s, people became aware about water pollution due to organic chemicals, including pesticide residues and industrial sludge. Many techniques such as aeration, flocculation, and activated carbon adsorption were used to combat water pollution. In the 1980s, membrane development for water treatment was added to the list.

There have been several new developments in the water treatment field in the last three decades. Conventional methods for water treatment can address the issues of disinfection, decontamination, and desalination. These water treatment methods are heavily dependent on large supplies of chemicals and energy. They also require huge operational complexity and are focused on large systems requiring considerable infusion of capital, engineering expertise, and infrastructure, all of which precludes their use in many parts of the world. Materials commonly used in these technologies are sediment filters, activated carbon, water softeners, ion exchangers, ceramics, activated alumina, organic polymers, and many hybrid materials.

Environmental considerations demand the development of strong, economically viable and eco-friendly replacements for conventional methods. Such interventions should be based upon renewable materials which are economical and which tend to degrade naturally if ever released in the environment. It is also important to develop technologies which consume less energy and have minimal effect on global warming. Green remediation is the practice of minimizing the environmental footprint of cleanup actions. It considers all environmental effects of cleaning up a contaminated site. Green and sustainable remediation of water is a rapidly growing field of interest to one and all: governmental agencies, corporations, academia, environmental consultants, public interest groups, and individuals. With the advancement of science in the 21st century, scientists are now able to create lighter and stronger materials for remediation of contaminated water which are not detrimental to our environment. Sources for such green materials include a wide range, from inorganic to organic to hybrid and from plant biomass to animal biomass, non-porous to porous, microbial to antimicrobial, and from solids to liquids.

In light of the above considerations, a focused set of articles covering a range of green materials for water remediation has been included in this book. Chapter 1 discusses the guidelines being followed for materials to be used for water remediation. It also discusses the directives given by the various world authorities in this regard. The information in this chapter provides the basic starting knowledge to new researchers in the field.

Chapter 2 presents a generalized and yet comprehensive view of available green technologies covering all biological and chemical methods, as well as their processes and applications for metal remediation. Chapter 3 gives a compilation of studies done by researchers on plant biomass-based materials as treatment agents for the removal of heavy metals from wastewater. The major advantages of biosorption over conventional treatment methods include low cost, minimization of chemical and/or biological sludge, no additional nutrient requirement, regeneration of the biosorbent, and the possibility of metal recovery. The author also discusses the types of mechanisms involved in the process. Chapter 4 evaluates the application of plant and animal polysaccharides as flocculants in effluent treatment.

Chapter 5 is a review of the application of zeolites in wastewater treatment. A brief overview on water softening and recent applications for removal of ammonia from wastewater are given in the chapter. Immobilization of organic complexing agents on the surface of an inorganic or organic solid support is usually aimed at modifying the surface with certain target functional groups that can be exploited for specific metal extraction. Chapter 6 presents functionalized silica gel, an organic–inorganic hybrid material, for metal remediation. Ease of synthesis of these green materials is discussed, along with methods based on solid-phase extraction using them for the separation and preconcentration of metal ions in polluted water resources.

Chapter 7 presents nanotechnologies developed rapidly in the past decade for water remediation. It gives an account of various types of nanomaterials evaluated/being evaluated as functional materials for water purification, *e.g.* metal-containing nanoparticles, carbonaceous nanoparticles, nanocrystalline zeolites, photocatalysts, magnetic nanoparticles, and dendrimers.

Chapter 8 emphasizes the potential of ionic liquids in many separation processes. Ionic liquids have been emerging as "green" solvents in separation processes due to many fascinating properties and having the potential to replace conventional solvents. Moreover, the required properties in extraction systems, *i.e.* hydrophobicity, polarity, efficiency, and selectivity, can be tailor-made using ionic liquids. They are used for simple biphasic liquid–liquid extractions, liquid-phase micro-extractions, ionic liquid-based solid-phase micro-extractions, thin layer chromatographic and high-performance liquid chromatographic methods, electro-migration methods, gas–liquid chromatographic methods, and supported ionic liquid membrane separations.

Chapter 9 describes the composition and structure of periphyton biofilms studied in recent years along with two aspects of their application, firstly in the purification of water and secondly in phosphorus release from sediments, cyanobacterial blooms, and periphyton biofilms. Periphyton communities are often used as monitors of ecosystem health and indicators of contamination in aquatic ecosystems. They are largely phototrophic benthic microbial biofilms. Owing to their microporous structure, complex composition, and extracellular polymeric substances, periphyton biofilms are applied in water and wastewater treatment. Chapter 10 describes the importance of microorganisms, especially green algae, in water remediation processes. Using green algae for the treatment of textile wastewater is slowly making a mark in the field of water treatment. The mechanism involved and factors affecting biosorption and the parameters used for predicting the efficacy of the use of viable green algae are discussed.

Chapter 11 describes green materials that exhibit ion-exchange properties and can undergo surface modification with positively charged surfactants. This property gives them efficiency for oxo anion removal from water. This process of surface modification shows high promise for a fraction of the cost of commercially available ion-exchange media.

All chapters of the book comprise fundamental information about the various types of promising green materials used for water remediation. The use of such materials will lead to a better and more sustainable way of treating polluted water. The book may be of use to students and researchers in this field.

Anuradha Mishra
James H. Clark

Contents

RSC Green Chemistry No. 23
Green Materials for Sustainable Water Remediation and Treatment
Edited by Anuradha Mishra and James H. Clark
© The Royal Society of Chemistry 2013
Published by the Royal Society of Chemistry, www.rsc.org

**Chapter 9   Periphyton Biofilms for Sustainability of Aquatic
Ecosystems   181**
Yonghong Wu

**Chapter 10   Remediation of Dye Containing Wastewater Using Viable
Algal Biomass   212**
Seema Dwivedi and Tanvi Vats

CHAPTER 1

Greening the Blue: How the World is Addressing the Challenge of Green Remediation of Water

ANURADHA MISHRA*[a] AND JAMES CLARK*[b]

[a] School of Vocational Studies & Applied Sciences, Gautam Buddha University, Greater Noida, Gautam Budh Nagar – 208310, India; [b] Green Chemistry Centre of Excellence, Department of Chemistry, University of York, UK
*Email: anuradha_mishra@rediffmail.com; james.clark@york.ac.uk

1.1 Introduction

The waters of the oceans, rivers and lakes have always been of vital importance for humanity. They are the very basis of life on the planet Earth and have enticed poets and artists. The availability of clean water is an essential requirement for humans and all other creatures. Good quality water is needed for direct consumption and for many types of industries.

There is a serious water quality crisis across the world and many factors are responsible for a continuous deterioration of water quality. These include rapid population growth, widespread urbanization, massive industrialization, and expanding and intensifying food production.[1,2] Worldwide, the need for drinkable water is increasing while the supply is decreasing. In certain places, water is very scarce, but in many other areas there is plenty of water that is not

RSC Green Chemistry No. 23
Green Materials for Sustainable Water Remediation and Treatment
Edited by Anuradha Mishra and James H. Clark
© The Royal Society of Chemistry 2013
Published by the Royal Society of Chemistry, www.rsc.org

drinkable. The situation tends to increase the unregulated or illegal discharge of contaminated water within and beyond national boundaries. This presents a global threat to human health and well-being, with both immediate and long-term consequences and a detrimental effect on poverty alleviation. Water supply and sanitation are key factors determining human well-being.[3] The Millennium Development Goals' report shows that, worldwide, 1.1 billion people lack access to safe drinking water, 2.6 billion people lack adequate sanitation, and 1.8 million people die every year from diarrheal diseases, 90% of which are children under the age of five.[4]

Water remediation can be described as the process to render water free from any contamination. Water remediation is applicable for groundwater, which is the predominant source of water used in cities as well as for farming, for wastewater from industries, which needs to be remediated to prevent contaminants entering the environment, and for several other types. Water remediation is important for several reasons. Firstly, water that is considered unsafe for human consumption must always be completely cleansed to meet well-established health criteria. Furthermore, water remediation is also important to keep the environment free from contamination. Impurities in wastewater can potentially damage the local topography and negatively affect agriculture and all types of farming. It can also adversely impact plant and animal life.

Water recycling is the reuse of treated wastewater for beneficial purposes, such as agricultural and landscape irrigation, industrial processes, toilet flushing, and replenishing the groundwater basin that is often referred to as groundwater recharge. Sometimes water is recycled and reused on site as, for example, when an industrial facility recycles wastewater for cooling processes. A common example of recycled water is water that has been reclaimed from municipal wastewater, or sewage. The term "water recycling" is often used synonymously with water reclamation and water reuse. If adequately treated to ensure water quality appropriate for the end use, recycled water can meet most but not all water demands. Recycled water is most commonly used for non-potable (not for drinking) purposes. Common uses of recycled water include agriculture, landscapes, public parks, and golf course irrigation. Other non-potable applications include cooling water for power plants and oil refineries, industrial process water for such establishments like paper mills and carpet dyers, toilet flushing, dust control, construction activities, concrete mixing, and artificial lakes. Figure 1.1 shows types of treatment processes and suggested uses at each level of treatment. In uses where there is a greater chance of human exposure to water, more treatment is required.

In order to live and reproduce, plants, wildlife, and fish depend on sufficient water flows to their habitats. Lack of adequate flow can result from diversion of water for agricultural, urban, and industrial purposes. Such diversions can cause deterioration of water quality and ecosystem health. Use of recycled water can significantly reduce diversion of freshwater from sensitive ecosystems. Human non-drinking water requirements can be supplemented by using recycled water, which can free considerable amounts of water for the environment and increase flows to vital ecosystems. In recent years, many

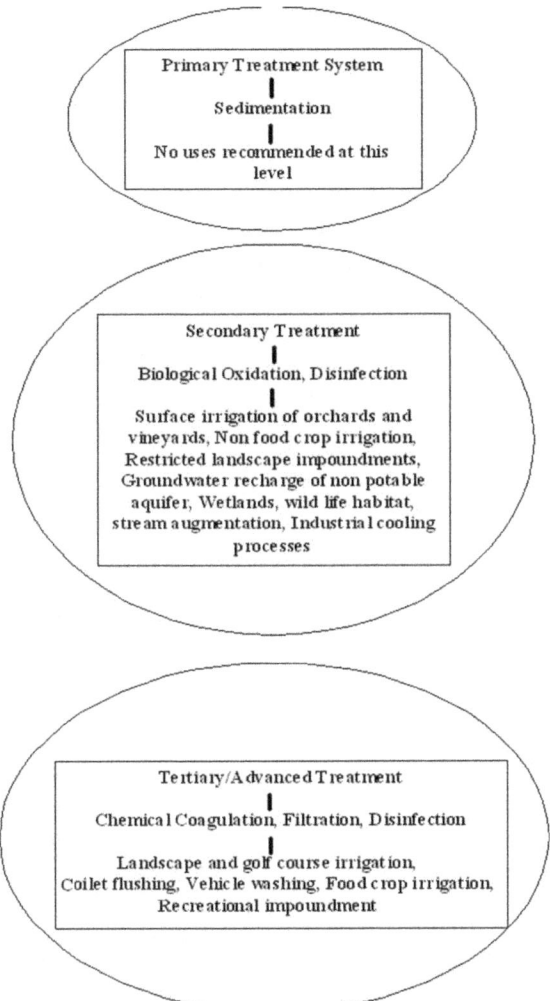

Figure 1.1   Various stages of wastewater treatment.

changes have been made in the processes of wastewater management. These changes are made because of tighter governmental regulations, on the one hand, and the fact that wastewater infrastructure needs major repairs, on the other. This has led to a new trend in wastewater management systems. Instead of considering wastewater as waste, it is now being increasingly considered as a carrier for raw materials for definite end uses.

Common water remediation techniques include phytoremediation, bio-augmentation, ozone and oxygen gas injection, and chemical precipitation. In most water remediation centers across the world, a combination of various methods is used. As such, no single water remediation method can completely rid the water of all contaminants and impurities.

1.2 Green Remediation (Greening the Blue)

Sustainability initiatives have addressed both the broader scope of applications as well as the selected elements of green remediation. The concept of sustainability has been derived from the realization that the Earth's natural resources are limited, and that human activity is depleting these resources at an alarming rate. This activity in turn has a significant impact on the environment. The concept of sustainability first found shape in the form of sustainable development. Sustainable development was defined in 1987 in the Brundtland Commission's report to the United Nations as "development that meets the needs of the present without the ability of future generations to meet their own needs". Issues such as climate change and resource conservation have brought increasing focus on protection of the environment. As a result, sustainability has evolved to become a holistic approach to environmental management. Sustainable practices are such practices that consider the preservation and augmentation of economic and natural resources, ecology, human health and safety, and quality of life.[5]

With advancing cleanup technologies and evolving incentives, green remediation, green materials, or technologies for making clean water offer significant potential for increasing the net benefit of cleanup (Figure 1.2).

Such strategies would tend to reduce project costs and expand the scope of long-term property use or reuse, without compromising the cleanup goals. Green remediation reduces the deleterious after-effects on the environment during cleanup processes, otherwise known as the "footprint" of remediation. It also avoids the potential for any collateral environmental damage. Green remediation promotes application of sustainable strategies at every site requiring environmental cleanup, whether conducted under federal, state, or local cleanup programs or by private parties. Green remediation requires close coordination between cleanup processes and reuse planning. Reuse goals influence the choice of remedial processes, cleanup standards, and the cleanup schedule. In turn, those decisions affect the approaches for investigating a site as well as selecting and designing a custom-made remedial process. They also

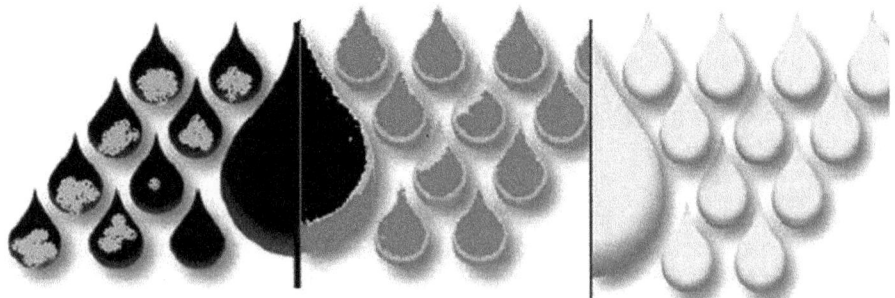

Figure 1.2   Polluted to clean water: a graphical presentation.

affect planning future operations and establishment of in-house remedial processes to ensure environmental protection.

The green solutions for water and wastewater remediation include bioremedial processes and chemical processes. Bioremedial processes are done either through plants (phytoremediation) or through microbes such as bacteria, algae, fungi, and yeast. Phytoremediation encompasses phytoextraction, phytostabilization, phytovolatilization, and rhizofiltration. The chemical solutions include chemical precipitation, ion exchange, liquid–liquid extraction, electrodialysis, and solid-phase extraction using natural materials or biodegradable synthetic materials.

1.3 Policy Directives for Water Remediation and Reuse

Different countries have specific policies for water remediation, and there are regulatory bodies to supervise and control water treatment plants. Additionally, global standards are maintained by international environment protection agencies and water regulatory bodies in accordance to which water remediation is carried out. The process of water remediation was initially emphasized only for potable water. Over the years, however, wastewater treatment has also become equally important due to environmental concerns.

The US Environmental Protection Agency (EPA) regulates many aspects of wastewater treatment and drinking water quality. Most states in US have established definite criteria and guidelines for the beneficial use of recycled water. In the year 1992, the EPA developed a technical document entitled "Guidelines for Water Reuse", which contains all the necessary information as a summary of state requirements and guidelines for the treatment and uses of recycled water.[6] State and federal regulations have been successful in providing a framework to ensure the safety of the many water recycling projects in the US. There is a wide range of EPA programs that tend to ensure sustainability of the cleanup processes along the categories of the built environment, water ecosystems and agriculture, energy and environment, and materials and toxics.[7] There are many programs, tools, and incentives available to help governments, business houses, communities, and individuals to serve as good environmental stewards, make sustainable choices, and effectively manage resources.

In its mission to protect human health and the environment, the EPA is dedicated to developing and promoting innovative cleanup strategies that can restore contaminated sites to productive use, along with a reduction of costs and promotion of environmental quality. The EPA strives for implementation of cleanup programs that tend to use natural resources and energy efficiently, reduce negative impacts on the environment, minimize or eliminate pollution at its source, and reduce waste to the greatest extent possible in accordance with the agency's strategic plan for compliance and environmental stewardship. The practice of "green remediation" uses these strategies to positively impact all environmental effects of the remedial processes for contaminated sites, and also incorporates options to maximize the net environmental benefit of cleanup actions. Strategies for green remediation also incorporate sustainability.

This essentially means that while, on the one hand, environmental protection does not preclude economic development, on the other, economic development is ecologically viable today and in the long run.

The UN envisioned and formulated the "Millennium Development Goals" (MDG), dedicated to reduce poverty and ensure sustainable development. Goal number 7 of target 10 of the MDG states: "Halve, by 2015, the proportion of people without sustainable access to safe water and basic sanitation". The mandate of the UN-Water Decade Programme on Capacity (UNW-DPC) is to strengthen the activities of UN-Water members and partners (more than two dozen UN organizations and programmes) and support them in their efforts to achieve the Millennium Goals related to water. UNW-DPC is directly working with members and partners towards improved human well-being through enhanced water and sanitation that is a core element of the green economy.

1.4 Eco-Labels and Standards

Eco-labels are often affixed to products by manufacturers to indicate to customers that the products meet certain environmental standards. These standards may be developed by private entities, by governmental or public agencies, or jointly by stakeholders and experts from the public and private sectors. As part of its mission, the EPA is working with a variety of non-governmental standards developers to promote the development of voluntary consensus towards standards for environmentally preferable goods and services. The National Technology Transfer and Advancement Act (NTTAA) and OMB Circular A-119 direct the US federal government to use and participate in the development of reference standards compatible with best environmental goals.[8–10] The consensus standards should meet government needs.

The number of standards for green products has increased in recent years due to a growth in market demand for "green" products. The Federal Trade Commission (FTC) has created its Green Guides[11] to help ensure that marketing claims regarding the environmental friendliness of products and production processes are truthful and documentarily substantiated.[12] These guidelines largely address the issues of when and how very specific and narrow environmental attributes can be claimed, and not how to construct a broad-based environmental standard or eco-labeling program. A green label demonstrates a product's sound environmental performance and the supplier's commitment to protect the environment. The society has become so environmentally conscious that now green labeling improves the corporate image, brand reputation, and recognition of high product quality. Many countries use the international standard ISO (International Organization for Standardization) 14021 on self-declared environmental claims as a basis to inform the aware consumers. The green label should include within it standards on water use and recycling.

The EU Ecolabel system[13,14] helps one to identify products and services that have a desirable environmental impact throughout their life cycle, right from the extraction of raw material through to production, use, and final disposal and would be expected to include aspects of water management. Recognized throughout Europe, the EU Ecolabel is a highly trustworthy label, promoting environmental excellence. The EU Ecolabel scheme is indeed a safeguard for environmental sustainability. The criteria have been developed and agreed upon by scientists, NGOs, and stakeholders to create a credible and reliable way to make environmentally responsible choices.

From the raw materials to manufacturing, packaging, distribution, and disposal, EU Ecolabel products are evaluated by independent experts to ensure they meet the predefined criteria that ensure their desired environmental impact. The EU Ecolabel is an easy way to make an informed choice about the products that one may be buying. Although the scheme is voluntary, hundreds of companies across Europe have joined up because of EU Ecolabel's competitive edge and firm commitment to the environment. Customers can rely on the logo as every product is checked by independent experts.

Defra's (UK) guidelines[15] have also drawn from ISO 14021 and in this respect tend to align with international practices. For clarity and ease of reference, the Defra guidelines refer to the relevant provisions of ISO 14021. In the year 2003, Defra published sector-specific guidelines where research showed that further guidance may be useful.

1.4.1 Globalization of Green Labels

Eco-labeling schemes have been widely used worldwide since the late 1970s. As of today, there are close to 30 different green label schemes worldwide. Most of them are run on a voluntary basis. They all provide opportunities to include the provision of clean water and to encourage water recycling.

Germany's "Blue Angel" eco-label, the first national scheme in the world, was introduced in 1977.

In Asia, countries such as China, Japan, Korea, India, Thailand, Malaysia, and Singapore have already established their own eco-labeling schemes. The Green Council (GC) of Hong Kong started the Hong Kong Green Label Scheme (HKGLS)[16] in December 2000. The scheme sets environmental standards and awards a "green label" to products that match the criteria regarding their environment performance. In establishing the standards, HKGLS takes inputs from relevant international standards. It is benchmarked with well-developed eco-labels to ensure credibility of the standards. As with the majority of eco-labeling programs, HKGLS is an ISO 14024[17] compliant Type 1 label, which involves a third-party certification requiring considerations of lifecycle impacts. Some of the key criteria contained in these standards also require compliance with applicable legislation.

"Eco Mark Program" is the first environmental label in Japan.[18] It was started by the Japan Environment Association (JEA) in 1989. The "Eco Mark" is the only Type I environment label defined by ISO that is implemented in Japan. In the "Eco Mark Program", the use of the "Eco Mark" is permitted only to those products that are certified by the JEA to have predefined lesser environmental impact compared to other similar products in each lifecycle. The lifecycle includes the total span from excavation of raw materials, manufacturing, distribution, usage, and up to final recycling/disposal of the product. The certificate ensures that the processes behind the product contribute to preservation of the environment. The certification criteria are described for each type or genre of products.

The "Eco-Leaf Environmental Declaration" is a newer eco-labeling program, whose full implementation was started in 2002 by the Japan Environmental Management Association for Industry (JEMAI).[19] This label belongs to the Type III labeling category as defined by the ISO. The "Eco-Leaf Environmental Declaration" uses the LCA method to quantitatively calculate the environmental impact of products through all lifecycle stages, from extraction of resources to processes of manufacturing and assembly, distribution, usage, and discarding/recycling. These qualitative data are disclosed to the public. It is up to users to decide how to interpret and evaluate the data.

The ISO is a private non-profit organization that was established for the objective of developing and promoting equivalent international standardization and its related activities. The ISO publishes the "Environmental Labels and Declarations" series as an international standard on environmental representation (eco-labeling).[20] The ISO provides three types of "environmental labels and declarations" (Types 1, 2, and 3), whose definitions and requirements are stipulated respectively. In addition, there are certain general principles that are common to all three types as well. The ISO series on environmental management are framed under ISO 14000. The various versions in this series include ISO 14020 (1998), ISO 14024 (1999), ISO 14021 (1999), and ISO 14025 (2000).

1.5   Ecological and Economic Considerations

Strategies for green remediation rely on sustainable development whereby environmental protection does not preclude economic development that is ecologically viable today and in the long run. A green economy can be thought of as one which is low carbon, resource efficient, and socially inclusive.[21] In other words, it stands for an economy that results in improved human well-being and social equity, while significantly reducing environmental and ecological scarcities.

Water in the green economy is based on proper water management to facilitate social and economic development, whilst also safeguarding freshwater ecosystems. In green economies, the role of water in both maintaining biodiversity and ecosystem services is now widely recognized as vital.

"The central role of wastewater management in sustainable development" not only identifies the threats to human and ecological health and the consequences of inaction, but also presents opportunities, where appropriate policy and management responses over the short and longer term can trigger employment, support livelihoods, boost public and ecosystem health, and contribute to more intelligent water management.

Water recycling provides tremendous environmental benefits by providing an additional source of water. Water recycling can help in decreasing the diversion of water from sensitive ecosystems. Benefits also include decreased wastewater discharges, thus reducing and preventing contamination in water resources. Wetlands and riparian habitats can also be created or enhanced with recycled water. Application of recycled water for agricultural and landscape irrigation can provide an additional source of nutrients and lessen the need to apply synthetic fertilizers.

1.6   Conclusions and Future Directions

There are no clear-cut directives from any governmental agency of any country in the world for green technologies for water treatment or green materials to be used in water remediation. Water recycling has proven to be effective and successful in creating a new and reliable water supply without compromising public health. Non-potable reuse is a widely accepted practice that will continue to grow. However, in many parts of the US, the uses of recycled water are expanding in order to accommodate the needs of the environment and growing water supply demands. Advances in wastewater treatment technology and health studies of indirect potable reuse have led many to predict that planned indirect potable reuse will soon become more common. Recycling waste and gray water requires far less energy than treating salt water using a desalination system.

Today, global conditions are such that the momentum for initiatives, such as the counter measures against water pollution and establishment of recycling-based society, should be picked up at a faster pace. In order to accomplish these tasks, it is important to promote the use of green remediation of water and wastewater. Innovation and new techniques rely on a strong science and technology base which are needed to eliminate the pollution of surface and ground water resources so as to improve water quality. Science and technology and training have to play important roles in water resources development and management in general. For effective and economical management of our water resources, the frontiers of knowledge need to be pushed forward in several directions by intensifying research efforts in various areas.

References

1. K. McLaughlin, *Water on the Road to Rio*, UN-Water Programme for Advocacy and Communication; waterinthegreeneconomyinpractice.files. wordpress.com/water-road-to Rio, 2011.
2. UNEP, ERCE, UNESCO, *Water Quality for Ecosystem and Human Health*, National Water Research Institute, Burlington, Ontario, Canada, 2nd edn, 2008.
3. *UN-Water Taskforce on Wastewater Management*; www.un.org/ waterforlifedecade/quality.shtml, 2009.

4. *Meeting the MDG Drinking Water and Sanitation Target: the Urban and Rural Challenge of the Decade*; www.who.int/water_sanitation_health/monitoring/jmp2006/en/index.html, 2006.
5. Wisconsin Initiative for Sustainable Remediation & Redevelopment, *A Practical Guide to Green and Sustainable Remediation in the State of Wisconsin*, Pub-RR-911, Wisconsin Department of Natural Resources, Madison, 2012.
6. *Guidelines for Water Reuse*, EPA/625/R-04/108, U.S. Environmental Protection Agency, Washington, 2004.
7. *Green Remediation: Incorporating Sustainable Environmental Practices into Remediation of Contaminated Sites*, EPA 542-R-08-002, U.S. Environmental Protection Agency, Washington, 2008.
8. *National Technology Transfer and Advancement Act*, Pub. L. 104–113, OMB Circular A-119, U.S. Food and Drug Administration, Washington, 1998.
9. *ANSI Essential Requirements: Due Process Requirement for American National Standards*, American National Standards Institute, New York, 2013.
10. *Toward Sustainability: Building a Better Understanding of Ecosystem Services*, U.S. Environmental Protection Agency, Washington, 2011.
11. ISO 14021:1999, *Environmental Labels and Declarations – Self-declared Environmental Claims (Type II Environmental Labelling)*, International Organization for Standardization, Geneva, 1999.
12. http://www.globalecolabelling.net/, 2004.
13. http://ec.europa.eu/consumers/cons_safe/news/green/guidelines_en.pdf, 2000.
14. http://ec.europa.eu/consumers/rights/docs/Guidance_UCP_Directive_en.pdf, 2009.
15. http://ecolabel.defra.gov.uk/pdfs/PitchingGreen, 2008.
16. http://www.greenlabel.com.hk/, 2013.
17. ISO 14024: *Environmental Labels and Declarations – Type I Environmental Labelling – Principles and Procedures*, International Organization for Standardization, Geneva, 1999.
18. Eco Mark Office, Japan Environment Association (JEA), Tokyo; http://www.ecomark.jp/english/index.html, 2007.
19. *Eco-Leaf Environmental Labeling Program*, Japan Environmental Management Association for Industry (JEMAI), Tokyo; http://www.env.go.jp/policy/hozen/ecolabel/index.html, 2009.
20. http://www.iso.org/iso/en/ISOOnline.frontpage, 2012.
21. *California Energy Commission's 2005 Report: California's Water-Energy Relationship*, CEC#700-2005-011-SF, California Energy Commission, Sacramento, 2005.

CHAPTER 2

Green Materials for Sustainable Remediation of Metals in Water

R. K. SHARMA,*[a] ALOK ADHOLEYA,[b] MANAB DAS[b]
AND ADITI PURI[a]

[a] Green Chemistry Network Centre, Department of Chemistry, University of
Delhi, Delhi – 110007, India; [b] Biotechnology and Management of
Bioresources Division, The Energy and Resource Institute,
New Delhi – 110007, India
*Email: rksharmagreenchem@hotmail.com

2.1 Introduction

Out of all natural resources, water is considered as the most essential and precious blessing of nature. It is both crucial and vital for sustainable and wholesome development of mankind. Also, in the universal framework, it is the fundamental element not solely in the social realm but also in the sphere of integral economic growth. However, degradation of aquatic wetlands due to an increase in water pollution has taken centre stage in global reflection as a serious and threatening phenomenon. This deterioration is mainly attributable to the metals which are currently a major source of ecological hazard. Interestingly, small amounts of these metals are common in our environment and a balanced concentration of a few of them (trace elements) is obligatory for good health in the human diet. However, excessive exposure to any of them may cause acute or chronic toxicity (poisoning) as they are non-degradable and often assimilate through the food chain.[1] Although they disperse both naturally and by anthropogenic input into the biosphere, accelerated globalization and

RSC Green Chemistry No. 23
Green Materials for Sustainable Water Remediation and Treatment
Edited by Anuradha Mishra and James H. Clark

industrial revolution are considered to be the main culprits responsible for metal-contaminated wastewater.[2] The huge volumes of generated raw effluents and sludge are a rich source of toxic metals and are often discharged or dumped to nearby sites and landfills that lead to degradation of environmental health. The conditions are further worsened in water-scarce regions in several parts of world where industrial wastewater is used for irrigation and the resulting metal-polluted lands serve as the transmission route for metals in the human food chain. In addition to this, the metal contaminants present could run off to pollute surface water and percolate through soil to groundwater, thereby posing serious health issues. Exposure to metals further continues with the direct and indiscriminate release of wastewater to nearby water bodies like lakes and estuaries, which in turn are taken up by roots of plants through the soil.[3,4] Prolonged human consumption of unsafe concentrations of metals may lead to the disruption of numerous biological and biochemical processes in the human body. There are reports highlighting health issues related to metal intoxication, including decreased immunological defenses, intrauterine growth retardation, impaired psycho-social faculties, disabilities associated with malnutrition and high prevalence of upper gastrointestinal cancer rates (Table 2.1).[1,5,6]

The increased concern of social elements and stringent national and international regulations imposed by federal and state governments on water pollution and the discharge of metals makes it essential to develop non-intrusive, low-cost, aesthetically pleasing, ecologically benign, and socially accepted technologies to remediate polluted ecosystems. Driven by their commitment for continual environmental improvement, development of green technologies is of major interest to the environmental scientist, particularly in the study of the fate of environmental pollutants. Undoubtedly, these green extraction methods have gained considerable momentum in recent times as they can recover significant amounts of mobile metallic pollutants.

Table 2.1 Health effects of various toxic metals.

Metal contaminant	Health hazards
Arsenic	Carcinogenic, producing liver tumors, skin and gastrointestinal effects
Mercury	Corrosive to skin, eyes and muscle membranes, dermatitis, anorexia, kidney damage and severe muscle pain
Cadmium	Carcinogenic, causes lung fibrosis, dyspnea and weight loss
Lead	Suspected carcinogen, loss of appetite, anemia, muscle and joint pains, diminishing IQ, causes sterility, kidney problems and high blood pressure
Chromium	Suspected human carcinogen, producing lung tumors, allergic dermatitis
Nickel	Causes chronic bronchitis, reduced lung function, cancer of lungs and nasal sinus
Zinc	Causes short-term illness called "metal fume fever" and restlessness
Copper	Long-term exposure causes irritation of nose, mouth, eyes, headache, stomach ache, dizziness, diarrhea

This chapter is an attempt to represent a simplistic view of the enthusiastic and collaborative efforts put forward by the scientific community so far in the form of "green" technologies and materials for remediation of existing waste sites and polluted water sources directly or indirectly. The target of the authors is not to overload the content with the burdensome literature presented so far but to provide a basic understanding of the green solutions to decontaminate the environment that have developed with time. The content will help to build up a clear mental view of the methods so far available to people not belonging to any partitioned compartment of science but belonging to the ecosystem for which they are concerned.

2.2   The Biological Solution

2.2.1   Phytoremediation

Holding the credit of the mother of humanity, Nature is performing its duty to decontaminate the environment for the benefit of mankind. It engages plants to cleanse the environment and to equilibrate and restore the ecological balance. The plants growing in metal-contaminated and polluted terrestrial ecosystems take up toxic metals and bind them, thus helping in both environmental decontamination and recovery of metals. The whole phenomena of uptake, accumulation, translocation, sequestration and detoxification of metals from soil and aqueous environments is known as phytoremediation. The increased inclination of researchers towards this green remediation process has further opened new dimensions in its enhanced environmental application for metal detoxification. It is now generally being permuted with genetic engineering and advanced agricultural practices. The generic term "phytoremediation" consists of the Greek prefix *phyto* (plant), attached to the Latin root *remedium* (to correct or remove an evil). It is broadly categorized as four main processes: phytoextraction, phytovolatilization, phytostabilization (Figure 2.1) and rhizofiltration.[7–11] Although these processes work in different environmental compartments, this plant-based environmental remediation technology has been widely pursued in recent years as a green, cost-effective strategy to trap metal contaminants that are in mobile chemical forms which are the most threatening to human and environmental health.

2.2.1.1   Phytoextraction

Phytoextraction is the removal of pollutants by the roots of plants, followed by translocation to above-ground plant biomass, which is subsequently harvested. Hence it aims at reducing the metal concentration in contaminated soils to regulatory levels within a reasonable time frame. Here the harvested mass of plants acts as a sink to concentrate metals from soils. Over time, a few species have evolved with a high tolerance ability to survive in metal contaminated lands and have developed the capability to uptake these metals to their shoot system.[12,13] This natural ability of plant materials to extract metal ions is

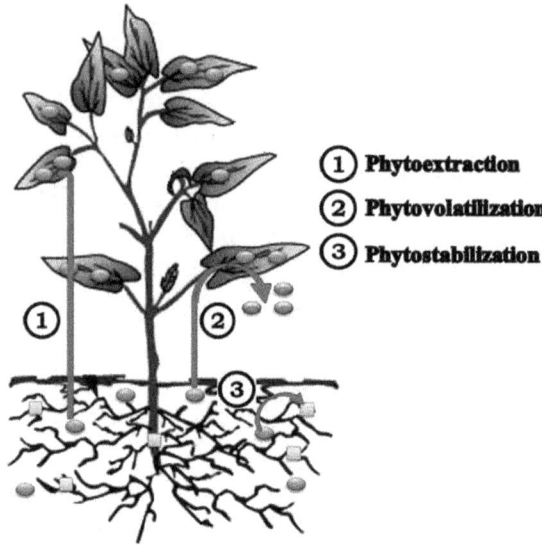

① **Phytoextraction**
② **Phytovolatilization**
③ **Phytostabilization**

Figure 2.1   Phytoremediation.

known as natural phytoextraction and plant species involved in this phenomenon are termed hyperaccumulating species. However, the small shoot and root
growth response, slow growing rate and less commercial availability of these
plants have stimulated study on alternative species which are faster growing
and can be easily cultivated using established agronomic practices. These are
called high biomass plants. Though they have fewer propensities to accumulate
metallic species compared to hyperaccumulating plants, this is easily compen-
sated by the huge amount of root and shoot biomass produced which enhances
their ability to accumulate metal ions. Their metal uptake capacity can further
be improved by adding a conditioning fluid containing a chelator or another
agent to the soil to upsurge metal solubility or mobilization so that the plants
can absorb them more easily. This is known as chemically induced/assisted
phytoextraction. The addition of chelators increases the metal bioavailability,
uptake and translocation of metals. A few of the available chelating agents are:
ethylenediaminetetraacetic acid (EDTA), 1,2-cyclohexylenedinitrilotetraacetic
acid (CDTA), diethylenetriaminepentaacetic acid (DTPA), (hydro-
xyethyl)ethylenediaminetriacetic acid (HEDTA) and nitriloacetic acid (NTA)
as synthetic chelators and humic acid as a natural chelator.[14,15]
   The main advantage of phytoextraction is environmental friendliness, but for
cost effective and sustainable phytoextraction processes it is essential to create
stabilizing plants which produce high levels of root and shoot biomass, high
tolerance and resistance to metals, fast growth kinetics, an extended root
system for exploring large soil volumes, an adaptability to specific environ-
ments/sites and easy agricultural management. This can be achieved by
mycorrhizal association. The mycorrhizal associated fungi are soil microorganisms which enhance plant growth on severely disturbed sites

contaminated with metals. They establish a mutual symbiosis with the plants, providing a direct physical link between soil and plant roots and play an important role in metal tolerance and accumulation. The mutualistic association provides the fungus with relatively constant and direct access to carbohydrates, *i.e.* glucose and sucrose supplied by the plant. The carbohydrates are translocated from the leaves to root tissue and finally on to fungal partners. In return, the plant gains the benefits of the mycelium's higher absorptive capacity for water and mineral nutrients (due to the comparatively large surface area of the mycelium:root ratio), thus improving the plant's mineral absorption capabilities. Hence, it plays an important role in successful survival and growth of plants in contaminated systems. This root colonization also exploits the ability of plants to translocate a great fraction of the metals taken up to harvestable biomass.[16]

Another important component of phytoextraction technology is plant growth-promoting rhizobacteria. The rhizosphere contains a large microbial population which enhances metal mobility and availability to the plant. Isolation of the indigenous and presumably stress-adapted fungi and rhizosphere can act as a potential biotechnological tool for inoculation of plants for successful restoration of degraded ecosystems.[17]

2.2.1.2   *Phytostabilization*

Phytostabilization does not remove but rather prevents exposure pathways of metals to ground water and to the aerial environment by stabilizing/inactivating the toxic metal contaminants physically and chemically and preventing them from mobilizing or leaching in a manner that would endanger public health. It uses perennial and non-harvested plants to suppress the vertical migration of pollutants through absorption and accumulation by roots or precipitation in the rhizosphere. So, the toxic sites are stabilized by plants that serve as ground cover, thereby reducing the distribution of the toxic metals to other areas. The tolerance ability of a plant plays a crucial role in this process, as disposal of the hazardous material or the biomass is not required. There are reports of integration of phytostabilization with physical processes to enhance the immobilization of toxic metals, like addition of soil amendments that increase the soil organic matter and pH (using lime), phosphate fertilizers, organic matter or bio-solids, iron or manganese oxyhydroxides, natural or artificial clay minerals or mixtures of these amendments or by binding certain constituents with phosphate or carbonate without using soil amendments.[18]

2.2.1.3   *Phytovolatilization*

Use of plant material to modify and transform the metal contaminants to less toxic species and subsequently releasing them into the atmosphere is known as phytovolatilization. As the process is highly dependent on the physical characteristics of the contaminant itself, so it is primarily used for the removal of selenium, mercury and the mercuric ion. Instead of accumulating, these metals move along the way as water travels along the plant vascular system and are enzymatically transformed into less polluting volatile substances, from root membranes to shoots and then through the stomata of leaves into the aerial compartment of the ecosystem. *Nicotiana tabacum* and *Arabidopsis thaliana* have been genetically engineered to insert bacterial Hg ion reductase genes into them for the purpose of Hg phytovolatilization. The work is highly encouraging to convert the mercuric ion to less toxic metallic mercury and volatilize it, as naturally occurring plants cannot accomplish this. *Arabidopsis* and Indian mustard have also been overexpressed with the gene encoding enzyme SMT (sterol methyl transferase), which is cloned from an Se hyper-accumulator (*Astragulus bisculatus*). The use of these genetically modified species causes high selenium tolerance, accumulation and volatilization in the plants.[19,20]

2.2.1.4 Rhizofiltration

This works on a parallel principle to phytoextraction; the only difference is that this process is concerned with the remediation of ground water and industrial waste streams directly rather than remediation of polluted soils (Figure 2.2). The root biomass of terrestrial or aquatic plants is used to sieve contaminants directly out of the aquatic environment and operate in either hydroponic floating rafts on ponds (*in situ*) or a constructed wetland setting (*ex situ*). Plant

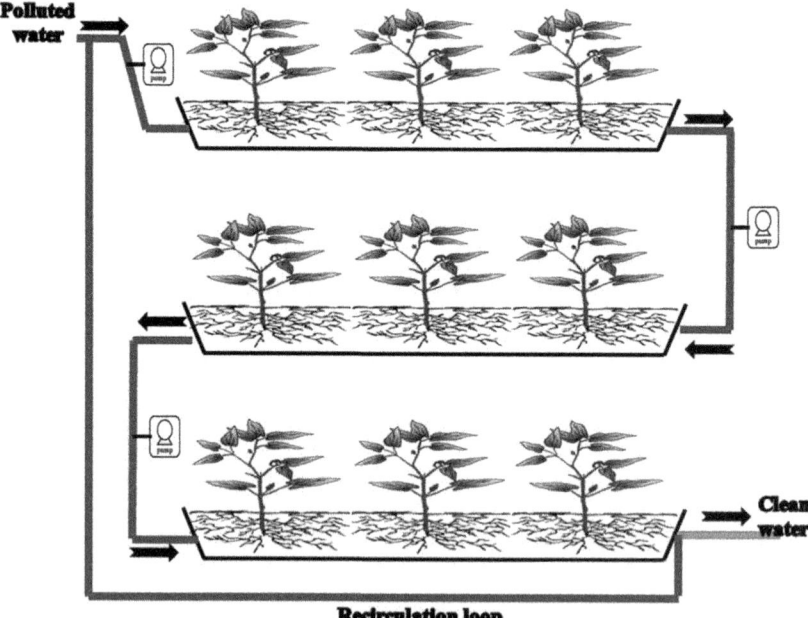

Figure 2.2   A rhizofiltration unit.

materials are harvested periodically for disposal or metal recovery from their roots and any other tissue-containing metals. Generally the contaminated water is pumped through a continuous flow system to the designed greenhouse units where plants having longer, substantial and rapid growth root systems with large surface areas are planted for the removal of the metals. The plants are harvested periodically after saturation of their roots.[19,21]

2.2.2 Bioremediation

Microorganisms are available everywhere and can adapt and grow at subzero temperatures, as well as in extreme heat, in desert conditions, in water, in the presence of excess oxygen, and in anaerobic conditions, with the presence of hazardous compounds or on any waste stream. Because of the wide range of adaptability of microbes, these can be used to degrade or remediate environmental contaminants, including organic and inorganic compounds; the process is known as bioremediation.[22]

Although microorganisms cannot destroy metals, they can react with them in natural and synthetic environments, altering their physical and chemical state through various mechanisms depending on the context and environment. The mechanisms by which microorganisms remove metals from solution are: (1) extracellular accumulation/precipitation; (2) cell surface sorption or complexion; and (3) intracellular accumulation.[23] A schematic diagram depicting metal–microbe interaction is presented in Figure 2.3. Extracellular accumulation/precipitation may be facilitated by using viable microorganisms, by cell-surface sorption or complexation which can occur with alive or dead microorganisms, while intracellular accumulation requires microbial activity. Although living and dead cells are both capable of metal accumulation, the

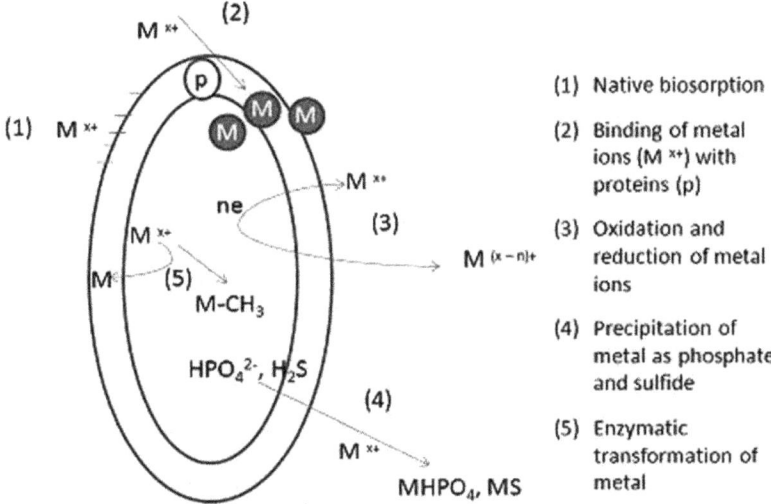

Figure 2.3 The mechanism of biosorption and bioaccumulation.[26]

Table 2.2   Characteristics of metal biosorption and bioaccumulation.[25]

Biosorption	*Bioaccumulation*
A reversible passive process	A partially reversible active process
Metals are bound with dead biomass	Metals are bound with cellular surface and accumulated in living biomass
Nutrient-independent adsorption process	Nutrient-dependent absorption process
Fast single-stage process	Slow double-stage process
No scope for toxic effects	Scope for toxic effect on cellular growth

difference lies in the mechanisms involved, which is given by the extent of metabolic dependence.[24] Metabolism-dependent intracellular uptake of metal ions may be a slower process than biosorption. It is inhibited by low temperatures, the absence of an energy source and metabolic inhibitors. Many metals are essential for growth and metabolism and organisms possess transport systems for their accumulation. Non-essential metals may be taken *via* such systems. However, when the metal ion concentration becomes too high or sufficient metal ions are absorbed by the microorganism, the organism's metabolism is disrupted, thus causing the organism to die. This disadvantage does not exist if non-living organisms (biomass) or biological materials derived from microorganisms are used to adsorb metal ions from solution/effluents. Besides, use of dead biomass (biosorption) also eliminates the economic component of maintenance, including nutrient supply. The fundamental differences in characteristics of biosorption using dead biomass and bioaccumulation by living cells are presented in Table 2.2. A range of effluents, including wastewaters from metallurgical industries, rinse waters from electroplating, metal finishing and printed circuit boards manufacturing, mining operations, leachates, surface and ground waters can treated by both biosorption and bioaccumulation. Metals that can be targeted include Ag, Al, Au, Cd, Co, Cr, Cu, Fe, Hg, Mn, Mo, Ni, Pb, Pt, Se, U, V and Zn.[25]

2.2.3   Types of Microbial Agents

2.2.3.1   *Bacteria*

Bacteria are the most abundant and versatile of microorganisms and constitute a significant fraction of the entire living terrestrial biomass.[27] Bacteria are useful as biosorbents because of their small size, their ubiquity, their ability to grow under controlled conditions, their resilience to a wide range of environment situations, and their heterogeneity among different species in relation to their number of surface binding sites, binding strength and mechanisms.[28] The walls of bacteria are efficient metal chelators, though a wide spectrum of uptake capacity may be exhibited. Metal binding may be at least a two-stage process, first involving interaction between metal ions and reactive groups, followed by inorganic deposition of increased amounts of metal. The carboxyl groups of the glutamic acid part of peptidoglycans are the major sites of metal deposition.

Table 2.3 Bacterial biomass used for metal removal.[27]

Metal ion	Bacteria
Pb	*Bacillus* spp., *Bacillus firmus, Corynebacterium glutamicum, Enterobacter* spp., *Pseudomonas aeruginosa, Pseudomonas putida, Streptomyces rimosus*
Zn	*Streptomyces rimosus, Bacillus firmus, Aphanothece halophytica, Pseudomonas putida, Streptomyces rimosus, Streptoverticillium cinnamoneum, Thiobacillus ferrooxidans*
Cu	*Bacillus* spp., *Bacillus subtilis, Enterobacter* spp., *Micrococcus luteus, Pseudomonas aeruginosa, Pseudomonas cepacia, Pseudomonas putida, Pseudomonas stutzeri, Sphaerotilus natans, Streptomyces coelicolor, Thiobacillus ferrooxidans*
Cd	*Ochrobactrum anthropic, Sphingomona spaucimobilis, Aeromonas caviae, Enterobacter* spp., *Pseudomonas aeruginosa, Pseudomonas putida, Staphylococcus xylosus, Streptomyces pimprina, Streptomyces rimosus*
Cr(VI)	*Bacillus coagulans, Bacillus megaterium, Zoogloea ramigera, Aeromonas caviae, Bacillus coagulans, Bacillus licheniformis, Bacillus megaterium, Bacillus thuringiensis, Pseudomonas* spp., *Staphylococcus xylosus*
Ni	*Bacillus thuringiensis, Streptomyces rimosus*
Pd	*Desulfovibrio desulfuricans, Desulfovibrio fructosivorans, Desulfovibrio vulgaris*
Pt	*Desulfovibrio desulfuricans, Desulfovibrio fructosivorans, Desulfovibrio vulgaris*
U	*Arthrobacter nicotianae, Bacillus licheniformis, Bacillus megaterium, Bacillus subtilis, Corynebacterium equi, Corynebacterium glutamicum, Micrococcus luteus, Nocardia erythropolis, Zoogloea ramigera*
Th	*Arthrobacter nicotianae, Bacillus licheniformis, Bacillus megaterium, Bacillus subtilis, Corynebacterium equi, Corynebacterium glutamicum, Micrococcus luteus, Zoogloea ramigera*

In some bacteria, metabolism-independent biosorption may be the most significant proportion of the total uptake. Although biosorption is independent of metabolism, it is possible that a metabolism-dependent microenvironment may enhance metal deposition.[29] Bacteria species such as *Bacillus, Pseudomonas, Streptomyces, Escherichia* and *Micrococcus* have been tested for uptake metals.[27] Table 2.3 summarizes some of the important results of metal removal using bacterial biomasses.

2.2.3.2 Fungi and Yeasts

Fungi have been utilized as efficient and economical biological agents for the removal of toxic metals from effluents because of having a high percentage of cell wall material capable of binding metal ions in fungal biomass.[30] Besides, fungi are easy to grow, produce high yields of biomass and can be manipulated genetically and morphologically. Although fungi are a large and diverse group of eukaryotic microorganisms, three groups of fungi have major practical importance: the molds, yeasts and mushrooms. Filamentous fungi and yeasts have been observed in many instances to bind metallic elements.[27] Metabolism-independent binding of metal ions to fungal and yeast cell walls is usually rapid and large amounts may be bound. The biosorption capacity of

Table 2.4 Fungal and yeast biomass used for metal removal.[27]

Metal ion	Fungi and yeasts
Pb	*Saccharomyces cerevisiae, Penicillium purpurogenum, Aspergillus, Trichoderma, Rhizopus, Mucor, Fusarium* (living cells), *Phanerochaete chryosporium* (living cells)
Cu	*Saccharomyces cerevisiae, Penicillium, Aspergillus, Trichoderma, Rhizopus, Mucor, Fusarium* (living cells), *Phanerochaete chryosporium* (living cells), *Penicillium italicum*
Zn	*Saccharomyces cerevisiae, Penicillium italicum, Aspergillus, Trichoderma, Rhizopus, Mucor, Fusarium* (living cells)
Cd	*Saccharomyces cerevisiae, Penicillium, Aspergillus, Trichoderma, Rhizopus, Mucor, Fusarium* (living cells), *Phanerochaete chryosporium* (living cells)
Ni	*Saccharomyces cerevisiae, Penicillium chrysogenum*
Cr(VI)	*Saccharomyces cerevisiae, Penicillium chrysogenum*
Pd	*Saccharomyces cerevisiae*
Pt	*Saccharomyces cerevisiae*
Ag	*Saccharomyces cerevisiae, Aspergillus niger, Mucor rouxii, Rhizopus arrhizus* (living cells), *Penicillium* spp. (living cells)
Au	*Aspergillus niger, Mucor rouxii, Rhizopus arrhizus* (living cells), *Penicillium* spp. (living cells)
U	*Saccharomyces cerevisiae, Aspergillus, Penicillium, Rhizopus, Saccharomyces, Trichoderma, Mucor, Rhizopus* (living cells)
Th	*Saccharomyces cerevisiae, Aspergillus, Penicillium, Rhizopus, Saccharomyces, Trichoderma, Mucor, Rhizopus* (living cells)

dead biomass may be greater, equivalent to, or less than that of living cells, which may depend on the killing process used. As in other microbes, a variety of ligands may be involved, including carboxyl, amino, phosphate, hydroxyl and thiol groups. In certain fungi, and especially yeasts, greater amounts of metals may be taken up by transport than by biosorption. However, for many filamentous fungi it appears that general biosorption accounts for a major portion of metal uptake.[29] Different fungi and yeasts used in the removal of metals from aqueous solution are presented in Table 2.4.

2.2.3.3 Algae

High sorption capacity and ready availability in practically unlimited quantities in the seas and oceans make algae efficient biological agents for removal and recovery of metals from solution.[27,31] Algae have low nutrient requirements and being autotrophic they produce a large biomass; unlike other microbes, such as bacteria and fungi, they generally do not produce toxic substances.[30] In common with other microbial groups, many potential metal-binding sites occur in algae cell walls, which include polysaccharides, cellulose, uronic acid and proteins. Metabolism-independent accumulation of metals is often rapid and usually completes in 5–10 min.[29] Plenty of algae have been reported as efficient tools for the removal of metals from solution (Table 2.5).

The biosorption capacity of a number of macro- and microalgae has also been reviewed by Brinza *et al.*[33] About 17 microalgal and 39 macroalgal species

Table 2.5 Algal biomass used for metal removal.[32]

Metal ion	Macro- and microalgal species
Pb	*Asccophyllum nodosum, Chlorella vulgaris, Cladophora glomerata, Chondrus crispus, Codium taylori, Fucus vesiculosus, Galaxaura marginata, Padina gymnospora, Padina tetrasomatica, Polysiphonia violacea, Sargassum fluitans, Sargassum hystrix, Sargassum natans, Sargassum vulgare, Ulfa lactuca, Undaria pinnatifida, Gracilaria corticata*
Cu	*Cholrella miniata, Scenedesmus obliquus, Sargassum* spp.
Zn	*Chlorella vulgaris*
Cd	*Asccophyllum nodosum, Chaetmor halinum, Chlorella vulgaris, Codium fragile, Corallina officinalis, Fucus vesiculosus, Gracilaria edulis, Gracilaria salicornia, Padina tetrasomatica, Porphira columbina, Sargassum bacculuria, Sargassum natans, Sargassum siliquosum*
Ni	*Asccophyllum nodosum, Cholrella miniata, Chlorella vulgaris, Chondrus crispus, Codium taylori, Fucus vesiculosus, Galaxaura marginata, Padina gymnospora, Sargassum fluitans, Sargassum natans, Sargassum vulgare, Scenedesmus obliquus*
Cr(VI)	*Chlorella vulgaris, Scenedesmus obliquus, Sargassum* spp.

are mentioned in the review. They have been reported to adsorb one or more metal ions, including K, Mg, Ca, Fe, Sr, Co, Cu, Mn, Ni, V, Zn, As, Cd, Mo, Pb, Se and Al.

2.3 The Chemical Solution

Although the scientific boom has been considered as a major culprit for polluting the biosphere, it has also provided various chemical ways for cleaning the existing waste sites and contaminated water. These chemical methods are the evidence of accelerated efforts being put forward by the scientific community for continual environmental improvement.

2.3.1 Chemical Precipitation

Chemical precipitation is a widely accepted, simple, easy to operate and proven technology for the removal of metals from wastewater. It involves the reduction of the metal content from wastewater to an appropriate and acceptable concentration by forming an insoluble solid precipitate (hydroxide, sulfide or carbonate). The formed precipitate can either be filtered or centrifuged from the liquid portion. The phenomenon is dependent on the solubility product (K_{sp}) of the metal involved, the pH of the wastewater and the concentration of relevant metal ions. Precipitation is further supported through the use of a coagulant which acts an agent causing smaller particles suspended in solution to gather into larger aggregates.

Additionally, co-precipitation is also an adsorptive phenomenon in the removal process in which metals can be co-precipitated with secondary minerals in wastewater. When added to metallic waste streams, these minerals are oxidized and contribute in the precipitation of other metallic contaminants. The process enhances metal removal and precipitates can then be separated from the treated water by chemical coagulation, flocculation and clarification processes. Based on the low cost and ease of handling, various hydroxides have been used to precipitate metals from wastewater. However, lime is the preferred choice as it can act as a coagulant itself, sweeping ions out of solution in formation and settling. The sulfide precipitation has an upper hand over hydroxide precipitation as metal sulfides have much lower solubilities than the corresponding metal hydroxides, thus allowing lower residual metal content extraction in the wastewater. Being a well established technology, it has been practiced as a prime method of treatment in industrial waters for many years. Nowadays, however, it is becoming uneconomic and causes more problems than the benefits it offers. It requires a large amount of chemicals to reduce metals to an acceptable level for discharge. Large volumes of sludge are generated during the operations, hence causing severe disposal problems. Moreover, it is not metal selective and requires corrosive chemicals to work with, thereby increasing safety concerns.[34–36]

2.3.2   Ion Exchange

Ion exchange is a reversible and versatile separation process having inherent advantages of fast kinetics, high treatment capacity and efficiency for removal of undesirable ionic impurities from wastewaters. It involves transfer of ionic species from one liquid phase to another *via* intermediate solid resins. The majority of the target ions to be replaced in the waste streams are toxic, thus affecting water purity and public health. The transfer takes place without any physical alteration to the ion exchange material. To achieve this, a group of ion species (the analyte ions) present in a solution is exchanged with another group of ion species belonging to a solid phase. The ion exchange resins for metal removal are either synthetic or natural solid cation exchange resins which concentrate the metal onto their surface. Synthetic ion exchange resins are basically organic compounds polymerized to form a porous tridimensional matrix, and their structure and porosity are determined principally by the conditions of polymerization of the backbone polymer. The most common cation exchangers are either strongly acidic resins with sulfonic acid groups ($-SO_3H$) or weakly acid resins with carboxylic acid groups ($-COOH$) having exchangeable hydrogen ions; the uptake behavior is also affected by certain variables such as pH, temperature, initial metal concentration and contact time. Although the use of synthetic resins is an attractive way for removal of metals from water and wastewaters, the process is limited due to the fact that wastewater must be extensively pretreated to avoid resin contamination and not all metals can be removed with the employed resins. Also, ion exchange resins must be regenerated by chemical reagents when they are exhausted and the regeneration can cause serious secondary pollution.

Naturally occurring zeolites have also been utilized as a worthy substitute to synthetic materials for the removal of metal cations owing to their high abundance and, more importantly, their relatively low prices. Zeolites are naturally occurring hydrated aluminosilicate minerals and their structures consist of three-dimensional frameworks of SiO_4 and AlO_4 tetrahedra. The net negative charge on the lattice due to the presence of Si^{4+} and Al^{3+} is balanced by the exchangeable cation (sodium, potassium or calcium). Natural zeolites generally have greater thermal stability and better resistance to acid environments than many common commercial synthetic adsorbents.[34,37–41]

2.3.3 Liquid–Liquid Extraction

This process involves a two-phase system in which recovery of the metals from an aqueous phase is normally achieved by contacting the aqueous phase with an organic phase containing a metal chelator. It can selectively extract a desired metallic cation from a feed solution containing a significant amount of metallic impurities. In this context, the link between interfacial activity and complexing properties has an important influence on the mass transfer kinetics and permits selective extraction against a thermodynamic equilibrium constant. The organic phases used in this extraction are generally flammable and volatile organic compounds (VOCs), and large consumption of organic solvents generally changes the atmospheric and aquatic environments. One more drawback associated with liquid–liquid extraction is its limited ability to concentrate the metals during the extraction process; in most cases, extraction produces no more than a ten-fold increase.[42]

Nowadays, these traditional organic solvents have been replaced by alternative green ones such as ionic liquids. They are made up of two components, an anion and a cation. Generally, one or both of the ions is large and the cation has a low degree of symmetry, which reduces the crystal lattice energy. Because these ions are poorly coordinated, the ionic liquids have a melting point below 100 °C or sometimes even at room temperature. Their unique features, such as their very low volatility, non-flammability and ability to dissolve both organic and inorganic compounds, are important advantages over conventional organic solvents. Because of this friendly environment perception, they have been increasingly investigated worldwide as alternatives to replace traditional organic solvents in metal separation.[43,44]

2.3.4 Electrodialysis

Electrodialysis is the process for the separation of ionic components (metal ions) through the use of semi-permeable ion-selective membranes. It implies the application of an electrical potential between two electrodes which causes migration of cations and anions towards the respective electrodes. Because of the alternate spacing of cation- and anion-permeable membranes, cells of concentrated and dilute salts are formed. This process creates a highly concentrated metal stream which is effective for recovery. This is one such technique that has many advantages, such as avoidance of hazardous sludge production, less area requirement, application of low voltage and operation under ambient pressure. However, it is not so economically competitive because of the high cost of electrodes and the relatively short lifetime of the membranes when working in a high-density electrical field. The treatment is also limited to very dilute and relatively clean wastewaters because of membrane fouling.[34,45,46]

2.3.5 Solid Phase Extraction

Solid phase extraction (SPE) is currently the most popular technique for the isolation of target analytes from a fluid phase. When compared with liquid–liquid extraction, SPE reduces the consumption of organic solvents, which has a dramatic effect on the reduction of waste. In addition, high enrichment factors and elevated recoveries are feasible with SPE using short times. It involves partitioning of solutes between two immiscible phases, *i.e.* a liquid sample solution and a solid sorbent phase. In many cases the process leads to high quality treated effluents and also offers flexibility in design and operation. Also, in some cases, after yielding metals in concentrated form *via* an elution process, it regenerates adsorbents with similar efficiency and adsorption capacity.[47,48]

2.3.5.1 Silica-Based Sorbents

Silica-based, chemically bonded sorbents are the most promising sorbents for the separation of trace metal ions from aqueous solution. Silica as a support material has good mechanical strength and swelling stability, which is a prime requirement for its use. Moreover, it can undergo heat treatment without being affected and is highly inexpensive. In addition to this, organic groups can be easily loaded on silica gel with high stability (Table 2.6). The modified silica gels generally exhibit higher sorption capacities than those of organic polymer-based resins. This is because organic functional groups increase the binding capacity and selectivity pattern of silica gel. The synthetic pathway of functionalized silica gel involves the reaction of an organosilane with silanol groups on the silica gel,

Table 2.6 Various chelating ligands immobilized on silica gel for metal extraction.[51–62]

Chelating ligand immobilized	Metal targeted
Diethylenetriamine	Hg^{2+} and Cu^{2+}
Diphenyl diketone monothiosemicarbazone	Pd^{2+}
4-Phenylacetophenone 4-aminobenzoylhydrazone	Cu^{2+}, Ni^{2+} and Co^{2+}
1,4-Bis(3-aminopropyl)piperazine	Cu^{2+}, Ni^{2+} and Co^{2+}
Poly(methacrylic acid)	Cd^{2+}
Dithiocarbamate	Pb^{2+}, Cd^{2+}, Cu^{2+} and Hg^{2+}
Diethylenetriamine and tetraethylenepentamine	U^{6+}
Tris(2-aminoethyl)amine	Cr^{3+}, Cd^{2+} and Pb^{2+}
Curcumin	Cu^{2+}, Fe^{3+} and Zn^{2+}
Gallic acid	Pb^{2+}, Cu^{2+}, Cd^{2+} and Ni^{2+}
Glycerol	Al^{3+}
4-(Dimethylamino)benzaldehyde	Cr^{3+}, Cu^{2+}, Ni^{2+}, Pb^{2+} and Zn^{2+}
Polypyrrole nanoparticles	Cr^{6+}

which is similar in every reaction but the difference is in the type of chelating molecules which selectively bind to a particular metal ion. Once accumulated, the metal can be recovered with suitable eluting media and the solid material can be easily regenerated and reused an ample number of times.[49,50]

2.3.5.2 Activated Carbon Adsorbents

Activated carbons (ACs) are extensively used in water treatment as they possess high porosity and a high surface area for potential removal of a variety of pollutants from effluent streams. They usually acquire a significant weak acidic ion exchange character, which results in interactions between the carbon surface and the adsorbate, to remove trace metal contaminants. However, the high cost of activated carbon due to the depleted source of commercial coal-based ACs and its loss during the regeneration restricts its widespread use and application.

To decrease treatment costs, attempts have been made to find inexpensive alternative AC precursors, such as waste materials. Nowadays, ACs are prepared from a wide variety of agricultural waste materials, which are abundant, cheap and have low economic value. They are a rich source for AC production due to their low ash content and reasonable hardness. Various kinds of wastes that have been used for this purpose are wood, coconut shells, nutshells, cotton stalks and waste rubber. At present, many investigations are directed towards modifying carbon surfaces to increase their acidic surface functional groups and hence increasing their efficiency.[34,41,63]

2.3.5.3 Carbon Nanotubes

Carbon nanotubes (CNTs) as a sorbent material exhibit great potential for the adsorption of metal ions from aqueous solution because of their unique physical and chemical properties. They possess a highly porous and hollow structure, a large specific surface area, a light mass density and a strong interaction between carbon and hydrogen atoms. CNTs are hydrophobic in nature and thus have low dispersibility in water. However, this aggregation is reduced by functionalization of the surface. The mechanism of metal sorption onto CNTs is complex and is affected by several pathways involving electrostatic attraction, sorption–precipitation and chemical interaction between the metal ions and the surface functional groups of the CNTs. However, CNTs are expensive and improvements in synthetic methods and in the control of conditions to develop a cost effective way of CNT production are required.[34,41,64–66]

2.3.5.4 Agricultural Waste Materials as Solid Sorbents

Metal extraction *via* waste materials holds the benefits of high abundance, renewability, efficiency and low cost and hence has emerged as a viable option for metal remediation. These materials possess an enriched platform of a variety of functional groups which facilitate metal complexation and help sequestration of the metals. The availability of these organic functionalities is due to the presence of basic components, including hemicellulose, lignin, extractives, lipids, proteins, simple sugars, water hydrocarbons and starch. It is amazing to know that various agricultural waste materials such as rice bran, rice husks, wheat bran, wheat husks, sawdust of various plants, tree bark, groundnut shells, coconut shells, black gram husks, hazelnut shells, walnut shells, cotton seed hulls, waste tea leaves, *Cassia fistula* leaves, maize corn cobs, jatropha deoiled cakes, sugarcane bagasse, apples, bananas, orange peels, soybean hulls, grapes stalks, water hyacinths, sugar beet pulp, sunflower stalks, coffee beans, cotton stalks, *etc.*, have been used successfully for metal removal. Not only this, these sorbents can be modified for better efficiency and multiple reuse to enhance their applicability on an industrial scale.[6,34,41]

2.4   Conclusion

Wastewater streams overburdened with metals are a worldwide concern. Also, the issue prevails more in developing countries owing to a lack of technical and financial resources and the non-regulatory control for the management of hazardous wastes in industrial developments. Therefore, permanent cost-effective solutions which treat multiple metals concurrently, even in the presence of other contaminants, are highly desirable. However, although there have been substantial scientific efforts (as have been highlighted in the chapter), the current signs are not encouraging due to the non-relativity of interdisciplinary research, inadequate funds to support the newly innovated ideas, and an irrational mindset of industrialists and the entire society to consider alternative technology and contemporary thoughts. Therefore, the cooperative and collective contributions of scientists, industrialists and engineers are needed to overcome these constraints and reservations. The integrated thinking of the scientific community to interconnect the appropriate biological and physicochemical processes through a green channel can bring revolutionary changes for environmental remediation. Also, continuous efforts should be made to transform the conservative thoughts of the entire community towards science by highlighting the respective strengths and constructive contributions for ecological conservation. Moreover, these solutions should have enough space to accommodate pioneering thoughts for their amelioration and betterment. Not only this, financial organizations should step forward to support the novel ideas to be implemented in a fruitful manner.

References

1. V. Mudgal, N. Madaan, A. Mudgal, R. B. Singh and S. Mishra, *Open Nutraceuticals J.*, 2010, **3**, 94.
2. J. O. Duruibe, M. O. C. Ogwuegbu and J. N. Egwurugwu, *Int. J. Phys. Sci.*, 2007, **5**, 112.
3. J. T. Srinivasan and V. R. Reddy, presented at the Indian Society for Ecological Economics (INSEE) Conference, Ahmedabad, India, 2009.

4. M. Qadir and C. A. Scott, in *Non-Pathogenic Trade-Offs of Wastewater Irrigation*, ed. P. Drechsel, C. A. Scott, L. Raschid-Sally, M. Redwood and A. Bahri, Earthscan, London, 2010, p. 101.
5. L. Jarup, *Br. Med. Bull.*, 2003, **68**, 167.
6. D. Sud, G. Mahajan and M. P. Kaur, *Bioresour. Technol.*, 2008, **99**, 6017.
7. V. Hooda, *J. Environ. Biol.*, 2007, **28**, 367.
8. T. Mahmood, *Soil Environ.*, 2010, **29**, 91.
9. S. K. Vishnoi and P. N. Srivastava, in *Proceedings of Taal 2007: The 12th World Lake Conference*, ed. M. Sengupta and R. Dalwani, 2008, p. 1016.
10. I. Raskin, R. D. Smith and D. E. Salt, *Curr. Opin. Biotechnol.*, 1997, **8**, 221.
11. O. B. Akpor and M. Muchie, *Int. J. Phys. Sci.*, 2010, **5**, 1807.
12. T. Vamerali, M. Bandiera and G. Mosca, *Environ. Chem. Lett.*, 2010, **8**, 1.
13. E. A. H. P. Smits and J. L. Freeman, *Front. Ecol. Environ.*, 2006, **4**, 203.
14. R. K. Sharma, A. Adholeya, A. Puri and M. Das, in *Bioextraction: The Interface of Biotechnology and Green Chemistry*, ed. C. Baskar, S. Baskar and R. S. Dhillon, Springer, Berlin, 2012, p. 435.
15. C. Williams and A. N. B. Xing, *Sci. Agric. (Piracicaba, Braz.)*, 2006, **63**, 299.
16. A. Gaur and A. Adholeya, *Curr. Sci.*, 2004, **86**, 528.
17. Y. Jing, Z. He and X. Yang, *J. Zhejiang Univ. Sci. B*, 2007, **8**, 192.
18. R. Jabeen, A. Ahmad and M. Iqbal, *Bot. Rev.*, 2009, **75**, 339.
19. M. N. V. Prasad, *Russ. J. Plant Physiol.*, 2003, **50**, 686.
20. C. D. Jadia and M. H. Fulekar, *Afr. J. Biotechnol.*, 2009, **8**, 921.
21. M. Ghosh and S. P. Singh, *Appl. Ecol. Environ. Res.*, 2005, **3**, 1.
22. M. Vidali, *Pure Appl. Chem.*, 2002, **73**, 1163.
23. T. R. Muraleedharan, I. Leela and C. Venkobachar, *Curr. Sci.*, 1991, **61**, 379.
24. Z. Asku, T. Kutsal, S. Gun, N. Haciosmanoglu and M. Gholminejad, *Environ. Technol.*, 1991, **12**, 915.
25. K. Chojnacka, *Environ. Int.*, 2010, **36**, 299.
26. M. Valls and V. de Lorenzo, *FEMS Microbiol. Rev.*, 2002, **26**, 327.
27. J. Wang and C. Chen, *Biotechnol. Adv.*, 2009, **27**, 195.
28. K. M. Paknikar, A. V. Pethkar and P. R. Puranik, *Indian J. Biotechnol.*, 2003, **2**, 426.
29. A. I. Zouboulis, E. G. Rousou, K. A. Matis and I. C. Hancock, *J. Chem. Technol. Biotechnol.*, 1999, **74**, 429.
30. N. Das, R. Vimla and P. Karthika, *Indian J. Biotechnol.*, 2008, **7**, 159.
31. J. Rincon, F. Gonzalez, A. Ballester, M. L. Blazquez and J. A. Munoz, *J. Chem. Technol. Biotechnol.*, 2005, **80**, 1403.
32. E. Romera, F. Gonzalez, A. Ballester, M. L. Blazquez and J. A. Munoz, *Crit. Rev. Biotechnol.*, 2006, **26**, 223.
33. L. Brinza, M. J. Dring and M. Gavrilescu, *Environ. Eng. Manage. J.*, 2007, **6**, 237.
34. F. Fu and Q. Wang, *J. Environ. Manage.*, 2011, **92**, 407.

35. EPA 832-F-00-018, *Wastewater Technology Fact Sheet: Chemical Precipitation*, U. S. Environmental Protection Agency, Washington, 2000.
36. M. A. Barakat, *Arab. J. Chem.*, 2011, **4**, 361.
37. S. D. Alexandratos, *Ind. Eng. Chem. Res.*, 2009, **48**, 388.
38. H. Leinonen, Academic dissertation, University of Helsinki, 1999.
39. E. Erdem, N. Karapinar and R. Donat, *J. Colloid Interface Sci.*, 2004, **280**, 309.
40. S. Babel and T. A. Kurniawan, *J. Hazard. Mater.*, 2003, **B97**, 219.
41. G. Zhao, X. Wu, X. Tan and X. Wang, *Open Colloid Sci. J.*, 2011, **4**, 19.
42. J. Rydberg, G. R. Chopin, C. Musikas and T. Sekine, in *Solvent Extraction Equilibria*, ed. M. Dekker, Wiley, New York, 1992, p. 101.
43. P. Sun and D. W. Armstrong, *Anal. Chim. Acta*, 2010, **661**, 1.
44. L. Fischer, T. Falta, G. Koellensperger, A. Stojanovic, D. Kogelnig, M. Galanski, R. Krachler, B. K. Keppler and S. Hann, *Water Res.*, 2011, **45**, 4601.
45. M. R. Jakobsen, J. Fritt-Rasmussen, S. Nielsen and L. M. Ottosen, *J. Hazard. Mater.*, 2004, **106B**, 127.
46. T. Xu and C. Huang, *AIChE J.*, 2008, **54**, 3147.
47. R. K. Sharma, A. Pandey, S. Gulati and A. Adholeya, *J. Hazard. Mater.*, 2012, **209–210**, 285.
48. R. K. Sharma and P. Pant, *J. Hazard. Mater.*, 2009, **163**, 295.
49. R. K. Sharma, S. Mittal, S. Azmi and A. Adholeya, *Surf. Eng.*, 2005, **21**, 232.
50. R. K. Sharma, S. Mittal and M. Koel, *Crit. Rev. Anal. Chem.*, 2003, **33**, 183.
51. Y. Zhang, R. Qu, C. Sun, H. Chen, C. Wang, C. Ji, P. Yin, Y. Sun, H. Zhang and Y. Niu, *J. Hazard. Mater.*, 2009, **163**, 127.
52. I. Hatay, R. Gup and M. Ersoz, *J. Hazard. Mater.*, 2008, **150**, 546.
53. R. K. Dey and C. Airoldi, *J. Hazard. Mater.*, 2008, **156**, 95.
54. W. Wang, *Process Saf. Environ. Prot.*, 2011, **89**, 127.
55. L. Bai, H. Hu, W. Fu, J. Wana, X. Cheng, L. Zhuge, L. Xiong and Q. Chena, *J. Hazard. Mater.*, 2011, **195**, 261.
56. A. M. Donia, A. A. Atia, A. M. Daher, O. A. Desouky and E. A. Elshehy, *Int. J. Miner. Process.*, 2011, **101**, 81.
57. X. Huang, X. Chang, Q. He, Y. Cui, Y. Zhai and N. Jiang, *J. Hazard. Mater.*, 2008, **157**, 154.
58. X. Zhu, X. Chang, Y. Cui, X. Zou, D. Yang and Z. Hu, *Microchem. J.*, 2008, **86**, 189.
59. F. Xie, X. Lin, X. Wu and Z. Xie, *Talanta*, 2008, **74**, 836.
60. A. Safavi, S. Momeni and N. Saghir, *J. Hazard. Mater.*, 2009, **162**, 333.
61. Y. Cui, X. Chang, X. Zhu, H. Luo, Z. Hu, X. Zou and Q. He, *Microchem. J.*, 2007, **87**, 20.
62. P. Mondal, K. Roy, S. P. Bayen and P. Chowdhury, *Talanta*, 2011, **83**, 1482.

63. J. M. Diasa, M. C. M. Alvim-Ferraza, M. F. Almeida, J. Rivera-Utrillab and M. Sanchez-Polo, *J. Environ. Manage.*, 2007, **85**, 833.
64. G. P. Rao, C. Lu and F. Su, *Sep. Purif. Technol.*, 2007, **58**, 224.
65. V. A. Lemos, L. S. G. Teixeira, M. A. Bezerra, A. C. S. Costa, J. T. Castro, L. A. M. Cardoso, D. Santiago de Jesus, E. Souza Santos, P. X. Baliza and L. N. Santos, *Appl. Spectrosc. Rev.*, 2008, **43**, 303.
66. P. Liang, Y. Liu, L. Guo, J. Zeng and H. Lu, *J. Anal. At. Spectrom.*, 2004, **19**, 1489.

CHAPTER 3

Role of Plant Biomass in Heavy Metal Treatment of Contaminated Water

RAJANI SRINIVASAN

Department of Chemistry, Geosciences and Physics, College of Science and Technology, Tarleton State University, Stephenville, TX 76401, USA
Email: srinivasan@tarleton.edu

3.1 Introduction

Heavy metals released into the environment have been increasing continuously as a result of industrial activities and technological developments all over the world. Owing to their extended persistence in biological systems and tendency to bioaccumulate as they move up the food chain, they represent important environmental and occupational hazards.[1] The presence of metals above critical levels may cause various types of acute and chronic disorders in human health and are toxic to the ecosystem.

Because of the rapid increase in metal concentrations as well as an increase in the awareness of the toxicological effects of metals released into the environment, a number of studies on metal recovery and removal of metal solutions have been carried out. Conventional methods for metal removal include chemical precipitation, lime coagulation, ion exchange, reverse osmosis and solvent extraction.[2] These conventional methods for the removal of heavy metals from wastewaters are often cost prohibitive, having inadequate efficiencies at low metal concentrations, particularly in the range of $1–100 \, mg \, L^{-1}$.

RSC Green Chemistry No. 23
Green Materials for Sustainable Water Remediation and Treatment
Edited by Anuradha Mishra and James H. Clark
© The Royal Society of Chemistry 2013
Published by the Royal Society of Chemistry, www.rsc.org

Some of these methods generate toxic sludge, the disposal of which is a burden on the techno-economic feasibility of treatment procedures.[3] The search for new technologies involving the removal of toxic metals from wastewaters has directed attention to biosorption, based on the metal-binding capacities of various biological materials. Biosorption can be defined as the ability of biological materials to accumulate heavy metals from wastewater through metabolically mediated or physicochemical pathways of uptake.[4] The major advantages of biosorption over conventional treatment methods include low cost, high efficiency of metal removal from dilute solutions, minimization of chemical and biological sludge, no additional nutrient requirement, regeneration of the biosorbent and the possibility of metal recovery.[5–7] Biosorption for the removal of heavy metal ions may provide an attractive alternative to physicochemical methods.[8] Only within the past decade has the potential of metal biosorption by biomass materials been well established. Biosorption is considered to be a fast physical or chemical process. A significant number of biosorption studies on the removal of heavy metals from aqueous solutions have been conducted worldwide. Nearly all of them have been directed towards optimizing biosorption parameters to obtain the highest removal efficiency, while the rest were concerned with the biosorption mechanism. The biosorption rate depends on the type of process. According to the literature, biosorption can be divided into two main processes: adsorption of ions on the cell surface and bioaccumulation within the cell.[9] Natural materials are available in large quantities and certain waste from agricultural operations has great potential to be used as low cost adsorbents. They represent unused resources, have wide availability and are environmentally friendly.[10]

Mechanisms involved in the biosorption process include chemisorption, complexation, adsorption–complexation on surfaces and pores, ion exchange, microprecipitation, heavy metal hydroxide condensation onto the bio-surface, and surface adsorption. Biosorption largely depends on parameters such as pH, the initial metal ion concentration, biomass concentration, the presence of various competitive metal ions in solution, and to a limited extent on the temperature.[11]

Very few reviews are available where readers can obtain an overview of the sorption capacities of agro-based biomasses used for metal remediation. This chapter includes a compilation of studies carried out by researchers using plant-derived biomass as treatment agents for metal contaminated wastewater.

3.2 Various Types of Biomass Used for Metal Removal

The removal of cadmium, arsenic and lead from drinking and irrigation water is a recurring challenge, especially in developing countries. Cost considerations can make it expedient to use local materials, produced in agricultural or industrial operations, as adsorbents for these toxins. The performance of these materials may not always be optimal, but their immediate availability often makes them attractive choices. The review by Yadanaparthi *et al.*[12] presents a compilation of adsorption studies, many of which are based on the use of low-value products derived from plants in the removal of the metals cadmium, arsenic and lead.

Datura innoxia biomass has been targeted as a promising sorbent for heavy metal reclamation and remediation. Although the metal sequestering capabilities of biomaterials have been known for a long time, a general inability to accurately predict their binding characteristics has been a barrier to their implementation in remediation schemes. Of interest in Lin and Rayson's[13] laboratory were the binding mechanisms as well as basic chemistry involved in the metal uptake of material derived from fragments of cell walls from cultured cells from the plant *D. innoxia*. Previous studies have suggested the dependency of lead-ion affinity distributions of the biomaterial upon solution variables (*i.e.*, ionic strength, pH) to be the result of two classes of binding sites: low-affinity sites involving sulfonates and carboxylates in an ion-exchange process and high-affinity sites resulting from the coordination of carboxylates to the metal. This hypothesis has been further investigated by the removal of the dominant functionality on the biomaterial (carboxylates) by carboxyl esterification by mildly acidic methanol. The resulting modified biomaterial was then studied in terms of affinity distributions toward protons and lead ions. The content of sulfonates in the biomaterial was $43\,\mu\text{mol}\,\text{g}^{-1}$. The total amount of carboxylates involved in Pb(II) binding was $560\,\mu\text{mol}\,\text{g}^{-1}$, with $147\,\mu\text{mol}\,\text{g}^{-1}$ involved in an ion-exchange mechanism and coordination sites including $212\,\mu\text{mol}\,\text{g}^{-1}$ biomaterial.

Goyal *et al.*[14] explored the effectiveness of *Saraca indica* leaf powder, a surplus low-value agricultural waste, in removing Pb ions from aqueous solution. Batch studies indicated that the maximum biosorption capacity for Pb was 95.37% at pH 6.5. The sorption process followed first-order rate kinetics. The adsorption equilibrium data fitted best to both Langmuir and Freundlich isotherms. Morphological changes observed in scanning electron micrographs of untreated and metal-treated biomass confirmed the phenomenon of biosorption. Fourier transform infrared (FTIR) spectroscopy of native and exhausted leaf powder confirmed lead biomass interactions responsible for the sorption. Acid regeneration was tried for several cycles with a view to recovering the sorbed metal ion and also to restore the sorbent to its original state. The findings showed that *S. indica* leaf powder can easily be envisaged as a new, vibrant, low-cost biosorbent for metal cleanup operations.

An adsorbent was developed from the mature leaves of the neem (*Azadirachta indica*) tree for removing Pb(II) from water by Bhattacharyya and Sharma.[15] Adsorption was carried out in a batch process and the uptake of the metal was very fast initially but gradually slowed down, indicating penetration into the interior of the adsorbent particles. Both first-order and second-order kinetics were tested and it was found that the latter gave a better explanation. The adsorbent had a very high Langmuir monolayer capacity of $300\,\text{mg}\,\text{g}^{-1}$. A small amount of the adsorbent ($1.2\,\text{g}\,\text{L}^{-1}$) could remove as much as 93% of Pb(II) in 300 min from a solution of concentration $100\,\text{mg}\,\text{L}^{-1}$ at 300 K. The adsorption continuously increased in the pH range 2.0–7.0, beyond which the adsorption could not be carried out owing to precipitation of the metal.

The adsorption was exothermic at ambient temperature and indicated the interactions to be thermodynamically favorable.

Rima *et al.*[16] reported the use of beetroot fibers to eliminate heavy metals from polluted water. This biomass was used to remove lead, copper and zinc ions. The kinetics and beetroot fiber fixation capacities of lead, zinc and copper according to various physicochemical parameters, such as pH and the concentration of the metal solutions, were studied. The quantity of the lead, zinc and copper cations retained by this biomass in simple and mixture solutions was estimated to be 23.6, 14.02 and 14.64 mg g^{-1}, respectively. The results indicated the usefulness of this biomass in the remediation of water contaminated by heavy metals by a rapid, practical and efficient method.

The bark of *Hemidesmus indicus*, an extensively available plant biomass commonly called Indian sarsaparilla, was used as biomaterial for removal of lead from aqueous streams by Chandrasekhar *et al.*[17] Batch experiments were carried out with immobilized biomass of *H. indicus* (IPBFIX) to optimize the experimental parameters like the effect of contact time, initial metal concentration, initial IPBFIX concentration and co-metal ion effect on biosorption of lead from contaminated waters. Column experiments were performed under flow conditions for the regeneration and recycle efficiency of IPBFIX and was found to be effective for three cycles. Elution experiments were carried out to remove lead ions from loaded IPBFIX and 100% elution was achieved with a 0.1 M HNO$_3$ solution. The effectiveness of the IPBFIX for biosorption of lead ions was demonstrated using wastewater samples emanating from a non-ferrous metal industry.

An attempt has also been made to remove lead from the lead polluted waters (both ground and surface) of industrially contaminated sites. Saka *et al.*[18] described various kinds of agricultural and forest waste adsorbents used to remove Pb(II) ions in wastewater treatment, and the technical feasibilities were reviewed mainly from the year 2000 to 2010. They all were compared with each other by metal binding capacities, metal removal performances, sorbent dose, optimum pH, temperature, initial concentration and contact time. It was shown that these alternative adsorbents had sufficient binding capacity to remove Pb(II) ions from wastewater.

Mawangi *et al.*[19] studied powdered maize tassels and found that they exhibited metal sorption properties due to the availability of metal absorbing functional groups. The tassels have a high amount of soluble organic substances that can dissolve in aqueous media, contributing to secondary pollution during a water treatment process. A chelating agent was chemically attached on the maize tassels with a view to increasing the sorption capacity, minimizing leaching and enhancing the tassels' stability. Thermogravimetric analysis confirmed that modification improved their thermal stability to withstand temperatures above 600 °C as well as reducing the secondary pollution. The modified sorbent was employed for the sorption of lead, copper and cadmium ions in both the model solutions and real samples. The contact time and pH were optimized, after which Langmuir and Freundlich isotherms were applied to the data. The sorption capacities for Cu(II), Cd(II) and Pb(II) improved from 3.4, 0.8 and 1.7 g kg^{-1}, respectively, to 6.3, 2.6 and 2.6 g kg^{-1} in the same order. The sorbent was shown to remove up to 95% of the metals in less than 10 min.

The adsorption of Cd(II) and Pb(II) onto natural dye waste (aqueous color extracted) of the *Hibiscus rosa-sinensis* flower has been studied by Vankar *et al.*[20] Batch adsorption studies were carried out to examine the effects of process parameters such as pH, temperature, agitating rate, adsorbent mass and particle size on metal uptake by hibiscus dye waste (HDW) for Cd(II) and Pb(II). The sorption process was fast and equilibrium was achieved in about 150 min of contact. The extent of adsorption of lead and cadmium increased with an increase of pH and adsorbent dose, while decreased with adsorbent particle size. The adsorption data followed the Langmuir model better than the Freundlich model. The adsorption equilibrium was described well by the Langmuir isotherm model, with adsorption maxima at equilibrium found to be 90.91 mg g^{-1} of lead ions and 103.09 mg g^{-1} of cadmium ions on HDW at 20 °C. The reversibility of the process was investigated and desorption was observed only at acidic pH values, and was generally high.

Garg *et al.*[21] reported the feasibility of using various agricultural residues, *viz.* sugarcane bagasse (SCB), maize corncob (MCC) or Jatropha oil cake (JOC), for the removal of Cd(II) from aqueous solution under different experimental conditions. Batch experiments were carried out at various pH values (2–7), adsorbent doses (250–2000 mg) and Cd(II) concentrations (5–500 mg L^{-1}) for a contact time of 60 min. The maximum cadmium removal capacity was shown by JOC (99.5%). The applicability of both Langmuir and Freundlich isotherms suggested the formation of a monolayer of Cd(II) ions onto the outer surface of the adsorbents. The maximum metal removal was observed at pH 6.0 with a contact time of 60 min at a stirring speed of 250 rpm and with an adsorbent dose of 20 g L^{-1} of the test solution. The maximum adsorption of Cd(II) metal ions was observed at pH 6 for all the adsorbents, *viz.* 99.5%, 99% and 85% for JOC, MCC and SCB, respectively. The order of Cd(II) removal by various biosorbents was JOC > MCC > SCB. JOC may thus be an alternative biosorbent for the removal of Cd(II) ions from aqueous so-lution. These results could be helpful in designing a batch mode system for the removal of cadmium from dilute wastewaters.

Biomass derived from the leaves of *Acacia nilotica* was used as an adsorbent material for the removal of cadmium and lead from aqueous solution by Waseem *et al.*[22] The effect of various operating variables, *e.g.* adsorbent dosage, contact time, pH and temperature, on the removal of cadmium and lead was studied. The maximum adsorption of cadmium and lead arises at concentrations of 2 g/50 mL and 3 g/50 mL and at pH values of 5 and 4, respectively. The sorption data favored the pseudo-second-order kinetic model. Langmuir, Freundlich and Dubinin–Radushkevich (DR) models were applied to describe the biosorption isotherm of the metal ions by *A. nilotica* biomass. Based on regression coefficients, the equilibrium data found were fitted better to the Langmuir equilibrium model than other models. Thermodynamic par-ameters such as free energy change ($\Delta G°$), enthalpy change ($\Delta H°$) and entropy change ($\Delta S°$) revealed the spontaneous, endothermic and feasible nature of the adsorption process.

Ahmad *et al.*[23] has presented a review on the role of oil palm biomass (trunks, fronds, leaves, empty fruit bunches, shells, *etc.*) as adsorbents in the removal of water pollutants such as acid and basic dyes, heavy metals, phenolic compounds, various gaseous pollutants, and so on. Studies have shown that an oil palm-based adsorbent, among the low-cost adsorbents mentioned, is the most promising adsorbent for removing water pollutants. Further, these bioadsorbents can be chemically modified for better efficiency and can undergo multiple reuses to enhance their applicability at an industrial scale.

A review by Kumar *et al.*[24] provides detailed information on various aspects of the utilization of agro-based biomasses for zinc metal ion removal. These biosorbents can be modified using various methods for better efficiency and multiple reuses to enhance their applicability at an industrial scale. The removal of Cr(III) ions from water by *Polyalthialongi folia* leaves was studied as a function of adsorbent dose, pH, contact time and agitation speed by Anwar *et al.*[25] The surface characteristics of the leaves were evaluated by recording IR spectra. The Langmuir, Freundlich and Temkin adsorption isotherms were employed to explain the sorption process. It was found that 1 g of leaves can remove 1.87 mg of trivalent chromium at pH 3.0. It was concluded that *P. folia* leaves can be used as cost-effective and benign adsorbents for removal of Cr(III) ions from wastewater.

Biosorption was carried out in a batch process to test the suitability of *Amaranthus hybridus* (African spinach) stalks and *Carica papaya* (pawpaw) seeds for removal of Mn(II) and Pb(II) ions from aqueous solution by Egila *et al.*[26] The amount of metal ions removed from solution depended on the metal ion–substrate contact time, ion concentration and ion type. The results indicated that the amount of adsorbed metal ions varied with substrate materials, and the order of removal of heavy metals by the biosorbents was Mn(II) > Pb(II). Furthermore, Mn(II) had higher removal percentages than Pb(II) ions in both substrates. In all cases, *C. papaya* seeds showed greater adsorptive capacity than *A. hybridus* stalks. The results from both biomass substrates well fitted the Freundlich adsorption isotherm.

Low-cost adsorbents were prepared from three dried plants, namely *Carpobrotus edulis*, *Launea arborescens* and *Withania frutescens*, for the removal of heavy metals, nitrate and phosphate ions from industrial wastewaters. The efficiency of these adsorbents was investigated using a batch adsorption technique at room temperature by Chiban *et al.*[27] The dried plant particles were characterized by N_2 at 77 K adsorption, scanning electron microscopy (SEM), energy-dispersive X-ray (EDAX) spectroscopy, FTIR spectroscopy and phytochemical screening. The adsorption experiments showed that the microparticles of the dried plants presented a good adsorption of heavy metals, phosphate and nitrate ions from real wastewaters. After the adsorption process, the Pb(II) concentrations, as well as those of Cd(II), Cu(II) and Zn(II), were below the European drinking water norm concentrations. The percent removal of heavy metals, nitrates and phosphates from industrial wastewaters by dried plants was 94% for Cd(II), 92% for Cu(II), 99% for Pb(II), 97% for Zn, 100% for NO_3^{2-} and 77% for PO_4^{3-} ions. For all heavy metal ions, the uptake efficiency of the studied plants ranged as *C. edulis* > *W. frutescens* > *L. arborescens*; however, the differences were rather small.

Dos Santos *et al.*[28] proposed the use of *Agave sisalana* (sisal fiber) as a natural adsorbent for Pb(II) and Cd(II) ion biosorption from natural waters. Flame atomic absorption spectrometry was used for the quantitative determination of Pb(II) and Cd(II) adsorption on the solid phase. The results showed that sisal had a surface area for adsorption of $0.0233\,m^2\,g^{-1}$ and that hydroxyl and carbonyl were the main functional groups involved in the biosorption. The best interpretation for the experimental data was given by a Freundlich isotherm that proposed monolayer sorption with a heterogeneous energetic distribution of active sites, accompanied by interactions between sorbed molecules. The maximum monolayer biosorption capacity was found to be $1.85\,mg\,g^{-1}$ for Cd(II) and $1.34\,mg\,g^{-1}$ for Pb(II) at pH 7 and 296 K. This adsorbent can be used for biosorption of cadmium and lead in polluted natural waters.

Lezczno *et al.*[29] examined the sorption capacity of a natural biomass collected from an irrigation pond. The biomass mainly consisted of a mixture of chlorophyte algae with caducipholic plants. Biosorption experiments were performed in monometallic and bimetallic solutions containing different metals commonly found in industrial effluents (Cd, Cu and Pb). The biosorption process was slightly slower in the binary system compared with the monometallic system, which was related to competition phenomena between metal cations in solution. The biosorbent behavior was quantified by the sorption isotherms fitting the experimental data to mathematical models. In monometallic systems the Langmuir model showed a better fit with a sorption order of Cu > Pb > Cd and a biomass–metal affinity order of Pb > Cd > Cu. In bimetallic systems, the binary-type Langmuir model was used and the sorption order obtained was Pb > Cu > Cd. In addition, the effectiveness of the biomass was investigated in several sorption–desorption cycles using HCl and $NaHCO_3$. The recovery of metals was higher with HCl than with $NaHCO_3$, although the sorption uptake of the biomass was sensitively affected by the former desorption agent in subsequent sorption cycles.

A study by Zu *et al.*[30] proved that the invasive alien plant *Iva xanthifolia* Nutt (IXN) can be used as an effective and low-cost absorbent for the treatment of wastewater containing Cd(II) and Pb(II). The Langmuir biosorption isotherm was applied to correlate the equilibrium data and fitted quite well. The ability of abundantly available heartwood of *Areca catechu* to adsorb Cd(II) ions from aqueous solution has been investigated through batch experiments at room temperature by Chakrabarty *et al.*[31] The adsorbent was found to be effective for quantitative removal of Cd(II) ions in acidic conditions and equilibrium was achieved in 30 min at pH 6.0. The equilibrium adsorption data were fitted to Langmuir, Freundlich and DR adsorption isotherm models and the model parameters were evaluated. The kinetic study showed that the pseudo-second-order rate equation better described the biosorption process. Chakraborty *et al.*[32] also studied the removal of Pb(II) from aqueous solution using the same biosorbent. The adsorbent was effective for the quantitative removal of Pb(II) ions in acidic conditions and equilibrium was achieved in 25 min. The equilibrium adsorption data were fitted to Langmuir and Freundlich adsorption isotherm models and the model parameters were evaluated. The kinetic study showed that the pseudo-second-order rate equation better described the biosorption process. The FTIR spectrum analysis revealed that hydroxyl, carboxyl, amide and amine groups were the major binding groups for Cd(II) and Pb(II).

The biosorption of Cd(II), Cu(II), Fe(II) and Zn(II) from aqueous solutions by an agro-waste, namely banana trunk fibers (BTF), was investigated by Sathasivam and Harris.[33] The effect of pH, contact time, metal ion concentration, adsorbent dose and change in $[M^{2+}]$/biomass ratio were studied at ambient temperature (25 °C). The equilibrium process was described well by the Freundlich isotherm model, with an adsorption capacity, K_f, of 8.49, 2.68, 6.58 and 1.74 mg g^{-1} for Cd(II), Cu(II), Fe(II) and Zn(II), respectively. Kinetic studies showed good correlation coefficients for a pseudo-second-order kinetic model. The BTF were subjected to different chemical modification methods (mercerization, acetylation, formaldehyde treatment, peroxide treatment, stearic acid treatment and sulfuric acid treatment) and the adsorption capacity (q_e) of each modified BTF for the metal ions was obtained.

In the study by Tan *et al.*,[34] corncob biomass was utilized as an adsorbent to remove Pb(II) from aqueous solution. Ground corncobs were modified with CH_3OH and NaOH to investigate the effect of chemical modification on Pb(II) binding capacity. The results showed that Pb(II) binding on the biomass is pH dependent and the kinetics can be well described by the Lagergren second-order model. The maximum Pb(II) binding capacity, q_{max}, calculated from the Langmuir isotherm was 0.0783 mmol g^{-1}. After base hydrolysis of the biomass, the Pb(II) binding capacity increased from 0.0783 to 0.2095 mmol g^{-1} (about 43.4 mg Pb g^{-1}). However, the Pb(II) binding capacity on the esterified corncobs decreased greatly from 0.0783 to 0.0381 mmol g^{-1}. FTIR spectroscopy showed that hydroxyl and carboxylic (COO^-) groups on the biomass play an important role in the Pb(II) binding process. The X-ray photoelectron spectroscopy (XPS) data further indicated that lead is adsorbed as Pb(II) and is attached to oxide groups on the biomass.

Biosorption of Pb(II) from aqueous solution by biomass prepared from *Moringa oleifera* bark (MOB), an agricultural solid waste, has been studied by Reddy *et al.*[35] The experimental equilibrium adsorption data were tested by four widely used two-parameter equations: the Langmuir, Freundlich, DR and Temkin isotherms. The results indicated that the data for Pb(II) adsorption onto MOB were best fitted by the Freundlich model. The adsorption capacity (Q_m) calculated from the Langmuir isotherm was 34.6 mg Pb(II) g^{-1} at an initial pH of 5.0. Adsorption kinetic data were analyzed using the pseudo-first-order and pseudo-second-order equations and intra-particle diffusion models. The results indicated that the adsorption kinetic data were best described by the pseudo-second-order model. IR spectral analysis revealed that the lead ions were chelated to hydroxyl and/or carboxyl functional groups present on the surface of the MOB. The biosorbent was effective in removing lead in the presence of common metal ions like Na^+, K^+, Ca^{2+} and Mg^{2+} present in water. Desorption studies were carried out with dilute hydrochloric acid for quantitative recovery of the metal ions as well as to regenerate the adsorbent. Based on the results obtained, such as good uptake capacity, rapid kinetics and low cost, MOB appears to be a promising biosorbent material for the removal of heavy metal ions from wastewater or effluents.

Mata *et al.*[36] have reported their work on the effectiveness of sugar-beet pectin xerogels for the removal of the heavy metals Cd(II), Pb(II) and Cu(II) after multiple batch sorption–desorption cycles, with and without a gel regeneration step. The metals were recovered from xerogel beads without destroying their sorption capability and the beads were successfully reused (nine cycles) without significant loss in both biosorption capacity and biosorbent mass. The metal uptake leveled off or increased after using a 1 M $CaCl_2$ regeneration step after each desorption. Calcium, as a regenerating agent, increased the stability and reusability of the gels, repairing the damage caused by the acid and removing the excess protons after each elution to provide new binding sites. Because of their excellent reusability, pectin xerogels are suitable for metal remediation technologies.

Waste materials like rice bran and pine sawdust were investigated by Chen *et al.*[37] to assess their potential for removal of lead and cadmium from water by the process of biosorption by batch adsorption experiments studies. The obtained results showed that chemical treatment of the biomass, pH and initial concentration of the heavy metals highly affected the overall metal sorption capacity of the biosorbent. The effect of chemical treatment of biomasses on the sorption of the heavy metals showed that modification by NaCl greatly enhanced the metal removal. The maximal biosorption removals of Pb(II) and Cd(II) were 99.25% and 98.25% on rice bran and 95.25% and 90.21% on pine sawdust at pH 6. The removal of Pb(II) and Cd(II) by rice bran was higher than that of pine sawdust. The two metal–biomass sorption processes were well fitted to the Langmuir equation. Both the biosorbents were also successfully applied to effluent treatment in the electroplating industry for Pb(II) and Cd(II) removal and were also potential sorbents for zinc, calcium and magnesium.

The efficacy of banana peel (*Musa sapientum*) biomass was tested for the removal of lead, copper, zinc and nickel metal ions using batch experiments in single and binary metal solutions under controlled experimental conditions by Muhammad *et al.*[38] It was found that metal sorption increases when the equilibrium metal concentration rises. At the highest experimental solution concentration used ($150\,mg\,L^{-1}$) the removal of metal ions were 92.52% for lead, 79.55% for copper, 63.23% for zinc and 68.10% for nickel, while at the lowest experimental solution concentration ($25\,mg\,L^{-1}$) the removal of metal ions was 94.80% for lead, 86.81% for copper, 84.63% for zinc and 82.36% for nickel. The values for the separation factor were between zero and one, indicating favorable sorption for the four tested metals on the biosorbent. The surface coverage values were approaching unity with increasing solution concentration, indicating the effectiveness of the biosorbent.

In a study by Ogali *et al.*,[39] the use of orange mesocarp residue biomass as a cost-effective and environmentally safe technique to remove Mg(II), Zn(II), Cu(II) and Pb(II) from aqueous solution was investigated. The results showed that unmodified orange mesocarp residue bound 56% of Mg(II), 81% of Zn(II), 71% of Cu(II), 73% of Pb(II) and 85.05% of Cd(II). These results show that orange mesocarp residue biomass can be effectively used to adsorb heavy metals from aqueous solution. Peanut hull, an agricultural byproduct abundant in China, was used as adsorbent for the removal of Cu(II) from aqueous solutions by Zhu *et al.*[40] The extent of adsorption was investigated as a function of pH, contact time, adsorbate concentration and reaction temperature. The Cu(II) removal was pH dependent, reaching a maximum at pH 5.5. The biosorption process followed pseudo-second-order kinetics and equilibrium was attained at 2 h. The rate constant increase with the increase of temperature indicates the endothermic nature of the biosorption. The activation energy (E_a) of Cu(II) biosorption was determined as $17.02 \, kJ \, mol^{-1}$ according to the Arrhenius equation, which shows that the biosorption may be an activated chemical biosorption. The equilibrium data were analyzed using the Langmuir, Freundlich and DR isotherm models, depending on the temperature. The mean free energy obtained from the DR isotherm indicated a chemical ion-exchange mechanism.

In Chile and Norway, countries with the highest salmon production in the world, salmon have developed chronic diseases due to toxicity problems from free metallic ions. Aluminum and iron were found to be present in freshwater used by the Chilean salmon industry. Different alternatives for aluminum and iron removal were compared to achieve the required standards in salmon culture by Aspe and Fernandes.[41] Manganese removal was also assessed since Fe and Mn removal can be accomplished in a single process. Since cellulose production (a principal economic activity in Chile) and the sawmill industry generate *Pinus radiata* bark as a waste product, this study analyzed its application as an adsorbent to precipitate Al, Fe and Mn in comparison with the traditional method of granular filtration. $Al(OH)_3$ precipitation was achieved by pH exchange. For precipitation of Fe and Mn oxides, two alternatives were analyzed: (i) oxidation by the presence of dissolved oxygen and pH exchange and (ii) pH exchange by CO_2 injection and oxidation produced by chemical filtration. In solution, Fe and Mn at low concentrations (less than $1 \, mg \, L^{-1}$) presented a maximum precipitation at pH 8.7, different from the value they presented individually. The separation efficiencies for the three processes, namely (a) oxidation and filtration in a column packed with *P. radiata* bark, (b) oxidation and granular filtration (smaller particle size) and (c) CO_2 injection and chemical and granular filtration, were 93, 97 and 98% for Fe and 97, 99 and 29% for Mn, respectively. In all the studied alternatives, Fe concentrations less than $0.1 \, mg \, L^{-1}$, compatible with salmon life, were obtained; in contrast, for Mn it was only possible to reach an adequate concentration for salmon life with the granular filter for smaller particle sizes. The optimum pH for Al precipitation was 6.4 and the column filled with *P. radiata* bark achieved Al concentration values less than $0.01\,mg\,L^{-1}$, a limiting value for salmon farms, obtaining removal efficiencies greater than 99.5%; in contrast, in the granular filter the average obtained for cycle efficiency was 80.3%. Only the column filled with *P. radiata* bark achieved Al and Fe concentrations compatible with salmon life.

Luo *et al.*[42] reported the feasibility of using rice husks to remove Ag(I) from synthetic wastewater. The results indicated that rice husks offered high removal efficiency, a fast adsorption rate and a high uptake capacity for Ag(I) ions. The equilibrium was attained within 20 min and the maximum removal efficiency of $11\,g\,L^{-1}$ rice husk at pH 2 was found to be 99.76%. The kinetic data were well fitted to the pseudo-second-order model. The isotherm adsorption data were well described by the Langmuir isotherm model and the maximum uptake capacity of Ag(I) ions onto rice husks was reported to be $42.43\,mg\,g^{-1}$.

Uranium(VI) biosorption by grapefruit peel from aqueous solutions was studied by Zhuo *et al.*[43] Batch experiments were conducted to evaluate the effect of contact time, initial uranium(VI) concentration, initial pH, adsorbent dose, salt concentration and temperature. The equilibrium process was well described by the Langmuir, Redlich–Peterson and Koble–Corrigan isotherm models, with maximum sorption capacity of $140.79\,mg\,g^{-1}$ at 298 K. The pseudo-second-order model and Elovish model adequately described the kinetic data in comparison to the pseudo-first-order model; the process involving the rate-controlling step is more complex, involving both boundary layer and intra-particle diffusion processes. The effective diffusion parameter D_i and D_f values were estimated at different initial concentrations and the average values were determined to be 1.167×10^{-7} and $4.078 \times 10^{-8}\,cm^2\,s^{-1}$. Thermodynamic parameters showed that the biosorption of uranium(VI) onto grapefruit peel biomass was feasible, spontaneous and endothermic under the studied conditions. The physical and chemical properties of the adsorbent were determined by SEM, themogravimetry–differential scanning calorimetry (TG-DSC), X-ray diffraction (XRD) and elemental analysis, and the nature of the biomass–uranium(VI) interactions was evaluated by FTIR analysis, which showed the participation of COOH, OH and NH_2 groups in the biosorption process. The adsorbents could be regenerated using $0.05\,mol\,L^{-1}$ HCl solution for at least three cycles, with up to 80% recovery.

Arsenic, as one of the heavy toxic elements, greatly threatens the health of humans. Studies on arsenic sorption have recently come to the focus of world attention. The review by Ma *et al.*[44] gives an overview of the latest developments around the world and the application of biomacromolecule-based adsorbents that has been reported in the treatment of drinking water, groundwater and industrial effluent for the removal of arsenic. Nigam *et al.*[45] examined the removal of arsenic from water by biosorption through the potential application of herbal dye wastes. Four different flower dye residues (after extraction of natural dye), *viz. Hibiscus rosa-sinensis*, *Rosa rosa*, *Tagetes erecta* and *Canna indica*, were utilized successfully for the removal of arsenic from aqueous solutions. Batch studies were carried out for various parameters, namely pH, sorbent dose, contact time, initial metal ion concentration and temperature. The data were utilized for isothermal, kinetic and thermodynamic studies. SEM, EDAX spectroscopy and FTIR analyses of biomasses were performed. The results showed that 1 g/100 mL for 5.0–5.5 h contact time at pH 6.0–7.5 with an agitation rate of 150 rpm provided 98, 96, 92 and 85% maximum absorption of arsenic by *R. rosa*, *H. rosa-sinensis*, *T. erecta* and *C. indica*, respectively, at an initial concentration of 500 μg L^{-1}. The data followed the Langmuir isotherm, showing sorption to be a monolayer on a heterogeneous surface of biosorbent. Negative values of $\Delta G°$ indicated a spontaneous nature, whereas $\Delta H°$ indicated the exothermic nature of the system followed by pseudo-first-order adsorption kinetics. FTIR results showed apparent changes in functional group regions after metal chelation. SEM and EDAX analyses showed the changes in surface morphology of all test biosorbents. Herbal dye wastes, used as biosorbent, exhibited significant (85–98%) removal of arsenic from aqueous solution.

A biomass derived from the plant *Momordica charantia* has been found to be very efficient in arsenic(III) adsorption. Pandey *et al.*[46] used this biomass for As(III) removal under different conditions. The parameters optimized were contact time (5–150 min), pH (2–11), concentration of adsorbent (1–50 g L^{-1}), concentration of adsorbate (0.1–100 mg L^{-1}), *etc.* It was observed that the pH had a strong effect on the biosorption capacity. The optimum pH obtained for arsenic adsorption was 9. The influence of common ions such as Ca(II), Mg(II), Cd(II), Se(IV), Cl$^-$, SO$_4^{2-}$ and HCO$_3^-$, at concentrations varying from 5 to 1000 mg L^{-1}, was also investigated. To establish the most appropriate correlation for the equilibrium curves, isotherm studies were performed for the As(III) ion using Freundlich and Langmuir adsorption isotherms. The pattern of adsorption fitted well with both models. The biomass of *M. charantia* was found to be effective for the removal of As(III), with 88% sorption efficiency at a concentration of 0.5 mg L^{-1} of As(III) solution. This biomass might be used as a palliative food item and thus might play a role in the toxic effects of ingested arsenic.

Sorption efficacy studies were carried out on selected biomaterials known for their potential usage as natural dyes for removal of Cr(VI). *Canna indica* flowers, *Portulaca olecera* flowers and stems, *Hibiscus rosa-sinensis* flowers and *Trapa natans* fruit skins (exocarp) were used for sorption studies by Vankar *et al.*[47] These plant parts, after the extraction of natural dye, were dried and evaluated for biosorption of heavy metals from effluents. Batch tests indicated that the Cr(VI) sorption capacity (q_e) followed the sequence $q_e(Trapa) > q_e(Hibiscus) > q_e(Portulaca) > q_e(Canna)$. Owing to the high sorptive capacity, *Trapa* fruit skins (exocarp) were selected. The optimization studies were carried out by taking the *Trapa* exocarp in powdered form and of a particular mesh size. Sorption kinetic data showed a first-order reversible kinetic model for all the sorbents; however, the biosorption of chromium by *T. natans* biomass occurred in two stages. In the first stage, 95% sorption was reported for Cr(VI) in 15 min, followed by a slower second stage. It reached equilibrium in 1 h, at which time 90–98% of the Cr(VI) was biosorbed by the

T. natans. The experimentally reported equilibrium data fitted well with both the Langmuir and Freundlich isotherms. The FTIR, XRD and XPS analyses showed that the main mechanism of Cr(VI) biosorption onto *Trapa* dried powder was through the binding of chromium ions with amide groups of the biomass.

The potential to remove Cr(VI) from aqueous solutions at varying pH values, contact times, initial concentrations and adsorbent dosages through biosorption using coffee husks was investigated by Ahalya *et al.*[48] The data obeyed Langmuir and Freundlich adsorption isotherms. The Langmuir adsorption capacity was found to be 44.95 mg g^{-1}. The Freundlich constants K_f and n were 1.027 and 1.493, respectively. Desorption studies indicated the removal of 60% of the hexavalent chromium. IR spectral studies revealed the presence of functional groups, such as hydroxyl and carboxyl groups, on the surface of the biomass, which facilitates biosorption of Cr(VI). Moussavi *et al.*[49] used pistachio hull powder (PHP) for the removal of hexavalent chromium from wastewater. The effects of pH (2–8), PHP concentration (0.5–8 g L^{-1}), Cr(VI) concentration (50–200 mg L^{-1}), temperature (5–50 °C) and contact time (1–60 min) were studied for the removal of Cr(VI) from aqueous solution. The results revealed that PHP adsorbs over 99% of chromium from solutions containing 50–200 mg L^{-1} of Cr(VI) at pH 2 and an adsorbent concentration of 5 g L^{-1} after 60 min of equilibration. The percent chromium adsorbed from solution increased with an increase in temperature from 5 to 40 °C. Kinetic and isotherm modeling studies demonstrated that the experimental data best fit a pseudo-second-order and Langmuir model, respectively. The maximum Langmuir adsorption capacity was 116.3 mg g^{-1}. In the second part of the study, the efficacy of PHP was examined by analyzing the removal of Cr(VI) from industrial wastewater. The results revealed that 2 g L^{-1} of PHP decreased the Cr(VI) concentration from 25 mg L^{-1} to less than 0.05 mg L^{-1} after 30 min of equilibration.

A study by Martin *et al.*[50] explored the ability of orange waste biomass to remove Cr(III) from aqueous solutions. Batch kinetic and isotherm studies were carried out on a laboratory scale to evaluate the adsorption capacity of orange waste. The effects of particle size, adsorbent dose and solution pH on Cr(III) removal were also studied. A kinetic study revealed that the adsorption of Cr(III) onto orange waste was a gradual process and equilibrium was reached within 3 days. A pseudo-second-order model was the most appropriate to describe the kinetic experimental data. Equilibrium assays displayed a maximum sorption capacity ranging from 0.57 to 1.44 mmol g^{-1} when the pH increased from 3 to 5, according to the Sips model, which along with the Redlich–Peterson equation is very suitable for correlating equilibrium data. The use of the studied adsorbent in the removal of chromium in continuous mode was successful and the breakthrough curves were adequately represented by the BDST model. Owing to the slow kinetics of chromium sorption onto orange waste, the sorption capacity in batch assays was higher than that in continuous assays.

Grainless stalks of corn (GLSC) were tested for removal of Cr(VI) and Cr(III) from aqueous solutions at different pH values, contact times, temperatures and chromium/adsorbent ratios by Bellu *et al.*[51] The results showed that the optimum pH for removal of Cr(VI) was 0.84, while the optimum pH for removal of Cr(III) was 4.6. The adsorption processes for both Cr(VI) and Cr(III) onto GLSC were found to follow first-order kinetics. The adsorption capacity of GLSC was calculated from the Langmuir isotherm as 7.1 mg g^{-1} at pH 0.84 for Cr(VI) and as 7.3 mg g^{-1} at pH 4.6 for Cr(III), at 20 °C. EPR spectroscopy showed the presence of Cr(VI)- and Cr(III)-bound GLSC at short contact times and adsorbed Cr(III) as the final oxidation state of Cr(VI)-treated GLSC. The results indicate that, at pH $\approx$ 1, GLSC can completely remove Cr(VI) from aqueous solution through an adsorption-coupled reduction mechanism to yield adsorbed Cr(III) and the less toxic aqueous Cr(III), which can be further removed at pH 4.6.

Corn stalk powder was used as a biosorbent in aqueous solution to investigate its Cr(VI) removal capacity by Li *et al.*[52] The results showed that reduction of Cr(VI) to Cr(III) occurred during the adsorption process. The adsorption capacity of the biosorbent was 14.46 mg g^{-1}. Thermodynamic parameters of the adsorption process, such as $\Delta G°$, $\Delta H°$ and $\Delta S°$, were calculated and the results indicated that the overall adsorption process was endothermic and spontaneous. Jain *et al.*[53] investigated the Cr(VI) removal efficiency of sunflower waste from aqueous system under different process conditions. Two adsorbents were prepared by pretreating the sunflower stem waste, one by boiling it and the second by treating it with formaldehyde. The removal of chromium was dependent on the physicochemical characteristics of the adsorbent, adsorbate concentration and other studied process parameters. The maximum metal removal was observed at pH 2.0. The efficiencies of boiled sunflower stem absorbent and formaldehyde-treated sunflower stem absorbent for the removal of Cr(VI) were 81.7% and 76.5%, respectively, for dilute solutions at 4.0 g L^{-1} adsorbent doses. The results revealed that the hexavalent chromium is considerably adsorbed on sunflower stems and it could be an economical method for the removal of Cr(VI) from aqueous systems.

Vankar *et al.*[54] assessed the uptake of zinc ions from aqueous solutions onto natural dye waste (aqueous color extracted) of *Hibiscus rosa-sinensis* flowers. The study investigated the effects of process parameters such as pH, temperature, agitating rate, adsorbent mass and particle size. The extent of adsorption increased with an increase in pH. Furthermore, the adsorption of zinc increased with increasing adsorbent doses and decreased with the adsorbent particle size. The adsorption data followed both Langmuir and Freundlich adsorption models; however, the Langmuir model was better suited than the Freundlich model and the adsorption maximum at equilibrium was 94.34 mg g^{-1} of zinc ions on hibiscus dye waste (HDW) sorbent at 20 °C. The adsorption of zinc ions increased with increasing temperature, indicating the endothermic nature of the adsorption process. This study indicated that the HDW can be used as an effective and environmentally safe biosorbent for the treatment of zinc-containing aqueous solutions.

An adsorbent was developed from mature leaves and stem bark of the neem (*Azadirachta indica*) tree for removing zinc from water by Arshad et al.[55] Adsorption was carried out in a batch process with several different concentrations of zinc by varying the pH. The uptake of the metal was very fast initially but gradually slowed down, indicating penetration into the interior of the adsorbent particles. The data showed that the optimum pH for efficient biosorption of zinc by neem leaves and stem bark was 4 and 5, respectively. The maximum adsorption capacity showed that the neem biomass had a mass capacity for zinc (147.08 mg Zn g^{-1} for neem leaves and 137.67 mg Zn g^{-1} for neem bark). The experimental results were analyzed in terms of Langmuir and Freundlich isotherms. The adsorption followed a pseudo-second-order kinetic model. The thermodynamic assessment of the metal ion–neem tree biomass system indicated the feasibility and spontaneous nature of the process and ΔG° values were evaluated as ranging from –26.84 to –32.75 kJ mol^{-1} (neem leaves) and –26.04 to –29.50 kJ mol^{-1} (neem bark) for zinc biosorption. Owing to its outstanding zinc uptake capacity, the neem tree was proved to be an excellent biomaterial for accumulating zinc from aqueous solutions.

The removal of zinc ions from aqueous solutions on the biomass of *A. indica* bark has been studied using a batch adsorption technique by King et al.[56] The equilibrium metal uptake was increased and percentage biosorption was decreased with an increase in the initial concentration and particle size of the biosorbent. The maximum zinc biosorption occurred at pH 6. Biosorption isothermal data were well interpreted by the Langmuir model, with a maximum biosorption capacity of 33.49 mg g^{-1} of zinc ions on *A. indica* bark biomass, and kinetic data were properly fitted with the pseudo-second-order kinetic model.

The removal of Ni(II) from aqueous solutions by using different *A. indica* biomass, namely neem leaves fresh (NLF), neem leaves activated (NLA), leaves ash (LA), neem bark fresh (NBF) and neem bark activated (NBA), was investigated by Alam et al.[57] Neem leaves and neem bark were activated by heat treatment and with the use of concentrated sulfuric acid. The adsorption efficiencies were found to be pH dependent, which increased by increasing the pH of the solution in the range from 2 to 7. The equilibrium time was attained after 2 h and the maximum removal was achieved at an adsorbent loading weight of 2 g. The maximum biosorption capacity for Ni(II) was 95% at pH 7. The equilibrium adsorption data were interpreted using both Langmuir and Freundlich isotherms.

The effect of adsorbent dose, pH and agitation speed on nickel removal from aqueous media using an agricultural waste biomass, sugarcane bagasse, has been investigated by Garg et al.[58] Batch mode experiments were carried out to assess the adsorption equilibrium. The influence of three parameters on the removal of nickel was also examined using a response surface methodological approach. The central composite face-centered experimental design in response surface methodology (RSM) was used for designing the experiments as well as for full response surface estimation. A study by Abdulsalam and Prasad[59] showed that tamarind bark powder can be efficiently used as a low-cost alternative for the removal of divalent nickel from the rinsing wastewater of a plating factory. Biosorption was pH dependent and the maximum removal of Ni(II) was obtained at pH 7.0. The Langmuir isotherm model was found to be better applicable to the equilibrium data. The maximum biosorption capacity was 15.34 mg g^{-1}. From the DR isotherm model the free energy was calculated as 15.81 kJ mol^{-1}, indicating that the biosorption of nickel occurred by chemisorption. Kinetic evaluation of the experimental data showed that the biosorption process followed pseudo-second-order kinetics.

A series of batch adsorption experiments was carried out by Mosavi *et al.*[60] to determine the biosorption capacities of alfalfa and datura (dried biomass) for Pb and Ni removal from contaminated solutions. The data showed that these adsorbents could remove about 94–99% Pb and 30–50% Ni that existed in polluted solutions and the binding speed to adsorbent was high and occurred within 5 min. However, datura has a greater affinity and capacity for metal binding than that for alfalfa. Raw agricultural waste is an affordable adsorbent for the removal of industrial contaminants.

A new biosorbent produced from castor leaves powder (*Ricinus communis* L.) was used to remove mercury(II) from aqueous solutions.[61] The maximum capacity (Q_{max}) of biomass was found to be 37.2 mg Hg(II) g^{-1} at pH 5.5. Biosorption equilibrium was established in approximately 1 h. The equilibrium data were described well by Langmuir and Freundlich models. The adsorbed mercury on biomass was desorbed using 10 mL of 4 M HCl solution. The biomass could be reused for other biosorption assays.

Krishnani and Ayyappan[62] in their review have described the usefulness of lignocellulosic agro-wastes as effective adsorbents. They have high adsorption properties, due to their ion-exchange capabilities. These properties render agricultural wastes as good sorbents for the removal of many metals, which add to their value, help reduce the cost of waste disposal, and provide a potentially cheap alternative to existing commercial carbons. Although the abundance and very low cost of lignocellulosic wastes from agricultural operations are real advantages that render them suitable alternatives for the remediation of heavy metals, further successful studies on these materials are essential to demonstrate the efficacy of this technology.

The biosorption capacity of a plant biomass, the root bark of the Indian sarsaparilla, was studied by Chandrasekhar *et al.*[63] for many toxic metals and the results revealed higher Pb removal followed by Cr and Zn among the 11 metals studied (As, Se, Zn, Fe, Ni, Co, Pb, Mn, Hg, Cr and Cu). Indian sarsaparilla (*Hemidesmus indicus*) has a wide distribution and is taxonomically classified as a member of the family Asclepiadacea.[64] The biomass was oven dried at 100 °C before use. The results of shake flask experiments revealed enhanced metal removal with 15 min agitation for Pb and 180 min for Zn and Cr removal. Metal removal was higher at lower pH for Cr and Zn and an increased pH decreased the percentage metal removal. Lead removal was unaffected by pH changes. The presence of co-ions (As, Se, Hg, *etc.*) did not affect Pb removal by biomass, but, on the other hand, Zn and Cr uptakes decreased. For the reuse of biomass, the used biomass was subjected to desorption studies using HNO$_3$. The retention capacity of the biomass was almost constant after three cycles of chelation–desorption, suggesting that the lifetime cycle was sufficiently long for continuous industrial application. The biosorption model developed was applied to a real life system successfully.

Schneegurt *et al.*[65] have investigated the metal-binding qualities of two biomass byproducts that are commercially available in quantity and at low cost, namely spillage, a dried yeast and plant mixture from the production of ethanol from corn, and ground corn cobs used in animal feeds. The biomass materials effectively removed toxic metals, such as Cu, Cs, Mo, Ni, Pb and Zn, even in the presence of competing metals likely to be found in sulfide mine tailing ponds. The effectiveness of these biosorbents was demonstrated using samples from the Berkeley Pit in Montana. The results of their experiments demonstrated that the biosorption of metals from wastewaters using biomass byproducts is a viable and cost-effective technology that should be included in process evaluations.

This chapter has given an overview of the recent developments in the use of plant-derived biomass for metal remediation from wastewater. The literature complied here have confirmed the effectiveness of biomass in removal of toxic metals and the advantages of these materials over conventional treatment agents. The type of biomass used, the source and the type of metals that were removed using these materials are summarized in Table 3.1.

Table 3.1 Summary of the plant source and their parts used as biomass for removal of various metals from contaminated water.

Name of the plant source	Biomass material	Type of metal removed	Ref.
Acacia nilotica	Leaves	Cd, Pb	12
Pinus radiata	Bark	Al, Fe, Mn	13
Grapefruit	Peel	U	14
Agriculture and forest	Waste	Pb	15
Maize	Tassels	Cu, Cd, Pb	16
Cana indica	Flower	Cr	17
Portulaca olecera	Flower and stem	Cr	17
Hibiscus rosa-sinensis	Flower	Cr	17
Trapa natans	Fruit skin	Cr	17
Hibiscus rosa-sinensis	Flower	As	18
Tagetes erecta	Flower	As	18
Canna indica	Flower	As	18
Hibiscus rosa-sinensis	Flower	Zn	19
Hibiscus rosa-sinenisis	Flower	Cd, Pb	20
Palm tree	Trunks, fronds, leaves, empty fruit bunches, shells	Heavy metals	21
Several agro-based biomass	–	Zn	22
Polyalthia longifolia	Leaves	Cr	23
Amaranthus hybridus (African spinach)	Stalk	Mn, Pb	24
Carica papaya (pawpaw)	Seed	Mn, Pb	24

Table 3.1 (*Continued*)

Name of the plant source	Biomass material	Type of metal removed	Ref.
Carpobrotus edulis	Dried plants	Cd, Cu, Pb, Zn	25
Withania frutescens	Dried plants	Cd, Cu, Pb, Zn	25
Launea arborescens	Dried plants	Cd, Cu, Pb, Zn	25
Agave sisalana	Sisal fibers	Pb, Cd	26
Chlorophyte algae with caducipholic plants	–	Pb, Cd, Cu	27
Rice	Husk	Ag	28
Iva xanthifolia Nutt	–	Cd, Pb	29
Coffee	Husk	Cr	30
Alfalafa and *Datura*	Dried biomass	Pb, Ni	31
Pistachio	Hull powder	Cr	32
Azadirachta indica	Leaves	Ni	33
Areca catechu	Heartwood	Cd	34
Areca catechu	Heartwood	Pb	35
Banana	Trunk fibers	Cd, Cu, Fe, Zn	36
Chitosan, cellulose and other plant-derived biomass	–	As	37
Corn	Cobs	Pb	38
Moringa olifera	Bark	Pb	39
Sugar beet	Pectin xerogels	Cd, Pb, Cu	40
Orange	Waste biomass	Cr	41
Rice and pine	Rice bran and pine sawdust	Pb, Cd	42
Plants	Low value products	Cd, As, Pb	43
Peanut	Hull	Cu	44
Corn	Stalk Powder	Cr	45
Sunflower	Waste	Cr	46
Momordica charantia	–	As	47
Saraca indica	Leaf powder	Pb	48
Azadirachta indica	Mature leaves and stem bark	Zn	49,51
Azadirachta indica	Mature leaves	Pb	50
Tamarind	Bark	Ni	52
Banana	Peel	Pb, Cu, Zn, Ni	53
Orange	Mesocarp	Mg, Zn, Cu, Pb	54
Corn	Grainless stalk	Cr	55
Sugarcane	Bagasse	Cd	56
Maize	Corncob	Cd	
Jathropa	–	Cd	
Castor	Leaves powder	Hg	57
Sugarcane	Bagasse	Ni	58
Lignocellulosic	Waste	Heavy metals	59
Hemidesmus	Bark	Pb	60,
Hemidesmus	Bark	Pb, Zn, Cr	61
Corn	–	Cu, Cs, Mo, Ni, Pb, Zn	63
Beetroot	Fibers	Pb, Cu, Zn, Fe	64
Datura innoxia	–	Pb	65

3.3 Conclusions

A plethora of work has been done on treatment of water and wastewater using almost all kinds of plant biomass at a laboratory scale. Some of the researchers have carried out field-scale research also, but the data are still insufficient. To take significant advantage of these plant-derived biomasses as alternative, economical and eco-friendly materials for metal removal from contaminated water, large-scale experiments need to be performed.

References

1. A. A. Mohammed and A. G. Devi Prasad, *Aust. J. Basic Appl. Sci.*, 2010, **4**, 3591.
2. G. Rich and K. Cherry, *Hazardous Wastes Treatment Technologies*, Pudvan, Northbrook, IL, 1987, pp. 201–210.
3. D. Kratochvil and B. Velosky, *Water Res.*, 1998, **32**, 2760.
4. E. Fourest and C. J. Roux, *Appl. Microbiol. Biotechnol.*, 1992, **37**, 399.
5. F. Veglio and A. G. Beolchini, *Process Biochem.*, 1997, **32**, 99.
6. B. Volesky, *Biosorption of Heavy Metals*, CRC, Boca Raton, FL, 1990, p. 3.
7. B. Volesky, *Biosorption of Heavy Metals*, CRC, Boca Raton, FL, 1990, p. 7.
8. A. Kapoor and T. Viraraghavan, *Bioresour. Technol.*, 1995, **53**, 195.
9. G. M. Gadd, *Experientia*, 1990, **46**, 834.
10. J. R. Deans and B. G. Dixon, *Water Res.*, 1992, **26**, 469.
11. J. Kumar, C. Balomajumder and P. Mondal, *Clean: Soil, Air, Water*, 2011, **39**, 641.
12. S. K. R. Yadanaparthi, D. Graybill and R. V. Wandruszka, *J. Hazard. Mater.*, 2009, **171**, 1.
13. S. Lin and G. D. Rayson, *Environ. Sci. Technol.*, 1998, **32**, 1488.
14. P. Goyal, P. Sharma, S. Srivastava and M. M. Srivastava, *Int. J. Environ. Sci. Technol.*, 2008, **5**, 27.
15. K. G. Bhattacharyya and A. Sharma, *J. Hazard. Mater.*, 2004, **113**, 97.
16. J. Rima, A. Ghauch, M. Ghaouch and M. Martin-Bouyer, *Toxicol. Environ. Chem.*, 2000, **75**, 89.
17. Chandrasekhar, C. T. Kamala, N. S. Chary, A. R. K. Sastry, T. Nageswara Rao and M. Vairamani, *J. Hazard. Mater.*, 2004, **108**, 111.
18. C. Saka, O. Şahin and M. M. Küçük, *Int. J. Environ. Sci. Technol.*, 2012, **9**, 379.
19. I. W. Mwangi, C. J. Ngila and J. O. Okonkwo, *Toxicol. Environ. Chem.*, 2012, **94**, 20.
20. P. S. Vankar, R. Saraswat and D. S. Malik, *Environ. Prog. Sustainable Energy*, 2010, **29**, 421.
21. U. Garg, M. P. Kaur, G. K. Jawa, D. Sud and V. K. Garg, *J. Hazard. Mater.*, 2008, **154**, 1149.
22. S. Waseem, M. I. Din, S. Nasir and A. Rasool, *Arab. J. Chem.*, 2012, in press; doi: org/10.1016/j.arabjc.2012.03.020.

23. T. Ahmad, M. Rafatullah, A. Ghazali, O. Sulaiman and R. Hashim, *J. Environ. Sci. Health, Part C*, 2011, **29**, 177.
24. J. Kumar, C. Balomajumder and P. Mondal, *Clean: Soil, Air, Water*, 2011, **39**, 641.
25. J. Anwar, U. Shafique, Z. Waheed-uz, Z. un Nisa, M. A. Munawar, N. Jamil, M. Salman, A. Dar, R. Rehman, J. Saif, H. Gul and T. Iqbal, *Int. J. Phytorem.*, 2011, **13**, 410.
26. J. N. Egila, B. E. N. Dauda, Y. A. Iyaka and T. Jimoh, *Int. J. Phys. Sci.*, 2011, **6**, 2152.
27. M. Chiban, A. Soudani, F. Sinan, S. Tahrouch and M. Persin, *Clean: Soil, Air, Water*, 2011, **39**, 376.
28. W. N. L. dos Santos, D. D. Cavalcante, E. G. P. da Silva, C. F. das Virgens and F. D. S. Dias, *Microchem. J.*, 2011, **97**, 269.
29. J. M. Lezcano, F. González, A. Ballester, M. L. Blázquez, J. A. Muñoz and C. García-Balboa, *J. Environ. Manage.*, 2011, **92**, 2666.
30. Y. Zu, Z. Ren, X. Zhao, P. Xiao and Y. Zhang, *Adv. Mater. Res.*, 2010, **113**, 1464.
31. P. Chakravarty, N. S. Sarma and H. P. Sarma, *Chem. Eng. J.*, 2010, **162**, 949.
32. P. Chakravarty, N. S. Sarma and H. P. Sarma, *Desalination*, 2010, **256**, 16.
33. K. Sathasivam and M. R. H. M. Haris, *J. Chilean Chem. Soc.*, 2010, **55**, 278.
34. G. Tan, H. Yuan, Y. Liu and D. Xiao, *J. Hazard. Mater.*, 2010, **174**, 740.
35. D. H. K. Reddy, K. Seshaiah, A. V. R. Reddy, M. M. Rao and M. C. Wang, *J. Hazard. Mater.*, 2010, **174**, 831.
36. Y. N. Mata, M. L. Blázquez, A. Ballester, F. González and J. A. Muñoz, *J. Hazard. Mater.*, 2010, **178**, 243.
37. Y. N. Chen, J. X. Nie, M. Chen and D. C. Zhang, in *Proceedings of the International Conference on Management and Service Science (MASS 2009)*, *Wuhan/Beijing, China, IEEE*, Piscataway, NJ, 2009.
38. A. M. Aqeel, W. Abdul, M. Karamat, M. Mohd, Jamil and Y. Ismail, *Sci. Res. Essays*, 2011, **6**, 4055.
39. R. E. Ogali, O. Akaranta and V. O. Aririguzo, *Afr. J. Biotechnol.*, 2008, **7**, 3073.
40. C. S. Zhu, L. P. Wang and W. Chen, *J. Hazard. Mater.*, 2009, **168**, 739.
41. E. Aspé, M. Roeckel and K. Fernández, *Aquacultural Eng.*, 2012, **49**, 1.
42. X. Q. Luo, X. Y. Lin, X. G. Luo, A. K. Luo and X. Liang, *Mater. Sci. Forum*, 2010, **658**, 45.
43. W. H. Zou, L. Zhao and L. Zhu, *J. Radioanal. Nucl. Chem.*, 2012, **292**, 1303.
44. F. Ma, R. Qu, C. Sun, Y. Zhang, C. Ji, X. Li and X. Ruan, *Ion Exch. Adsorpt.*, 2010, **26**, 187.
45. S. Nigam, P. S. Vankar and K. Gopal, *Environ. Sci. Pollut. Res.*, 2013, **20**, 1161.
46. P. K. Pandey, S. Choubey, Y. Verma, M. Pandey and K. Chandrashekhar, *Bioresour. Technol.*, 2009, **100**, 634.

47. P. S. Vankar, R. Sarswat, A. K. Dwivedi and R. S. Sahu, *J. Clean. Prod.*, doi: org/10.1016/j.jclepro.2011.09.021
48. N. Ahalya, R. D. Kanamadi and T. V. Ramachandra, *Int. J. Environ. Pollut.*, 2010, **43**, 106.
49. G. Moussavi and B. Barikbin, *Chem. Eng. J.*, 2010, **162**, 893.
50. A. B. S. P. Marín, M. I. Aguilar, V. F. Meseguer, J. F. Ortuño, J. Sáez and M. Lloréns, *Chem. Eng. J.*, 2009, **155**, 199.
51. S. Bellú, S. García, J. C. González, A. M. Atria, L. F. Sala and S. Signorella, *Sep. Sci. Technol.*, 2008, **43**, 3200.
52. R. Li, Z. Zhang, Z. Meng and H. Li, *Acta Sci. Circumstantiae*, 2009, **29**, 1434.
53. M. Jain, V. K. Garg and K. Kadirvelu, *J. Hazard. Mater.*, 2009, **162**, 365.
54. P. S. Vankar, R. Saraswat and R. Sahu, *Environ. Prog. Sustainable Energy*, 2012, **31**, 89.
55. M. Arshad, M. N. Zafar, S. Younis and R. Nadeem, *J. Hazard. Mater.*, 2008, **157**, 534.
56. P. King, K. Anuradha, S. B. Lahari, Y. Prasanna Kumar and V. S. Prasad, *J. Hazard. Mater.*, 2008, **152**, 324.
57. M. Alam, S. Rais and M. Aslam, *Desalin. Water Treat.*, 2010, **21**, 220.
58. U. K. Garg, M. P. Kaur, V. K. Garg and D. Sud, *Bioresour. Technol.*, 2008, **99**, 1325.
59. A. M. Abdulsalam and A. G. Devi Prasad, *Aust. J. Basic Appl. Sci.*, 2010, **4**, 3591.
60. S. B. Mosavi, S. Karimi and V. Feiziasl, *Asian J. Chem.*, 2010, **22**, 1700.
61. S. W. Al Rmalli, A. A. Dahmani, M. M. Abuein and A. A. Gleza, *J. Hazard. Mater.*, 2008, **152**, 955.
62. K. K. Krishnani and S. Ayyappan, *Rev. Environ. Contam. Toxicol.*, 2006, **188**, 59.
63. K. Chandrasekhar, C. T. Kamala, N. S. Chary and Y. Anjaneyulu, *Int. J. Miner. Process.*, 2003, **68**, 37.
64. P. K. Warrier, *Indian Medicinal Plants, A Compendium of 500 Species*, Orient Longman, Madras, 1995, **vol. 3**, pp.141–145.
65. M. A. Schneegurt, J. C. Jain, J. A. Menicucci, S. A. Brown, K. M. Kemner, D. F. Garofalo, M. R. Quallick, C. R. Neal and C. F. Kulpa, *Environ. Sci. Technol.*, 2001, **35**, 3786.

CHAPTER 4

Natural Polysaccharides as Treatment Agents for Wastewater

RAJANI SRINIVASAN

Department of Chemistry, Geosciences and Physics, College of Science and Technology, Tarleton State University, Stephenville, TX 76401, USA
Email: srinivasan@tarleton.edu

4.1 Introduction

Environmental pollution due to developments in technology is one of the most significant problems of this century. "Pollution" is derived from the Latin word *Pollutionem*, meaning defilement. Pollution can be defined as an undesirable change in physical, chemical and biological characteristics of air, soil and water. In the present times, water has been the most exploited natural resource. The steady increase in water pollution is due to rapid population growth, industrial proliferation, urbanization, increasing living standards and wide spheres of human activities. The time has come when pure and clean water in densely populated industrial areas has become inadequate even for maintaining normal living standards. Almost all the sources of water supply around the world are highly polluted due to a heavy flux of sewage, industrial effluents and domestic and agricultural wastes discharged into them, which consist of substances varying from simple nutrients to highly toxic hazardous chemicals. The deterioration of aesthetic and life supporting qualities of rivers and lakes

RSC Green Chemistry No. 23
Green Materials for Sustainable Water Remediation and Treatment
Edited by Anuradha Mishra and James H. Clark
© The Royal Society of Chemistry 2013
Published by the Royal Society of Chemistry, www.rsc.org

are affecting the various flora and fauna and we human beings are also suffering from various problems.

The aqueous effluents from chemical and related industries contain organic pollutants such as phenol, which is toxic and causes considerable damage and threats to the ecosystem in water bodies and human health. Heavy metals are discharged in considerable quantities into the environment through numerous industrial activities. Heavy metals such as Cr, Cu, Pb, Ni, *etc.*, in wastewater are hazardous to the environment and health. The specific problem associated with heavy metals in the environment is their accumulation in the food chain and their persistence therein. Water pollution due to toxic metals and organic compounds remains a serious environmental and public problem. Moreover, faced with more and more stringent regulations, water pollution has also become a major source of concern and a priority for most industrial sectors. Heavy metal ions, aromatic compounds (including phenolic derivatives and polycyclic aromatic compounds) and dyes are often found in the environment as a result of their wide industrial uses. They are common contaminants in wastewater and many of them are known to be toxic or carcinogenic. For example, chromium(VI) is found to be toxic to bacteria, plants, animals and people.[1]

Mercury and cadmium are known as two of the most toxic metals that are very damaging to the environment.[2] In addition, heavy metals are not biodegradable and tend to accumulate in living organisms, causing various diseases and disorders. Therefore, their presence in the environment, in particular in water, should be controlled. Chlorinated phenols are also considered as priority pollutants since they are harmful to organisms even at low concentrations. They have been classified as hazardous pollutants because of their harmful potential to human health.[3,4] TNT (2,4,6-trinitrotoluene) is a nitroaromatic molecule that has been widely used by the weapons industry for the production of explosives. This compound is recalcitrant, toxic and mutagenic to various organisms.[5,6] Many synthetic dyes, which are extensively used for textile dyeing, paper printing and as additives in petroleum products, are recalcitrant organic molecules that strongly color wastewater. Strict legislation on the discharge of these toxic products makes it necessary to develop various efficient technologies for the removal of pollutants from wastewater. Different technologies and processes are currently used. Biological treatments,[7–9] membrane processes,[10–13] advanced oxidation processes,[14–17] chemical and electrochemical techniques[18–20] and adsorption procedures[21–49] are the most widely used for removing metals and organic compounds from industrial effluents. Amongst all the treatments proposed, adsorption using sorbents is one of the most popular methods, since proper design of the adsorption process will produce high-quality treated effluents. Adsorption is a well-known equilibrium separation process. It is now recognized as an effective, efficient and economic method for water decontamination applications and for separation analytical purposes. The adsorbents may be of mineral, organic or biological origin: activated carbons,[21–26] zeolites,[27,28] clays,[29–33] silica beads,[34,35] low-cost adsorbents (industrial byproducts,[36–41] agricultural wastes,[42,43] biomass[44,45]) and polymeric materials (organic polymeric resins,[46,47] macroporous hyper-crosslinked polymers[48,49]) are significant examples. Recently, numerous approaches have been studied for the development of cheaper and more effective adsorbents containing natural polymers. Among these, polysaccharides such as chitin[50–52] and starch,[53–55] and their derivatives (chitosan,[56,57] cyclodextrin[58–60]), deserve particular attention. These biopolymers represent an interesting and attractive alternative as adsorbents because of their particular structure, physicochemical characteristics, chemical stability, high reactivity and excellent selectivity towards aromatic compounds and metals, resulting from the presence of chemical reactive groups (hydroxyl, acetamido or amino functions) in polymer chains. Moreover, it is well known that polysaccharides, which are abundant, renewable and biodegradable resources, have a capacity to associate by physical and chemical interactions with a wide variety of molecules.[61,62] Hence, adsorption/flocculation on polysaccharide derivatives from natural sources can be a low-cost procedure of choice in water decontamination for extraction and separation of compounds, and a useful tool for protecting the environment.[63]

Advanced wastewater treatment is defined as the level of treatment required beyond conventional secondary treatment to remove constituents of concern, including nutrients, toxic compounds and increased amounts of organic material and suspended solids. These industries utilize flocculation as one of the advanced treatments for separation of suspended and dissolved colored and colorless solids from the effluents.

4.2 Flocculation

Flocculation can be described as the agent-induced aggregation of particles suspended in liquid media into larger particles called "flocs". It is a destablization process of a stable colloidal dispersion caused by addition of a chemical known as a flocculant.

All types of industrial wastewater treatment and the dewatering of mining municipal sludges, *etc.*, involve flocculation as an important wastewater cleanup stage for removal of suspended matter. This process has also received considerable study in the literature and several reviews have been published recently.[64–67] The objectives of flocculation are to allow rapid separation of a liquid phase from a suspended solid phase, minimize solids remaining in the liquid phase and maximize the solids content in the solid phase. The next sections are the mechanisms, followed in Section 4.3 by the flocculants for destabilization of aqueous suspensions.

4.2.1 Charge Neutralization Flocculation

Increasing the ionic strength of the suspending medium reduces the thickness of the double layer,[68] thereby reducing the range of interparticle repulsions. Ions of similar charge to the surface are repelled by the surface and counterions are attracted to its vicinity. As the counterion concentration increases, the volume

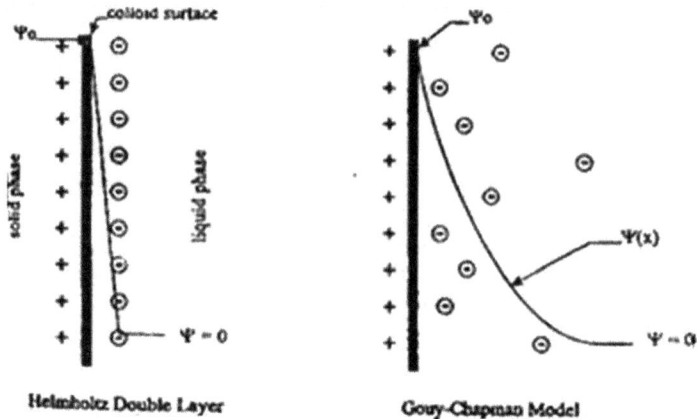

Figure 4.1 Charge distribution details of electrostatic double-layer models.

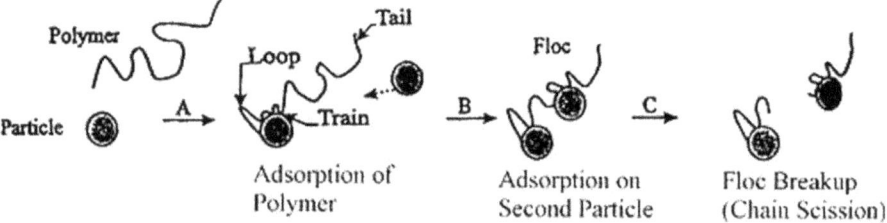

Figure 4.2 Bridging flocculation–deflocculation pathways.

of the diffuse ionic layer necessary for electro-neutrality decreases. This type of flocculation follows the Schulze–Hardy rule (Figure 4.1).[69]

4.2.2 Bridging Flocculation

When very long polymer molecules are absorbed into the surface of particles, they tend to form loops that extend some distance from the surface into the aqueous phase, and the ends may also dangle. These loops and ends may come into contact with and attach to another particle, forming a bridge between the two particles (Figure 4.2).[70]

A high molecular weight cationic polyelectrolyte has been used to flocculate a colloidal dispersion of anionic polystyrene latex particles. The polymer used had a high charge and the flocculation occurred at a solution pH where both the polymer and the particles were fully charged. Under these conditions, flocculation is expected to occur through a bridging mechanism.[71]

The rheological properties of suspensions flocculated by polymer bridging are studied as a function of adsorption and affinity of the polymer for the particles' surfaces. When both the particle and polymer concentration exceed some critical levels, the suspensions elastically respond to small deformations at low frequencies.[72]

4.2.3 Electrostatic Patch Mechanism

Cationic polymers having high charge density along the chain provide greater density of charge in the polymer molecule than is present on the surface of a particle. When this type of polymer molecule is adsorbed on the surface, it not only neutralizes the negative charge within the geometric area where it is attached, but also provides excess cationic charge to compensate for other negatively charged sites on the surface. The entire negative charge on a particle can be neutralized by a relatively small geometric surface coverage by a polymer.[73] Thus the polymer adsorbs onto a particle surface in patches of molecular size, forming positive patches surrounded by regions of negative surface sites. This mosaic of positive and negative surface areas allows direct electrostatic attraction between particles (Figure 4.3).

4.2.4 Sweep Floc Mechanism

Hydrolyzing metal salts are usually applied as flocculants in a concentration/ pH region that is oversaturated with respect to the neutral metal hydroxide. The rate of precipitation of the hydroxide, which is a function of temperature, other ions present and nucleation sites, may be slow unless a certain degree of supersaturation is exceeded. Under proper conditions, fluffy amorphous

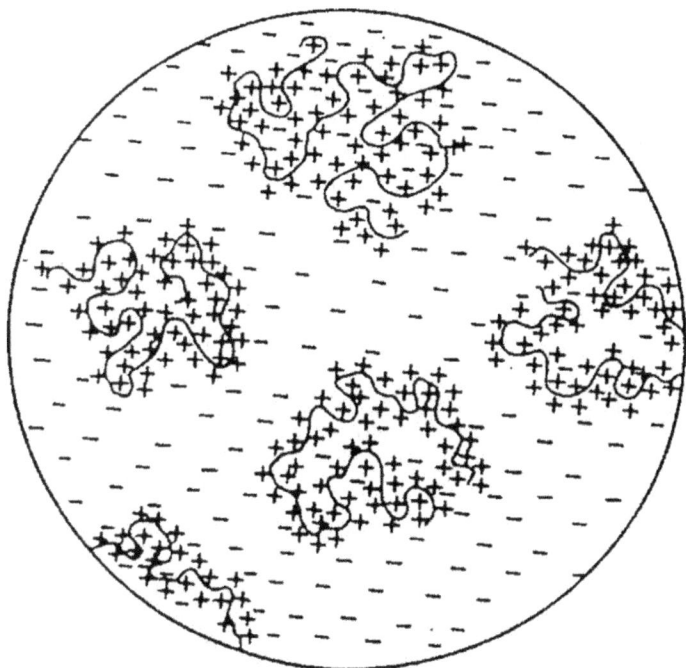

Figure 4.3 Arrangement of adsorbed polycations on a particle with low negative surface charge density.

hydroxides form rapidly, trapping and enmeshing a colloidal particle as it settles and giving rise to the term "sweep flocculation".[74,75]

4.3 Flocculants

The materials and the chemicals used for destabilization of colloidal particles in suspensions are called flocculants. These flocculants can be of two types: (i) inorganic salts; (ii) organic polymeric materials. Inorganic flocculants are more frequently referred to as coagulants and are based upon the hydrolyzable salts of aluminum ions, mainly aluminum sulfate [alum, $Al_2(SO_4)_3$], and ferric chloride ($FeCl_3$). Synthetic polymeric flocculants have replaced inorganic flocculants in most applications, except in water clarification where alum continues to be used. Polymeric flocculants are water-soluble organic macromolecules capable of destabilizing and flocculating suspended mater. Polymeric flocculants are of three types: natural, synthetic and modified natural polymeric flocculants.

Flocculants derived naturally (from plants, animals, bacteria, fungi, *etc.*) are called natural flocculants. The molecular weights depend upon the source and processing methods, but may be as high as several million. Guar gum and its derivatives are examples of natural polymer flocculants. Current flocculants derived from natural products involve starch, starch derivatives, plant gums and mucilages, seaweed extracts, cellulose derivatives, proteins and tannins.[76–83] Synthetic organic flocculants are water-soluble polymeric substances with weight-average molecular weights ranging from about 10^3 to greater than 5×10^6. If some subunits of the polymer molecule are charged, it is termed a polyelectrolyte.[84] Polymeric flocculants, natural or synthetic, are of three types: anionic, cationic and neutral. An anionic charge on a flocculant is obtained by the incorporation of carboxyl groups and sulfonic acid groups, *e.g.* 2-acrylamido-2-methypropanesulfonic acid (AMPS) (Figure 4.4).

A cationic charge is obtained exclusively by the incorporation of amine-type functionalities, usually quaternized amines such as poly[*N*-(dimethylaminopropyl)methacrylamide] (Figure 4.5).

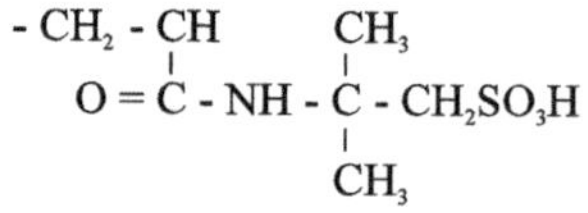

$$- CH_2 - CH \qquad CH_3$$
$$O = C - NH - C - CH_2SO_3H$$
$$CH_3$$

Figure 4.4   Structure of 2-acrylamido-2-methypropanesulfonic acid.

$$CH_3$$
$$(H_2C - C)_n$$
$$C - NHCH_2CH_2CH_2N \langle {CH_3 \atop CH_3}$$
$$O$$

Figure 4.5   Structure of poly[*N*-(dimethylaminopropyl)methacrylamide].

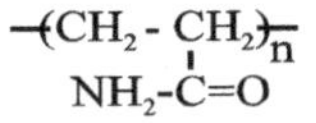

$$-(CH_2 - CH_2)_n$$
$$NH_2-C=O$$

Figure 4.6 Structure of polyacrylamide.

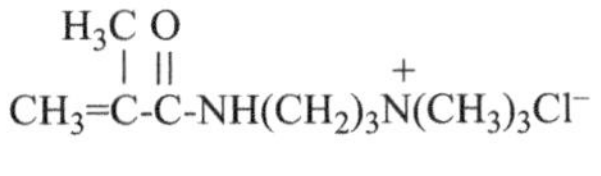

$$H_3C \quad O$$
$$| \quad ||$$
$$CH_3=C-C-NH(CH_2)_3\overset{+}{N}(CH_3)_3Cl^-$$

Figure 4.7 Structure of *N*-(3-methylacrylamidopropyl)-*N*,*N*,*N*-trimethylammonium chloride.

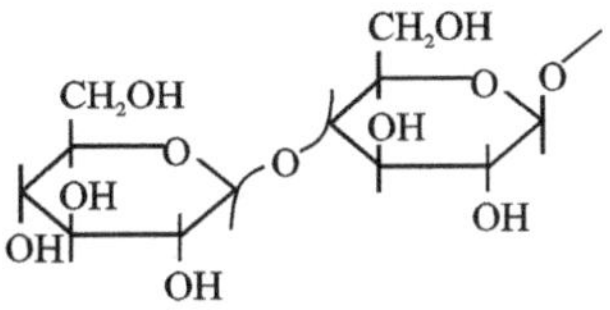

Figure 4.8 Structure indicating a glycosidic linkage.

Flocculants with no charge or with less than 1% monomer charged are considered non-ionic or neutral. Polyacrylamide is most prominent member of this class, followed by poly(ethylene oxide). They usually have a very high molecular weight (Figure 4.6).

Polymers containing both cationic and anionic functional groups in the same polymer molecule are polyampholytes. Zwitterion flocculants containing both cationic and anionic functional groups in the same polymer molecule are examples of polyampholytes, but little is known about them. One of the examples of this type of flocculant is *N*-(3-methylacrylamidopropyl)-*N*,*N*,*N*-trimethylammonium chloride (MAPTAC) (Figure 4.7).

4.3.1 Polysaccharide-Based Flocculants

Polysaccharides are naturally occurring carbohydrate polymers in which monosaccharide residues are linked directly through a glycosidic linkage (Figure 4.8).[85] They are found in the plant, animal and microbial kingdoms.

The generic name for polysaccharides is glycans. Polysaccharides composed of only one kind of monosaccharide residue are denoted by homoglycans, whereas those containing two or more monomer units are referred to as heteroglycans. Depending upon the source, polysaccharides can be grouped under plant, algal, bacterial or mammalian polysaccharides.

4.3.1.1 Plant Polysaccharides

This group includes two major industrial polysaccharides, starch and cellulose, together with hemicelluloses, pectins and plant exudates.

Starch is the principal food reserve polysaccharide in plants and serves as a main source of carbohydrate in the diet of man and animals. It is a mixture of two polysaccharides, amylose and amylopectin. Both are D-glucans wherein sugar residues are joined by α-(1→4) linkages, but the latter also contains some α-(1→6) bonds. Most uses of starch capitalize on the high viscosity of its solution and gelling characteristics. These properties may be altered by physical and chemical modifications.

Cellulose is a β-(1→4)-D-glucan commonly obtained from cotton or wood.[86,87] Associated with cellulose in many plants is a group of structurally similar polysaccharides called hemicelluloses. One component of hemicellulose is D-gluco-D-mannan together with other D-glucose sugars also linked by (1→4)-α-D- bonds; the sequence of the two sugars is random. The composition of a hemicellulose mixture depends on the source. The chemistry of hemicellulose has been extensively reviewed[88] and it is clear that given a certain fundamental structural type, an infinite range of variations may be achieved by altering the nature, frequency and pattern of substitution.

Pectins: pectic substances comprise a group of plant polysaccharides in which D-galacturonic acid is the principal component. The term pectin is used for water-soluble, gel-forming polysaccharides of this group; pectinic acids have a portion of uronic acid groups, esterified usually as methyl esters, and pectic acids are acidic materials devoid of ester groups.[89,90] Reference is often made to the pectic triad of D-galacturan, D-galactan and L-arabinan, and although a homopolymer of each type has been isolated, they are most commonly encountered as heteropolymers; certainly, the D-galactomannans usually encompass 10–25% of neutral sugars in their structures. Galactans are mainly linear chains of (1→4)-linked D-galactopyranosyl units, whereas arabinans are principally chains of (1→5)-linked L-arabinofuranosyl residues with substituents (at C-3) of additional units of the same sugar or of D-galactopyranose.

Pectin can also be categorized as one of the anionic polysaccharides in the cell wall that consists of smooth α-D-galacturonic acid with 1→4 linkages in the monomer region.[91] Pectin can be extracted from the outer cell layer of oranges, apples, grapes, peaches, *etc*. Pectin is extracted on an industrial scale from plant (waste) materials, *e.g.* apple pomace and citrus peel, and used by the food industry because of its ability to form gels under certain circumstances and to increase the viscosity of drinks. It is also widely applied as stabilizers in acidic milk products, and also has other pharmaceutical uses.[92]

Plant exudates: a gum is any material that swells or dissolves in water and exhibits gelling characteristics, acts as an emulsifier or possesses adhesive properties.[92–95] This wide-ranging term covers not only plant gums but also materials such as alginates, cellulose derivatives and modified starches. It is convenient to distinguish between plant exudates and gums obtained from seeds or bark since the chemical structures of the two are different. Historically, the term mucilage was associated with seeds and gums. Mucilages were considered chemically different from gums, with the former containing D-galacturonic acid and the latter containing D-glucuronic acid. Now that some gums and mucilages have been found to contain both acids, this distinction is chemically untenable, but the custom of referring to seed mucilages rather than seed gums has continued. Gums are used in industry because their aqueous solutions or dispersions possess suspending and stabilizing properties. Owing to their gel-forming capabilities, gums act as emulsifiers, flocculants, adhesives, *etc.*

From the point of view of chemical structure, plant exudates may be divided into three main groups and one minor group:

Group A has the main chain 3)-β-D-Gal-(1→ with side chains consisting of -D-GlcA-(1→6)-β-D-Gal-(1→ with some -L-Ara*f* and/or -L-Ara*p* (*e.g.* gum arabic).

Group B has a main chain of 4)-β-D-GlcA-(1→2)-D-Man-(1→ with side chains containing L-Ara*p* and/or D-Gal and/or D-GlcA (*e.g.* gum ghatti).[93]

Group C has the main chain 4)-α-D-GalA-(1→ or 4)-β-D-GalA-(1→2)-L-Rha-(1→ with side chains of D-glucuronic acid.

4.3.1.2 Algal Polysaccharides

Seaweeds represent a source of many different polysaccharides, several of which are of commercial importance, *e.g.* alginic acid (a polyuronide) and agar and carrageenan (sulfated galactans). There are three main families of seaweeds: Phaeophycae (brown), Rhodophycae (red) and Chlorophycea (green).[96] Alginates, because of their water holding, gelling and emulsifying properties, are also used in paper coating, explosive manufacture and latex emulsions.

4.3.1.3 Bacterial Polysaccharides

The bacterial kingdom is a source of many types of polysaccharides, several of which have attained commercial importance, and the majority are of medical interest because of their behavior as antigens.[96–101]

4.3.1.4 Mammalian Polysaccharides

Two homopolymers are found in animals, glycogen and chitin, but the majority of mammalian polysaccharides are heteroglycans (glycosamino-glycans). Glycogen has both α-(1→4) and α-(1→6) linkages.[101] The structure is similar to amylopectin. The greater degree of branching gives glycogen a higher degree of solubility in cold water than amylopectin. Chitin, a homopolymer, has units of 2-acetamido-2-deoxy-D-glucose (*N*-acetylglucosamine) 1→4 linked in the β-configuration; it is therefore an amino sugar analog of cellulose.[102]

The following gives an extensive review of natural polysaccharides derived from plants, animals and microorganisms as flocculants/adsorbents in the removal of contaminants from polluted water.

4.4 Plant-Derived Polysaccharides

Plant polysaccharides such as starches and pectin, guar gum, xanthan gum, alginates and various mucilages find extensive application as flocculants. Recently, food-grade polysaccharides, okra, fenugreek, tamarind, psyllium, *etc.*, have been used for treatment of wastewater. Polysaccharides, by virtue of being biodegradable, non-toxic, shear stable and easily available, are becoming popular in domestic and industrial effluent treatment. Starch has been used itself as a flocculant; however, its flocculation efficiency is low. Starch is usually modified in order to obtain products with good flocculation efficiency.

Ho *et al.*[103] compared the efficiency of a plant-derived pectin flocculant with the organic synthetic polyacrylamide (PAM) flocculant which is known to have high reduction in turbidity treatment. The objectives were (i) to determine the characteristics of both flocculants and (ii) to optimize the treatment processes of both flocculants in synthetic turbid wastewater. The treatment efficiency for PAM on flocculating activity was found to be equally excellent as the treatment efficiency using the biopolymeric flocculant. The results showed that when using the biopolymeric flocculant the optimum pH was 3, the cation concentration was 0.55 mM and the flocculant concentration was 3 mg L^{-1}, while when using PAM as the flocculant the process gave maximum response at a pH around 4, a cation concentration greater than 0.05 mM and a PAM concentration between 13 and 30 mg L^{-1}. The main concern for application in industry is to use low flocculant concentrations to achieve the maximum results. Thus, the use of a biopolymeric flocculant could achieve this goal as it is proven to be effective at a low concentration of 3 mg L^{-1}.

Mata *et al.*[104] reported the feasibility of using sugar-beet pectin gels for the removal of heavy metals from aqueous solutions. Sugar-beet pectin hydro- and xerogels were tested in the batch biosorption and desorption of cadmium, lead and copper. Pectins were successfully extracted and demethylated from the sugar-beet pulp, an agricultural residue, and gelled in the presence of CaCl$_2$. The stability of the hydro- and xerogel pectin beads made them suitable for biosorption of heavy metals under different conditions. Biosorption data were fitted to the pseudo-second-order kinetic model and the Langmuir isotherm model, obtaining the corresponding parameters. The main binding mechanisms involved were ion exchange with calcium of a gel structure and chelation or complexation with carboxyl groups. After biosorption, calcium in the gels was substituted by metal cations, reorganizing the structure of the gel matrix in a way that was visible using scanning electron microscopy. The use of 0.1 M HNO$_3$ was the best eluant for the reutilization of the gels and recovery of all the adsorbed metal, unlike HCl and H$_2$SO$_4$. The kinetic metal uptakes and rates of biosorption with pectin gels followed the order Cu > Pb > Cd, and the order of the Langmuir constant was Pb > Cu > Cd. Lead formed the most stable bonds with the pectin binding sites due to its higher affinity for both hydro- and xerogels. Compared to alginate, sugar-beet pectin gels presented higher metal affinities and could be an interesting alternative to the former biosorbent.

Pectins are polysaccharides of the middle lamella and primary cell wall in which they are crosslinked with cellulose and hemicellulose fibers. The structure of pectin is complex and can vary, depending on the source and the extraction method. It is a polysaccharide composed of galacturonic acid units with α-(1 → 4) bonds, which constitute the "smooth regions". In the "hairy regions", the rhamnose units in the carbon skeleton are branched with secondary chains, mainly arabinans, normally lost during extraction. Other residues are methanol, acetic acid, phenol and amides. Ferulic acid is a characteristic group of sugar-beet pectins. Therefore, the main functional groups of pectin are hydroxyl, carboxyl, amide and methoxyl. These functional groups have been traditionally associated with heavy metal binding, especially carboxyl groups with great biosorption and heavy metal removal potential.[105] Additionally, pectins in granulated form such as gel beads can be used for continuous applications, *e.g.* in fixed-bed columns. This makes sugar-beet pectins interesting alternatives to similar polysaccharides such as alginate that are already widely used and accepted. Most of the pectins present in sugar-beet pulp are high in methoxyl groups and have more than 50% of methoxylated residues. They gel at low pH values and in the presence of a high concentration of soluble solids. The resulting gels dissolve quickly in water and have a soft consistency and therefore have no application in the biosorption of metals or in the immobilization of biomass. Low-methoxyl pectins have less than 50% of methoxylated residues and can be obtained from high-methoxyl pectins by demethylation. These pectins are sensitive to gelation with divalent cations such as calcium according to the "egg-box model" proposed by Rees, but methoxyl groups are an impediment for the formation of the calcium bridges.[106] Their gels are stable in aqueous solutions and can be used in similar applications to those of alginate, including biomass immobilization and heavy and precious metal biosorption, among others.[107]

The use of new food-grade polysaccharides (mucilages) obtained from *Hibiscus esculentus* and *Trigonella foenum graceum*, commonly called okra and fenugreek, respectively, as flocculants was described by Srinivasan and Mishra.[108] These polysaccharides were used for removal of solids [suspended solids (SS) and total dissolved solids (TDS)] and dyes from real textile effluents and aqueous solutions of different classes of synthetic dyes. Okra gum obtained from the seedpods of *H. esculentus* is an anionic polysaccharide. It was used as a flocculant for removal of solid wastes from tannery effluent in a study by Agarwal *et al.*[109] The jar test method has been used for flocculation studies. Influences of varying the okra mucilage concentration, contact time and pH on removal of the pollutant from the textile wastewater were also investigated.[108] It was found that okra gum acts as a very effective flocculant, capable of removing more than 95% of SS and 69% TDS from the effluent. The flocculation efficiency was almost independent of pH variation. The mechanism of the flocculation was described as bridging. X-ray diffraction (XRD) patterns of solid waste material obtained before and after treatment with polysaccharides were used as supportive evidence to explain the mechanism of flocculation. However, in this experiment the extent of changes observed in the patterns suggests that, apart from secondary bonding between flocculant and solid waste, there may also be involvement of primary bonding like chelation[110] between waste crystalline matter and the polysaccharides.

The flocculant behavior of *Malva sylvestris* (mallow) and *H. esculentus* (okra) mucilages was assessed for the removal of turbidity from synthetic and bio-logically treated effluent by Anastasakis *et al.*[111] The mucilage was extracted from seed pods and lobes of mallow and seed pods of okra. Both are water soluble polysaccharides. A series of flocculation experiments was conducted to assess the optimal concentration of each species. The results showed that mallow and okra mucilages have significant flocculation properties. It was determined that okra was as efficient as mallow in removing turbidity, but at much lower doses. However, at higher dosages the dissolved organic carbon of both synthetic wastewater and effluent increased, probably due to the organic substances present in the okra and mallow mucilages. The flocculation efficiencies of okra and mallow when compared to commercial flocculants, such as Purifloc C-31,[112] Hercofloc and Nalco 610,[113] showed better flocculation behavior. In the case of the biologically treated effluent, okra and mallow mucilages achieved slightly higher turbidity removals (61–74% compared to 50–62% for the commercial flocculants) at doses similar or smaller. They further reported that all the studies took place under laboratory scale conditions. The development of extraction processes, the characterization of the natural active ingredients, the application of pilot and full-scale systems, as well as cost analysis, are necessary steps for the eventual application of natural flocculants.

Fenugreek mucilage, a food-grade natural polysaccharide, has been reported as a flocculant for tannery effluent treatment by Mishra *et al.*[114,115] The flocculation efficiency of the mucilage has been tested under various conditions by the standard jar test method. The effects of polysaccharide concentration, contact time and pH on percent solid removal were studied. The maximum SS and TDS were nearly 85% and 40%, respectively, using a flocculant dose of 0.08 mg L^{-1}. The suitable pH range was neutral for maximum efficiency of the mucilage as a flocculant. The time required for maximum TDS removal was 3 h, whereas it was only 1 h for SS removal. Wide-angle X-ray diffraction patterns of powder samples of the solids, before and after treatment, have also been reported and suggest interaction of the tannery waste with the mucilage. The flocculation efficiency of this mucilage was found to be at a par with that of a commercial flocculant.

Mishra and co-workers[116–121] have extensively studied the flocculation of different types of wastewater using *Plantago psyllium* mucilage. In the first of their studies,[116] they investigated the flocculation of textile wastewater by *P. psyllium* mucilage and their results showed that the natural anionic polysaccharide of the species was very effective. The efficiency of the mucilage in flocculating textile wastewater is dependent on the pH of the medium under the test conditions. Mishra *et al.*[122] assessed the same species (*P. psyllium*) for the removal of dyes from model textile wastewater containing golden yellow (C.I. Vat Yellow 4) and reactive black (C.I. Reactive Black 5). The mucilage reduced the dye concentration by flocculation and settling. The maximum removal of golden yellow and reactive black observed was 71% and 35%, respectively. The optimum time for the removal was 2 and 1 h, at an optimum mucilage dose of 10 mg L^{-1} for golden yellow and reactive black, respectively. The optimal flocculant concentration required to affect flocculation is independent of dye concentration within the range examined. The dye removal obtained was influenced by the salt concentrations in the wastewater sample. The flocculation efficiency was sensitive to pH when pure aqueous solutions of dyes were used, but it was relatively unaffected by pH change when salts were added to the dye solutions. The experimental results show that the mucilage is more effective for removal of solubilized vat dye than for reactive black. From the present set of experiments, flocculation using *P. psyllium* mucilage for dye removal was shown to be a simple and efficient treatment from an economic and technical point of view.

The mucilage extracted from tamarind (*T. indica*) seeds was used for various types of wastewater treatment.[123–125] It has been used for the removal of vat and direct dye in aqueous solution by Mishra and Bajpai.[124] The maximum removal obtained was 60% for golden yellow after 2 h and 25% for direct fast scarlet after 1 h. The optimum mucilage dose was 10 and 15 mg L^{-1} for golden yellow and direct fast scarlet, respectively. The same group has reported many studies on water treatment using this polysaccharide.

Mishra and Bajpai[126] investigated the efficiency of mucilage isolated from the fruits of *Coccinia indica* for the treatment of simulated textile wastewater samples containing two direct dyes, direct fast scarlet and direct fast yellow, and two vat dyes, golden yellow and nyanthrene yellow. Statistical analysis showed that the change in percent removal with pH was highly significant in the case of the direct dyes while it was less significant in the case of vat dyes removal.

M. oleifera Lam is a multipurpose tree native to Northern India that now grows widely throughout the tropics. It is generally known in the developing world as a vegetable, a medicinal plant and a source of vegetable oil. Several studies have shown that *M. oleifera* seeds possess effective coagulation properties. Recently, the flocculation properties of this species were tested on palm oil mill effluent. Bhatia *et al.*[127–129] used *M. oleifera* to reduce the SS, chemical oxygen demand (COD) and biological oxygen demand (BOD) of such effluent. Pilot-scale treatment resulted in 99.7% SS removal, 71.5% COD reduction, 68.2% BOD reduction and 100% oil and grease removal. Finally, they concluded that the volume of sludge produced using *M. oleifera* as coagulant is considerably less compared to alum.

A review of plant-based coagulant sources, processes, effectiveness and relevant coagulating mechanisms for treatment of water and wastewater has been presented by Yin.[130] Plant-based coagulants included in the review are nirmali seeds (*Strychnos potatorum*), *M. oleifera*, tannin and cactus. These coagulants are, in general, used as point-of-use technology in less-developed communities since they are relatively cost-effective compared to chemical coagulants, can be easily processed in a usable form and are biodegradable. These natural coagulants, when used for treatment of waters with low-to-medium turbidity range (50–500 NTU), are comparable to their chemical counterparts in terms of treatment efficiency. Their application for industrial wastewater treatment is still in its infancy, though they are technically promising as coagulants for dyeing effluent as afforded by Yoshida intermolecular interactions. These natural coagulants work by an adsorption mechanism, followed by charge neutralization or a polymeric bridging effect.

Verma *et al.*,[131] in their extensive review on chemical coagulation/flocculation technologies for removal of color from textile wastewaters, have included the experimental results from various researchers who have used plant-derived natural polysaccharides in removing the color from textile effluents. Various plant-extracted polymers such as starch, guar gum, gum arabic, tannin, mucilage from nirmali seeds, *M. oleifera* and cacti, *etc.*, are generally well known as coagulants within the scientific community. These polymers have a large number of industrial applications as these polysaccharides find use in the paper industry, as food additives, *etc.*

Adinolfi *et al.*[132] have reported that polysaccharides extracted from *Strychnos potatorum* (nirmali) seeds can effectively reduce up to 80% turbidity of a kaolin solution. *M. oleifera*, known as drumstick, is widely found throughout India and other parts of Asia, some parts of Africa and America. The tree's bark, root, fruit, flowers, leaves, seeds and gum are also used as medicines. The seed of these trees is also used as a coagulant and/or flocculant in water and wastewater treatment. Beltrán-Heredia *et al.*[133] have investigated the use of *M. oleifera* seed extract for the removal of anthraquinone dye and reported 95% dye removal at a coagulant dose of $100\,\mathrm{mg\,L^{-1}}$ at pH 7. Further, Lea[134] has investigated the effectiveness of *M. oleifera* seed extract for the treatment of turbid water and found 99.5% turbidity removal at a dosage of $400\,\mathrm{mg\,L^{-1}}$. Typically, an increased dosage of seed extract does not enhance the dye removal after maximum adsorption is reached. This might be due to the fact that no more new sites for adsorption remain available at the surface of the seed extract. *M. oleifera* seeds are also considered as an excellent biofuel source for making biodiesel.

Natural polysaccharides, galactomannans, occur in large amounts in the endosperm of the seeds of many Leguminoseae[135] and consist of a $(1\rightarrow4)$-linked β-D-mannopyranose backbone with branch points from their O-6 positions linked to α-D-galactose (*i.e.*, $1\rightarrow6$-linked α-D-galactopyranose);[136] they are used as adsorbents for removal of dye molecules from effluent, as discussed in a review by Blackburn.[137] In his review, adsorbents discussed were chitin, chitosan (low, medium and high M_r), locust bean gum, guar gum, corn starch, wheat starch, pectin, carrageenan (ι- and κ-forms), dextrin, alginic acid, tamarind gum and cassia gum. Chitin and chitosan were reported as highly effective and non-ionic galactomannans (locust bean gum, guar gum, cassia gum) were also effective in removing dye from effluent, whereas other nonionic polysaccharides, such as starch, were not effective. This was attributed to the structure of the polysaccharides and the relative degree of inter- and intramolecular interactions between separate polymer chains. The pendant galactose residues of galactomannans prevented strong interaction, allowing greater hydrogen bonding with dye; comparatively, starch has extensive chain interactions and as such has limited potential for hydrogen bonding with the dye molecules at the temperature of application. In addition, hydrophobic interactions between the hydrophobic parts of the dye and the α-face of the pendant galactose residues might have contributed to the superior performance. Repulsion between anionic polysaccharides and the dye anions prevented any hydrogen bonding and therefore pectin, carrageenans and alginic acid were ineffective in dye removal from effluent. The observation of most interest within this work was the adsorption performance of the galactomannans, locust bean gum, guar gum, and cassia gum.

Sanghi *et al.*[138] have investigated the use of *Ipomeoa dasysperma* seed gum and guar gum as coagulant aids for removal of acid dye. Investigations were also carried out by the same group for possible exploitation of *Cassia javahikai* seeds as a potential source of commercial gum for textile wastewater treatment.[139] Graft copolymerization with acrylamide was done to modify the seed gum for the favorable properties. *C. javahikai* seed gum, and its copolymer grafted with acrylamide, were synthesized in the presence of oxygen using a potassium persulfate/ascorbic acid redox system. Both *C. javahikai* seed gum and its grafted polyacrylamide were found to be good working substitutes as coagulant aids.

Gum arabic, also known as gum acacia, is a water- and fat-soluble polysaccharide; it is highly branched with a β-galactose backbone and having a high molecular weight of 250 000–750 000 Da. Coagulation studies with this novel natural coagulant are yet to be established. Although the mechanism of coagulation with natural coagulants has not been extensively investigated, the presence of hydroxyl groups along the polysaccharide chain provides a large number of available adsorption sites that might lead to interparticle bridging between the polysaccharide and the dye molecule.

Biosorbents, especially those derived from seaweed (macroscopic algae) and alginate derivatives, exhibit high affinity for many metal ions. Because biosorbents are widely abundant (usually biodegradable) and less expensive than industrial synthetic adsorbents, they hold great potential for the removal of toxic metals from industrial effluents.[140] Biosorption of ions from simulated aqueous media were studied using calcium alginate beads by Singh *et al.*[141] Experiments were designed and performed according to the Box–Behnken matrix of response surface methodology. The effects of four vital operating variables on the metal-ion sorption characteristics of calcium alginate beads were studied: alginate dosage, initial copper concentration, pH and agitation time. A high regression coefficient between the variables and response ($R^2 = 0.9974$) supported excellent evaluation of the experimental data by a second-order polynomial regression model. The experimental data obtained have been fitted well with Langmuir and Freundlich isotherm models and also exhibited very high correlation coefficients, which confirmed the suitability of the model and biosorption process. The study revealed that the alginate beads could be used as an ideal material for the removal of Cu^{2+} ions at about 85.3% from aqueous media and it would be applicable in process development to treat industrial effluents.

A study by Kalidhasan *et al.*[142] focused towards the preliminary study of the interaction of the diphenylthiocarbazone (DTZ) complex of chromium(VI) in acidic medium with a cellulose biopolymer. The chromium–DTZ complex could be quantitatively adsorbed on a cellulose column in the pH range 1.0–2.5 and the effect of various experimental parameters such as stability of the column and the complex, column breakthrough volume, and interfering ions have been studied in detail. The probable mechanism of adsorption of the complex on the cellulose biopolymer was corroborated using FTIR, SEM, EDX and solid-state ^{13}C NMR techniques. The pores formed due to the hydrogen bond between the cellulose layers, and then the ensuing occupation of the complex between these layers and on the surface of the biopolymer layer through electrostatic attractive forces and π-interaction of the aromatic ring with cellulose, are expected to play a vital role in the interaction. The cellulose column could be regenerated using environmentally benign poly(ethylene glycol)-400 in an acidic medium. Crini and Badot[143] in their extensive review have summarized the recent developments in polymeric materials derived from starch in removal of dyes from textile effluents.

4.5   Animal-Based Polysaccharides

Chitin is the second most common natural polysaccharide in the world, after cellulose. The main sources of chitin are two marine crustaceans, shrimp and crab. Chitosan is a linear copolymer of D-glucosamine (deacetylated unit) and *N*-acetyl-D-glucosamine (acetylated unit) produced by the deacetylation of chitin. Chitosan possesses several properties such as nontoxicity, biodegradability and outstanding chelation behavior that make it an effective coagulant and/or flocculant for removal of contaminants from water. This section provides an extensive review of the research done by various groups using chitosan as a flocculant. This is the most extensive animal-derived natural polysaccharide used as a flocculant for water treatment.

Various studies for treatment of industrial wastewater using chitosan were carried out during the late 1970s by Bough and co-workers.[144,145] They investigated the effectiveness of chitosan for coagulation and recovery of SS in the processing of waste from a variety of food processing industries and found that this novel coagulant was very effective for the efficient reduction of COD as well as removal of SS and turbidity. In order to better understand the peculiarities of using a product of natural origin in municipal wastewater treatment, laboratory testing (jar tests) was carried out with chitosan as a coagulant aid, as well as full-scale testing in a medium-sized physicochemical wastewater treatment plant by Babineau *et al.*[146] The full-scale test was performed in two parallel, identical systems, treating the same wastewater under the same conditions. The one using a combination of alum with a synthetic polymer (AL/SP) was compared with the other which used alum and chitosan (AL/CH). Removals for COD, SS and total phosphorus reached 87, 95 and 93%, respectively, for the AL/CH combination. These results were similar to those obtained for COD and SS with the AL/SP combination. Some results showed a coagulant dosage (alum) up to 24.8% lower with chitosan as the coagulant aid. For total phosphorus, however, the results showed that removals were higher with the AL/SP combination because of a higher coagulant dosage. Chitosan as coagulant has been promoted by Verma *et al.*[131] in their review. Numerous results claim that chitosan is involved in a dual mechanism, including coagulation by charge neutralization and flocculation by a bridging mechanism.[147,148]

The kinetics of aggregation of aqueous protein dispersions under the action of natural polysaccharide flocculants (cationic chitosan and anionic sodium alginate) were studied turbidimetrically in relation to the protein/polysaccharide ratio in the solution and the pH value. Sodium alginate and chitosan were tested in the treatment of wastewater from the flotation processing of hydrobiotics by Konovalova and Stepanova.[149] An overview of possible applications of chitosan on the wastewater treatment released from food-processing plants, typically seafood, dairy or meat processing industries, contains appreciable amounts of protein which can be recovered with the use of chitosan; this protein, after drying and sterilization, makes a great source of feed additives for farm animals.[150]

Marketa and Roman[151] presented a short review of adsorptive materials proposed and tested for removing phthalates from an aqueous environment. The major focus of the review was chitosan as an innovative absorbent for removal of phthalates from water. Chen and Chung[152–154] selected chitosan as the biosorption material, produced by *N*-acetylation of chitin, and it was employed to remove inorganic and organic pollutants, *e.g.* chlorophenols and nitrophenols. The phthalates under study were dimethyl phthalate (DMP), diethyl phthalate (DEP), di-*n*-propyl phthalate (DPP), di-*n*-butyl phthalate (DBP), diheptyl phthalate (DHpP) and di-2-ethylhexyl phthalate (DEHP). For studying adsorption, the authors selected a static arrangement of tests. Actual wastewater was used in a dynamic arrangement of tests. If actual wastewater mostly contains DEHP, the regeneration of chitosan is only recommended 15 times by the authors. The adsorption of phthalic acid esters (PAEs) and their degradation products, such as phthalate monomers (MPEs) and phthalic acid (PA), from the aqueous environment on chitosan was also subject to a study by Salim *et al.*[155] The results of their kinetic experiments indicated that DBP was adsorbed to a greater extent than DEHP or DMP. The adsorption capacity for the tested MPEs declined in the order monobuthyl phthalate > monomethyl phthalate > monoethylhexyl phthalate, with the highest adsorption capacity recorded for the sorption of PA. The results of a study by Salim *et al.*[155] indicated that chitosan adsorbed PAEs primarily because of hydrophobic interactions, and interacted with PA mainly due to interaction between polar active groups. For the monomers, especially monomethyl phthalate and monoethylhexyl phthalate, lower hydrophobicity than the PAEs and higher hydrophilicity than PA made them less adsorbable.[156] In terms of utilization, crawfish chitosan as a coagulant for recovery of organic compounds in wastewater was demonstrated to be equivalent, or superior to, the commercial chitosans from shrimp and crab waste shell and synthetic polyelectrolytes in turbidity reduction.[157]

Chitin and chitosan have been found to have extremely high affinity for dyes which may contribute to aquatic toxicity. It was also found that chitosan is effective for conditioning municipal and industrial sludge, mainly due to its effectiveness in sludge conditioning, rapid biodegradability in soil environments and economic advantages in centrifugal sludge dewatering.[157] A comprehensive review by Crini[158] on nonconventional low-cost absorbents on dye removal includes natural polysaccharides like starch, chitosan and others, and their efficiency. Chitosan-based sorbents have demonstrated outstanding removal capabilities for certain dyes in comparison to activated carbon. In a study by Blackburn,[137] natural polysaccharides were used as adsorbents for removal of dye molecules from effluent. The results showed that naturally cationic polysaccharides such as chitin and chitosan gave excellent levels of color removal, and this was attributed to a combination of electrostatic attraction, van der Waals forces and hydrogen bonding. The dyes used in this study were widely used dyes along with low, medium and high molecular weight chitosan. Zhang et al.[159] have used carboxymethyl chitosan for printing and dyeing wastewater treatment. The experimental results showed that carboxymethyl chitosan in wastewater decolorization and COD reduction is superior over other commonly used polymer flocculants.

Szygula et al.[160] reported approximately 99% color removal from the simulated textile wastewater containing Acid Blue. In continuation of this, Mahmoodi et al.[161] investigated the effectiveness of chitosan for removal of Acid Green 25 and Direct Red 23 and reported approximately 75% and 95% dye removal, respectively, in 10 min. Szygula et al.[160] examined the coagulation–flocculation process of a model dye (Acid Blue 92, AB92) using chitosan. The experimental procedure considered the effect of the initial pH and the dye concentration in order to evaluate the optimum molar ratio between the amine groups (of chitosan) and the sulfonic groups (of AB92) at biopolymer saturation. In their experiment they also described the removal of the dye from flocs using alkaline solutions (0.001–1 M NaOH solutions). The biopolymer was redissolved in acetic acid solution to be reused further. Gupta and Suhas[162] highlighted in their review on low-cost adsorbents that chitin and chitosan, in addition to having wide availability, also showed fast kinetics and appreciable adsorption capacities.

The review by Gerente et al.[163] described the significant developments in the new range of applications of chitosan as a potential water treatment material for removing metal ions from wastewaters. It also presented the developments in this area and identified the deficiencies in existing chitosan research by reviewing the equilibrium studies carried out to determine the capacity of chitosan for various metal ions. Then the kinetic studies were reviewed, as well as the solution methodologies adopted by various researchers to explain and model the rate of adsorption of the metal ions from solution. Moreover, chitosan was compared with two ion-exchange resins, Dowex A-1 and Zerolit 225, using copper effluent concentrations from 10 to 100 mg L^{-1}; chitosan had an uptake of greater than 10 times either ion exchanger.[164]

A review on recent developments in mercury(II) removal in wastewater treatment by Miretzky and Cirelli[165] gives a comprehensive report on using chitosan and its derivatives. Mercury is one of the most toxic heavy metals commonly found in the global environment. Its toxicity is related to the capacity of its compounds to bioconcentrate in organisms and to biomagnify through the food chain. Mercury(II) adsorption on chitosan is now assumed to occur through several single or mixed interactions: chelation or coordination on amino groups in a pendant fashion or in combination with vicinal hydroxyl groups, electrostatic attraction in acidic media or ion exchange with protonated amino groups. The adsorption capacity of chitosan under different experimental conditions is reported to help comparison of the efficacy of the Hg(II) removal process. A comparison with the adsorption capacity of other low-cost adsorbents is also tabulated in the review. Sahoo *et al.*[166] in their review have described chitosan as the prime commercial coagulant in wastewater treatment. Chitosan carries a partial positive charge and binds to metal ions, thus making easier metal ion removal from waste streams or contamination sites.[150]

Chromium(VI) is a well-known highly toxic metal, considered a priority pollutant. Industrial sources of Cr(VI) include leather tanning, cooling tower blowdown, plating, electroplating, anodizing baths, rinse waters, *etc.* A review by Owlad *et al.*[167] provides recent information about the most widely used techniques for Cr(VI) removal. It was reported that chitosan chelates five to six times greater amounts of metals than chitin, due to the free amino groups exposed during deacetylation.[168] It is widely known that the excellent adsorption behavior of chitosan for heavy metal removal is attributed to the high hydrophilicity due to a large number of hydroxyl groups, a large number of primary amino groups with high activity and the flexible structure of the polymer chain making a suitable configuration for adsorption of metal ions. Udaybhaskar *et al.*[77] investigated the interaction between chitosan and Cr(VI) and found that an adsorption capacity of 273 mg of Cr(VI) g^{-1} chitosan was achieved at pH 4.0. Schmuhl *et al.*[169] studied the crosslinking effects of chitosan. Crosslinking is the process of chemically joining two or more molecules by a covalent bond. Crosslinking reagents contain reactive ends to specific functional groups (primary amines, thiols, *etc.*) on molecules. Chitosan is commonly crosslinked with chemical agents such as diisocyanate, epoxy compounds, carbodiimides, glutaraldehyde and other reagents. It was found that non-crosslinked chitosan has a potential to adsorb 30 mg more of Cr(VI) g^{-1} than that of crosslinked chitosan. This is consistent with the fact that crosslinking reduces the adsorption capacities of chitosan, but this loss of capacity may be necessary to ensure the stability of the chitosan.

Lee *et al.*[170] prepared chitosan-based polymeric surfactants (CBPSs) and applied these to the removal of Cr(VI) commonly found in wastewater; a batch test was conducted to evaluate the adsorption capacity. The removal efficiency of Cr(VI) by the CBPSs depended on several factors, including the solution pH, CBPS dose and ionic strength. The results show that the CBPSs exhibited a greater adsorption capacity for Cr(VI) than have other modified chitosans reported in the literature. Sankararamakrishnan *et al.*[171] used chitosan crosslinked with glutaraldehyde, the xanthate group and chemically modified chitosan beads and flakes for the recovery of toxic Cr(VI). Spinelli *et al.*[166] synthesized a quaternary chitosan salt (QCS) and applied it to adsorb Cr(VI). It was reported that the adsorption capacity for Cr(VI) at pH 9.0 was 30.2 mg g^{-1} (0.58 mmol g^{-1}), while at pH 4.5 the capacity was 68.3 mg g^{-1}

(1.31 mmol g^{-1}). Chromium(VI) ions can be eluted from crosslinked QCS by treatment with a 1 mol L^{-1} solution of NaCl/NaOH, showing an efficiency of more than 95%.

The sorption of precious metals such as gold onto four chitosan derivatives yielded capacities in the range 400–600 mg Au g^{-1} chitosan, depending on derivative type and pH.[172] Niu and Volesky[173] found lower values on the order of 34 mg Au g^{-1} chitosan, but as Au(CN)$_2^{-}$. Silver ion sorption on chitosan and on commercial chelating resins has been studied by Lasko and Hurst[174] in terms of Ag, Ag(NH$_3$)$_2^{2+}$, Ag(S$_2$O$_3$), Ag(SCN)$_3^{2-}$ and Ag(CN)$_2^{-}$. All species showed good capacities in the range 18–24 mg Ag g^{-1} chitosan except the last one; they were in the same order for the tested resins. Platinum sorption of chitosan derivatives has also been studied by Guibal,[175,176] and the obtained capacities ranged from 290 to 390 mg g^{-1} chitosan flakes.

4.6　Microorganism-Based Polysaccharides

This section describes the microorganism-derived natural polysaccharides' efficiency in the treatment of water and wastewater. Their efficiencies in terms of water treatment capability, cost effectiveness, availability and ease of use are compared with the commercially available synthetic flocculants. Because of their biodegradability, harmlessness and lack of secondary pollution, flocculants derived from microorganisms have gained wide attention and research to date.

Wang *et al.*[177] isolated a bioflocculant producing bacterium named M09 from activated sludge. The major component of bioflocculant XM09 produced by strain M09 was identified as a polysaccharide rich in hydroxyl and carboxyl functional groups. The treatment effect on domestic wastewater by adding XM09 alone, adding poly(aluminum chloride) (PAC) alone and combining XM09 and PAC was compared. The results showed that the flocculating efficiencies of each flocculant alone were 87.74% and 89.96%, respectively; the combination of XM09 and PAC not only improved the flocculating efficiency, but also reduced the dosage of each flocculant greatly. In addition, to evaluate the safety of composite flocculants, the residual aluminum in the wastewater after treatment by compound flocculants was investigated to prove XM09 has an important role in removing residual aluminum, thereby decreasing the risk of secondary pollution brought by chemical flocculants. The flocculant potential of a bioflocculant exopolysaccharide (EPS) produced from an *Azotobacter indicus* ATCC 9540 strain was investigated at different pH values, temperatures and cation concentrations by Patil *et al.*[178] The flocculant activity at different concentrations of EPS in the absence of cations was reanalyzed by a slightly modified flocculant assay. It revealed that flocculant activity increased in a concentration-dependent manner up to a certain limit, with a maximum flocculation of 72% at 500 mg L^{-1} EPS concentration, even in the absence of cations. A differential scanning calorimetry study and flocculant assay revealed high-temperature stability of EPS up to 97 °C. Investigation of flocculation efficacy of the characterized EPS for wastewater treatment of dairy, woolen, starch and sugar industries suggested it to be effective and stable at wide pH range of 5–10.

The biopolymer produced by the *A. indicus* ATCC 9540 strain was found to have a high molecular mass, thermostability, pH receptivity and high flocculant activity. It possesses all the characteristics of an ideal flocculant agent required for industrial waste treatments. Recovering suspended solids not only decreases the BOD and COD load of pollutants for discharge, but also influences the economy with an environmentally friendly aspect as the recovered solids can be used as feed additives for animals. The *A. indicus* strain in the study is of nonpathogenic origin; therefore, EPS produced by the strain has the potential to be used for the treatment of drinking water and downstream processes, as well as industrial waste processing, because it works at a wide pH range (5.0–10). The EPS was able to significantly reduce BOD and COD above 60% in a range of industrial wastes and also showed an average reduction in suspended solids above 35%. Therefore, this study finds significance in the utilization of *Azotobacter* exopolysaccharides as potential natural bioflocculants.

The efficacy of the *A. indicus* ATCC 9540 strain for production of the EPS bioflocculant was investigated by Patil *et al.*,[179] using flower extract of *Madhuca latifolia* L. and yeast extract. EPS production was increased in the presence of nitrogen. The extracted polymer was characterized by different chemical tests, FTIR spectroscopy and TLC, which showed the presence of uronic acids, *O*-acetyl groups and orcinol, with suggestive indication of an alginate-like polymer. The isolated EPS showed cation-dependent flocculating activity.

Zhang *et al.*[180] derived the bioflocculant TJ-F1 from *Proteus mirabilis* from a mixed activated sludge. The flocculation efficiency was tested in kaolin solution and was found to have a high flocculation capacity of 93.1%. Physicochemical characterization of the bioflocculant showed that it mainly consisted of protein and acid polysaccharide. It contained carboxyl, hydroxyl and amino groups and hydrogen bonds, preferred for the flocculation process.

Verma and co-workers[131] extensively reviewed the various methods and technologies used for removal of dyes from the water discharged from textile industries. They described the potential of xanthum gum, a natural polysaccharide derived from the bacterial coat of *Xanthomonas campestris* and used as a food additive and rheology modifier,[181] as an efficient alternative for dye removal from textile wastewater. Xanthum gum is produced by the fermentation of glucose by the *X. campestris* bacterium. The possible mechanism of coagulation by interparticle bridging as observed for guar gum can also be observed for xanthan gum.

The biosorption behavior and mechanisms of a novel EPS, which is secreted by a mesophilic bacterium [namely *Wangia profunda* (SM-A87)] isolated from deep-sea sediment, for the heavy metals Cu(II) and Cd(II) have been studied by Zhou *et al.*[182] De Philippis *et al.*,[183] in their extensive review on exopolysaccharides derived from cyanobacteria and their metal removal efficiency, focused on defining the molecular mechanisms of the metal binding to the polysaccharidic exocellular layers and in the use of EPS-producing cyanobacteria for metal biosorption at a pilot scale with real wastewaters.

Their review also describes the main positive issues and the drawbacks so far emerging from these experiments.

The capability of microorganisms to remove heavy metals from the surrounding environment has long been recognized, and a large number of studies are available on the basic and applied aspects of this process, recently reviewed by Inthorn,[184] Mehta and Gaur,[185] Ahluwalia and Goyal,[186] Gadd[187] and Chojnacka.[188] A wide range of microorganisms, both eukaryotes and prokaryotes, has been investigated for the removal of heavy metals from water solutions, and promising results have been obtained for a number of them, including cyanobacteria. Cyanobacteria possess a cell wall capable of passively adsorbing high amounts of solubilized metals. Cyanobacterial cell walls possess unique characteristics in comparison with other Gram-negative bacteria. In many cases, cyanobacteria release a fraction of polysaccharidic material, which is solubilized in the culture medium and has been defined as released polysaccharide.[189]

Cyanobacterial EPSs exhibit two typical features that distinguish them from the polymers synthesized by other bacteria, namely more than 90% of those so-far characterized are complex hetero polysaccharides and about 75% of them show the presence of six or more different types of monosaccharides.[190] Owing to the presence of a large number of negative charges on the external cell layers, EPS-producing cyanobacteria have been considered very promising as chelating agents for the removal of positively charged heavy metal ions from water solutions,[190] and an increasing number of studies on their use in metal biosorption have been published in recent years. A few studies on the performances of EPS-producing cyanobacteria in heavy metal sorption showed that, owing to the presence of a large number of negative charges on the external cell layers, EPS-producing cyanobacteria could be considered very promising as chelating agents for the removal of positively charged heavy metal ions from water solutions.[190] Indeed, the above-mentioned research points out that metal biosorption depends on: (1) the chemical and morphological features of microbial cells, (2) the chemical characteristics of the EPS surrounding the cells and present in solution, (3) the chemical and physical properties of metals, (4) their interactions with the other compounds in solution and, finally, (5) the device and (6) the operational conditions utilized in the treatment. For these reasons, it is very difficult to define a general mechanism for the biosorption of metals by EPS-producing cyanobacteria, and the best way for assessing the potential of new cyanobacterial strains is the direct determination of their metal uptake capability under the intended operational conditions.

The most important positive aspects emerging from the studies on the use of EPS-producing cyanobacteria for the removal of metals from water solutions are the very good metal uptake observed in many cases and the possibility to recover the metals from the biomass at the end of the biosorption process. On the other hand, the main drawbacks are the costs of producing the biomass and the much slower growth rates of phototrophic microorganisms in comparison with chemoheterotrophs. However, the current studies on the genes involved in the biosynthesis of EPS in cyanobacteria by Pereira *et al.*[189]

might give the opportunity, in the future, to introduce specific alterations on the composition/structure of the polymers, thus producing tailored polysaccharides with much higher specificity and/or metal biosorption capacity.

Metal sorption by microorganisms can occur by following two different ways: an active, metabolically driven process, named bioaccumulation, or a passive, not metabolically driven process, named biosorption.[191] In bioaccumulation, the sorption is due to the transport of the metal across the cell membrane, with a consequent intracellular metal accumulation mediated by the cell metabolism. Consequently, bioaccumulation can occur only when cells are viable, and it is often associated with a defense mechanism activated by the microorganism in the presence of a toxic metal.[192] On the other hand, biosorption is a complex physicochemical process,[193] generally involving more than one mechanism (*e.g.* complexation, adsorption, ion exchange, precipitation, *etc.*) and characterized by faster kinetics in comparison with bioaccumulation, which requires the activation of the transport of the metal inside the cells.

The research done so far using natural polysaccharides derived from plants, animals and microorganisms is tabulated in Table 4.1.

Table 4.1   A list of polysaccharides used as water remediation materials.

Polysaccharide	Source	Type of contaminant removed	Ref.
Pectin	Plant	Turbidity	103
Okra	Plant	SS, TDS, color	104
Fenugreek	Plant seeds	SS, TDS, color	104
Kundoor	Plant fruits	Dyes	106
Fenugreek	Plant seeds	SS, TDS	107
Okra	Plant pods	SS, TDS	108
Ipomeoa dasysperma	Plant seeds	Dyes	110
Strychnos potatorum (nirmali)	Plant seeds	Turbidity	111
Moringa oleifera	Plant	Kaolin	111
Moringa oleifera	Plant	Dye	112
Moringa olifera	Plant	Turbidity	113
Sodium alginate	Plant	Cu(II)	114
Cellulose	Plant	Cr(VI)	115
Malva sylvestris	Plant	Dyes	117
Okra	Plant	Dyes	117
Plantago psyllium	Plant	SS, TDS	120,122,123
Plantago psyllium	Plant	Dyes	122
Tamarindus indica	Plant	Dyes	123
Moringa olifera	Plant	SS, COD, BOD	124–126
Moringa olifera	Plant	Dyes	127
Nirmali	Plant	Dyes	127
Tannin	Plant	Dyes	127
Cactus	Plant	Dyes	127
Pectin	Plant	Heavy metals	128
Alginates	Plant	Heavy metals	132
Alginates	Plant	Mn, Ni, Cr	132

Table 4.1 (*Continued*)

Polysaccharide	Source	Type of contaminant removed	Ref.
Cassia javahikai	Plant	Dyes	134
Locust bean gum	Plant	Dyes	137
Guar gum	Plant	Dyes	137
Cassia gum	Plant	Dyes	137
Tamarind gum	Plant	Dyes	137
Corn starch	Plant	Dyes	137
Wheat starch	Plant	Dyes	137
Dextrin	Plant	Dyes	137
κ-Carrageenan	Plant	Dyes	137
ι-Carrageenan	Plant	Dyes	137
Pectin	Plant	Dyes	137
Alginic acid	Plant	Dyes	137
Chitosan	Animal	COD, SS, total phosphorus	140
Chitosan	Animal	Turbidity	141
Chitosan	Animal	Phthalates	142
Chitosan	Animal	Chlorophenols, nitrophenols, phthalates	147
Chitosan	Animal	Color, COD	154
Chitosan	Animal	Dyes	155
Chitosan	Animal	Dyes	156
Chitosan	Animal	Hg(II)	157
Chitosan	Animal	Turbidity	158
Chitosan	Animal	Dye	160
Chitosan	Animal	Cr(VI)	161,164–166
Chitosan	Animal	Cu and Au ions	167
Chitosan	Animal	Au	169,170
Chitosan	Animal	Ag	171–173
Chitosan	Animal	Dyes	109,137,174
XM09	M09 bacterium	Al	175
EPS	*Azotobacter indicus* ATCC 9540	BOD, COD, SS	176
TJ-F1	*Proteus mirabilis*	Zeta potential	178
Bioflocculant	*Xanthomonas campestris*	Dyes	109
Bioflocculant	Cyanobacteria	Metal	181–186
SM-A87 EPS	*Wangia profunda* bacterium	Cu(II), Cd(II)	182

4.7 Conclusions

From the scientific literature available on the use of natural polysaccharides derived from plants, animals and microorganisms, it can be concluded that these materials, owing to their abundant availability, biodegradability, cost effectiveness and efficiencies, are better water treatment alternatives than their synthetic counterparts. Another important perspective from the review shows that for these natural polysaccharides to be used as water treatment agents commercially, large-scale and real-time field experiments need to be performed to obtain the maximum out of the available rich fauna and flora in every part of the world. These materials could be the inexpensive option for clean water availability around the world. Furthermore, the increase in prices of chemical products, the interest in using biosorption for the recovery of precious metals and the increasing public perception of the importance of green biotechnologies in the treatment of wastewaters might give the opportunity to develop new industrial processes based on the use of these materials.

References

1. C. A. Kozlowski and W. Walkowiak, *Water Res.*, 2002, **36**, 4870.
2. S. Rio and A. Delebarre, *Fuel*, 2003, **82**, 153.
3. R. Van der Oost, J. Beyer and P. E. Vermeulen, *Environ. Toxicol. Pharmacol.*, 2003, **13**, 57.
4. H. Bagheri and M. Saraji., *J. Chromatogr. A*, 2001, **910**, 87.
5. K. B. Lee, M. B. Gu and S. H. Moon, *Water Res.*, 2003, **37**, 983.
6. H. M. Hwang, L. F. Slaughter, S. M. Cook and H. Cui, *Bull. Environ. Contam. Toxicol.*, 2000, **65**, 228.
7. C. I. Pearce, J. R. Lloyd and J. T. Guthrie, *Dyes Pigments*, 2003, **58**, 179.
8. G. McMullan, C. Meehan, A. Conneely, N. Kirby, T. Robinson and P. Nigam, *Appl. Microbiol. Biotechnol.*, 2001, **56**, 81.
9. Y. Fu and T. Viraraghavan, *Bioresour. Technol.*, 2001, **79**, 251.
10. B. Van der Bruggen and C. Vandecasteele, *Environ. Pollut.*, 2003, **122**, 435.
11. A. Cassano, R. Molinari, M. Romano and E. Drioli, *J. Membr. Sci.*, 2001, **181**, 111.
12. V. V. Goncharuk, D. D. Kucheruk, V. M. Kochkodan and V. P. Badekha, *Desalination*, 2002, **143**, 45.
13. R. Y. Ning, *Desalination*, 2002, **143**, 237.
14. J. M. Lee, M. S. Kim, B. Hwang, W. Bae and B. W. Kim, *Dyes Pigments*, 2003, **56**, 59.
15. T. Kurbus, Y. M. Slokar, A. Majcen Le Marechal and D. B. Voncina, *Dyes Pigments*, 2003, **58**, 171.
16. F. Torrades, M. Pérez, H. D. Mansilla and J. Peral, *Chemosphere*, 2003, **53**, 1211.
17. F. Al-Momani, E. Touraud, J. R. Degorce-Dumas, J. Roussy and O. Thomas, *J. Photochem. Photobiol. A*, 2002, **153**, 197.
18. U. Von Gunten, *Water Res.*, 2003, **37**, 1443.
19. X. Chen, G. Chen and P. L. Yue, *Environ. Sci. Technol.*, 2002, **36**, 778.
20. C. Y. Hu, S. L. Lo and W. H. Kuan, *Water Res.*, 2003, **37**, 4513.
21. Z. Hu, L. Lei, Y. Li and Y. Ni, *Sep. Purif. Technol.*, 2003, **31**, 13.
22. J. M. Chern and Y. W. Chien, *Water Res.*, 2003, **37**, 2347.
23. M. F. R. Pereira, S. F. Soares, J. M. J. Orfao and J. L. Figueiredo, *Carbon*, 2003, **41**, 811.
24. J. Rivera-Utrilla, I. Bautista-Toledo, M. A. Ferro-Garcia and C. Moreno-Castilla, *Carbon*, 2003, **41**, 323.

25. S. Nouri, F. Haghseresht and G. Q. M. Lu, *Adsorption*, 2002, **8**, 215.
26. M. W. Jung, K. H. Ahn, Y. Lee, K. P. Kim and J. T. Rhee Park, *Microchem. J.*, 2001, **70**, 123.
27. S. T. Bosso and J. Enzweiler, *Water Res.*, 2002, **36**, 4795.
28. V. J. Inglezakis, M. D. Loizidou and H. P. Grigoropoulou, *J. Colloid Interface Sci.*, 2003, **261**, 49.
29. O. Abollino, M. Aceto, M. Malandrino, C. Sarzanini and E. Mentasti, *Water Res.*, 2003, **37**, 1619.
30. S. Al-Asheh, F. A. Banat and L. Abu-Aitah, *Sep. Purif. Technol.*, 2003, **33**, 1.
31. Y. H. Shen, *Water Res.*, 2002, **36**, 1107.
32. R. Celis, M. Carmen Hermosin and J. Cornejo, *Environ. Sci. Technol.*, 2000, **34**, 4593.
33. O. Yavuz, Y. Altunkaynak and F. Güzel, *Water Res.*, 2003, **37**, 948.
34. M. Ghoul, M. Bacquet and M. Morcellet, *Water Res.*, 2003, **37**, 729.
35. A. Krysztafkiewicz, S. Binkowski and T. Jesionowski, *Appl. Surf. Sci.*, 2002, **199**, 31.
36. F. A. Lopez, M. I. Martin, C. Pérez, A. Lopez-Delgado and F. J. Alguacil, *Water Res.*, 2003, **37**, 3883.
37. V. K. Garg, R. Gupta, A. B. Yadav and R. Kumar, *Bioresour. Technol.*, 2003, **89**, 121.
38. S. Netpradit, P. Thiravetyan and S. Towprayoon, *Water Res.*, 2004, **38**, 71.
39. V. K. Gupta, C. K. Jain, I. Ali, M. Sharma and V. K. Saini, *Water Res.*, 2003, **37**, 4038.
40. Z. Reddad, C. Gerente, Y. Andres, J. F. Thibault and P. Le Cloirec, *Water Res.*, 2003, **37**, 3983.
41. N. Calace, E. Nardi, B. M. Petronio and M. Pietroletti, *Environ. Pollut.*, 2002, **118**, 315.
42. T. Robinson, B. Chandran and P. Nigam, *Bioresour. Technol.*, 2002, **84**, 299.
43. T. Robinson, B. Chandran and P. Nigam, *Water Res.*, 2002, **36**, 2824.
44. P. Vasudevan, V. Padmavathy and S. C. Dhingra, *Bioresour. Technol.*, 2003, **89**, 281.
45. M. X. Loukidou, K. A. Matis, A. I. Zouboulis and M. Liakopoulou-Kyriakidou, *Water Res.*, 2003, **37**, 4544.
46. A. Zhang, T. Asakura and G. Uchiyama, *React. Funct. Polym.*, 2003, **57**, 67.
47. A. Atia, A. M. Donia, S. A. Abou-El-Enein and A. M. Yousif, *Sep. Purif. Technol.*, 2003, **33**, 295.
48. B. C. Pan, Y. Xiong, Q. Su, A. M. Li, J. L. Chen and Q. X. Zhang, *Chemosphere*, 2003, **51**, 953.
49. V. V. Azanova and J. Hradil, *React. Funct. Polym.*, 1999, **41**, 163.
50. J. Synowiecki and N. A. Al-Khateeb, *Crit. Rev. Food Sci. Nutr.*, 2003, **43**, 145.
51. M. N. V. Ravi Kumar, *React. Funct. Polym.*, 2000, **46**, 1.

52. S. E. Bailey, T. J. Olin, R. M. Bricka and D. D. Adrian, *Water Res.*, 1999, **33**, 2469.
53. *Starch and Starch Containing Origins*, ed. V. P. Yuryev, A. Cesaro and W. J. Bergthaller, Nova, New York, 2002.
54. O. B. Wurzburg, in *Modified Starches: Properties and Uses*, ed. O. B. Wurzburg, CRC, Boca Raton, 1986, p. 55.
55. P. A. Sandford and J. Baird, in *Industrial Utilization of Polysaccharides*, ed. G. O. Aspinall, Academic, New York, 1983, p. 411.
56. S. Babel and T. A. Kurniawan, *J. Hazard. Mater.*, 2003, **B9**, 219.
57. A. J. Varma, S. V. Deshpande and J. F. Kennedy, *Carbohydr. Polym.*, 2004, **55**, 77.
58. M. Singh, R. Sharma and U. C. Banerjee, *Biotechnol. Adv.*, 2002, **20**, 341.
59. G. Crini and M. Morcellet, *J. Sep. Sci.*, 2002, **25**, 789.
60. E. M. M. Del Valle, *Process Biochem.*, 2004, **39**, 1033.
61. W. Ciesielski, C. Y. Lii, M. T. Yen and P. Tomasik, *Carbohydr. Polym.*, 2003, **51**, 47.
62. E. Polaczek, F. Starzyk, K. Malenki and P. Tomasik, *Carbohydr. Polym.*, 2000, **43**, 291.
63. B. A. Bolto, *Prog. Polym. Sci.*, 1995, **20**, 987.
64. G. R. Rose and M. R. St. John, in *Encyclopedia of Polymer Science and Engineering*, Wiley, New York, 2nd edn, 1986, vol. 7, p. 211.
65. C. L. McCormick, J. Bock and D. N. Schulz, in *Encyclopedia of Polymer Science and Engineering*, Wiley, New York, 2nd edn, 1986, vol. 17, p. 730.
66. J. Gregory, in *Chemistry and Technology of Water Soluble Polymers*, ed. C. A. Finch, Plenum, New York, 1983, p. 307.
67. Y. A. Attia, *Coll. Chem. Miner. Processes*, 1992, **12**, 227.
68. P. Senett and J. P. Olivier, *Ind. Eng. Chem.*, 1965, **57**(8), 32.
69. W. J. Weber, *Physiochemical Processes for Water Quality Control*, Wiley-Interscience, New York, 1972, p. 68.
70. J. A. Kitchnener, *Br. Polym. J.*, 1972, **4**, 217.
71. S. Briggs, M. Habgood, G. J. Jameson and Y. D Yan, *Chem. Eng. J.*, 2000, **13**, 80.
72. Y. Ostubo, *Adv. Colloid Interface Sci.*, 1994, **53**, 1.
73. A. Rembaum, in *Recycling and Disposal of Solid Wastes*, ed. T. F. Yen, Ann Arbor Science, Ann Arbor, MI, 1974, p. 262.
74. W. Stumm and C. R. O'Melia, *J. Am. Water Works Assoc.*, 1968, **60**, 514.
75. R. F. Packham, *J. Colloid Sci.*, 1965, **20**, 81.
76. V. Tare and S. Choudhary, *Water Res.*, 1987, **22**, 1109.
77. P. Udaybhaskar, L. Iyenger and R. A. V. S. Prabhakar, *J. Appl. Polym. Sci.*, 1990, **39**, 739.
78. M. Javed and V. Tare, *J. Appl. Polym. Sci.*, 1991, **42**, 317.
79. W. Chan and C. Chiang, *J. Appl. Polym. Sci.*, 1995, **58**, 1721.
80. R. P. Singh, in *Polymers and Other Advanced Materials: Emerging Technologies and Business Opportunities*, ed. P. N. Prasad, J. E. Mark and T. F. Fai, Plenum, New York, 1995, p. 227.
81. M. I. Khalil and S. Farag, *J. Appl. Polym. Sci.*, 1998, **69**, 45.

82. S. Rajani, P. S. Vankar and A. Mishra, *Colourage*, 2001, **48**, 29.
83. M. Agarwal, S. Rajani and A. Mishra, *Macromol. Mater. Eng.*, 2001, **286**, 560.
84. A. Jha, S. Agarwal, A. Mishra and J. S. P. Rai, *Iran. Polym. J.*, 2001, **10**, 85.
85. W. Pigman and D. Horton, ed., *The Carbohydrates*, Academic, New York, 2nd edn, 1970.
86. K. Ward Jr. and P. A. Seib, in *The Carbohydrates*, ed. W. Pigman and D. Horton, Academic, New York, 2nd edn, 1970, p. 413.
87. R. H. Marchessault, and P. R. Sundarajan, in *The Polysaccharides*, ed. G. O. Aspinall, Academic, New York, 1983, vol. 2, p. 12.
88. K. C. B. Wilkie, *Adv. Carbohydr. Chem. Biochem.*, 1979, **36**, 215.
89. A. M. Stephen, *The Polysaccharides*, ed. G. O. Aspinall, Academic, New York, 1983, vol. 2, p. 98.
90. G. O. Aspinall, *The Carbohydrates*, ed. W. Pigman and D. Horton, Academic, New York, 2nd edn, 1970, p. 515.
91. H. A. Schols and A. G. J. Voragen, *Prog. Biotechnol.*, 1996, **14**, 3.
92. A. M. Stephen, *Food Polysaccharides and Their Applications*, Dekker, New York, 1995.
93. F. Smith and R. Montgomery, *The Chemistry of Plant Gums and Mucilages*, Van Nostrand Reinhold, New York, 1959.
94. R. L. Whistler and J. N. Bemiller, ed., *Industrial Gums*, Academic, Orlando, FL, 2nd edn, 1973.
95. E. Percival and R. H. McDowell, *Chemistry and Enzymology of Marine Algal Polysaccharides*, Academic, London, 1987.
96. R. C. W. Berkeley, G. W. Gooday and D. C. Elwood, *Microbial Polysaccharides and Polysaccharaces*, Academic, London, 1979.
97. P. Sandford, *Adv. Carbohydr. Chem. Biochem.*, 1979, **36**, 266.
98. C. T. Bishop and H. J. Jennings, in *The Polysaccharides*, ed. G. O. Aspinall, Academic, New York, 1982, vol. 1, p. 292.
99. P. A. Sandford and J. Baird, in *The Polysaccharides*, ed. G. O. Aspinall, Academic, New York, 1983, vol. 2, p. 412.
100. L. Kenne and B. Lindberg, in *The Polysaccharides*, ed. G. O. Aspinall, Academic, New York, 1983, vol. 2, p. 287.
101. R. Geddes, in *The Polysaccharides*, ed. G. O. Aspinall, Academic, New York, 1985, vol. 3, p. 283.
102. R. A. A. Muzzarelli, in *The Polysaccharides*, ed. G. O. Aspinall, Academic, New York, 1985, vol. 3, p. 417.
103. Y. C. Ho, I. Norli, A. F. M. Alkarkhi and N. Morad, *Bioresour. Technol.*, 2010, **101**, 1166.
104. Y. N. Mata, M. L. Blázquez, A. Ballester, F. González and J. A. Muñoz, *Chem. Eng. J.*, 2009, **150**, 289.
105. B. Volesky, *Sorption and Biosorption*, BV-Sorbex, St. Lambert, Quebec, 2003.
106. G. T. Grant, E. R. Morris, D. A. Rees, P. J. C. Smith and D. Thom, *FEBS Lett.*, 1973, **32**, 195.

107. P. Harel, L. Mignot, J. P. Sauvage and G. A. Junter, *Ind. Crop. Prod.*, 1998, **7**, 239.
108. R. Srinivasan and A. Mishra, *Chin. J. Polym. Sci.*, 2008, **26**, 679.
109. M. Agarwal, S. Rajani, A. Mishra and J. S. P. Rai, *Int. J. Polym. Mater.*, 2003, **52**, 1049.
110. D. W. Kang, R. C. Choi and D. K. Kweon, *J. Appl. Polym. Sci.*, 1999, **73**, 469.
111. K. Anastasakis, D. Kalderis and E. Diamadopoulos, *Desalination*, 2009, **249**, 786.
112. M. Rebhun, N. Narkis and A. M. Wachs, *Water Res.*, 1969, **3**, 345.
113. O. Ogedengbe, *Water Res.*, 1975, **10**, 343.
114. A. Mishra, A. Yadav, M. Agarwal and M. Bajpai, *React. Funct. Polym.*, 2004, **59**, 99.
115. A. Mishra, M. Agarwal and A. Yadav, *Colloid Polym. Sci.*, 2003, **281**, 164.
116. A. Mishra, A. Yadav, M. Agarwal and S. Rajani, *Colloid Polym. Sci.*, 2004, **282**, 300.
117. S. Rajani, M. Agarwal and A. Mishra, *Water Qual. Res. J. Can.*, 2002, **37**, 371.
118. A. Mishra, S. Rajani and R. Dubey, *Macromol. Mater. Eng.*, 2002, **287**, 592.
119. A. Mishra, M. Agarwal, M. Bajpai, S. Rajani and R. P. Mishra, *Iran. Polym. J.*, 2002, **11**, 381.
120. A. Mishra, R. Srinivasan, M. Bajpai and R. Dubey, *Colloid Polym. Sci.*, 2004, **282**, 722.
121. A. Mishra, A. Yadav, M. Agarwal and R. Srinivasan, *Chin. J. Polym. Sci.*, 2005, **23**, 113.
122. A. Mishra and M. Bajpai, *J. Hazard. Mater.*, 2005, **B118**, 213.
123. A. Mishra, M. Bajpai, S. Pal, M. Agarwal and S. Pandey, *Colloid Polym. Sci.*, 2006, **285**, 161.
124. A. Mishra and M. Bajpai, *Bioresour. Technol.*, 2006, **97**, 1055.
125. A. Mishra, *J. Biobased Mater. Bioenergy*, 2013, **7**, 12.
126. A. Mishra, M. Bajpai and S. Pandey, *Sep. Sci. Technol.*, 2006, **41**, 583.
127. S. Bhatia, Z. Othman and A. L. Ahmad, *J. Chem. Technol. Biotechnol.*, 2006, **81**, 1852.
128. S. Bhatia, Z. Othman and A. L. Ahmad, *Chem. Eng. J.*, 2007, **133**, 205.
129. S. Bhatia, Z. Othman and A. L. Ahmad, *J. Hazard. Mater.*, 2007, **145**, 120.
130. C. Y. Yin, *Process Biochem.*, 2010, **45**, 1437.
131. A. K. Verma, R. R. Dash and P. Bhunia, *J. Environ. Manage.*, 2012, **93**, 154.
132. M. Adinolfi, M. M. Corsaro, R. Lanzetta, M. Parrilli, G. Folkard, W. Grant and J. Sutherland, *Carbohydr. Res.*, 1994, **263**, 103.
133. J. Beltrán-Heredia, J. Sánchez-Martín, A. Delgado-Regalado and C. Jurado-Bustos, *J. Hazard. Mater.*, 2009, **170**, 43.
134. M. Lea, *Curr. Protoc. Microbiol.*, 2010, chap. 1, unit 1G.2; doi: 10.1002/9780471729259.mc01g02s16.

135. D. Brown, *Ecotoxicol. Environ. Saf.*, 1987, **13**, 139.
136. J. R. Daniel, R. L. Whistler, A. G. J. Voragen and W. Pilnik, in *Ullmann's Encyclopaedia of Industrial Chemistry*, VCH, Weinheim, 5th edn, 1994, vol. A25, p. 1.
137. R. S. Blackburn, *Environ. Sci. Technol.*, 2004, **38**, 4905.
138. R. Sanghi, B. Bhatttacharya, A. Dixit and V. Singh, *J. Environ. Manage.*, 2006, **81**, 36.
139. R. Sanghi, B. Bhattacharya and V. Singh, *Bioresour. Technol.*, 2006, **97**, 1259.
140. J. F. Fiset, J. F. Blais and P. A. Riveros, *Rev. Sci. Eau*, 2008, **21**, 283.
141. L. Singh, A. R. Pavankumar, R. Lakshmanan and G. K. Rajarao, *Ecol. Eng.*, 2012, **38**, 119.
142. S. Kalidhasan, A. S. Krishnakumar, V. Rajesh and N. Rajesh, *Spectrochim. Acta, Part A*, 2011, **79**, 1681.
143. G. Crini and P. M. Badot, *Int. J. Environ. Technol. Manage.*, 2010, **12**, 129.
144. W. A. Bough, *Process Biochem.*, 1976, **11**, 13.
145. W. A. Bough, W. L. Salter, A. C. M. Wu and B. E. Perkins, *Biotechnol. Bioeng.*, 1978, **20**, 1931.
146. D. Babineau, D. Chartray and R. Leduc, *Water Qual. Res. J. Can.*, 2008, **43**, 219.
147. H. K. No and S. P. Meyers, *Rev. Environ. Contam. Toxicol.*, 2000, **163**, 1.
148. E. Guibal and J. Roussy, *React. Funct. Polym.*, 2007, **67**, 33.
149. I. N. Konovalova and N. V. Stepanova, *Russ. J. Appl. Chem.*, 2005, **78**, 2006.
150. T. Asano, N. Havakawa and T. Suzuki, in *Proceedings of the 1st International Conference on Chitin/Chitosan*, ed. R. A. A. Muzzarelli and E. R. Pariser, MIT Sea Grant Program, Cambridge, MA, 1978, p. 80.
151. J. Marketa and S. Roman, *J. Environ. Manage.*, 2012, **94**, 13.
152. C. Y. Chen and Y. C. Chung., *J. Environ. Sci. Health, Part A*, 2006, **41**, 235.
153. C. Y. Chen and Y. C. Chung, *Environ. Eng. Sci.*, 2007, **24**, 534.
154. C. Y. Chen, C. C. Chen and Y. C. Chung, *Bioresour. Technol.*, 2007, **98**, 2578.
155. C. J. Salim, H. Liu and J. F. Kennedy, *Carbohydr. Polym.*, 2010, **81**, 640.
156. W. A. Bough, *Poult. Sci.*, 1975, **54**, 1904.
157. P. Miretzky and A. F. Cirelli, *J. Hazard. Mater.*, 2009, **167**, 10.
158. G. Crini, *Bioresour. Technol.*, 2006, **97**, 1061.
159. Q. H. Zhang, X. Tang and T. C. Luo, *J. Environ. Pollut. Control*, 1995, **17**, 7.
160. A. Szygula, E. Guibal, M. A. Palacin, M. Ruiz and A. M. Sastre, *J. Environ. Manage.*, 2009, **90**, 2979.
161. N. M. Mahmoodi, R. Salehi, M. Arami and H. Bahrami, *Desalination*, 2011, **267**, 64.
162. V. K. Gupta and Suhas, *J. Environ. Manage.*, 2009, **90**, 2313.
163. C. Gerente, V. K. C. Lee, P. Le Cloirec and G. McKay, *Crit. Rev. Environ. Sci. Technol.*, 2007, **37**, 41.

164. W. S. W. Ngah and I. M. Isa, *J. Appl. Polym. Sci.*, 1998, **67**, 1067.
165. P. Miretzky and A. F. Cirelli, *J. Hazard. Mater.*, 2009, **167**, 10.
166. D. Sahoo, S. Sahoo, P. Mohanty, S. Sasmal and P. L. Nayak, *Des. Monomers Polym.*, 2009, **12**, 377.
167. M. Owlad, M. K. Aroua, W. A. W Daud and S. Baroutian, *Water, Air, Soil Pollut.*, 2009, **200**, 59.
168. T. C. Yang and R. R. Zall, *Ind. Eng. Chem. Prod. Res. Dev.*, 1984, **23**, 168.
169. R. Schmuhl, H. M. Krieg and K. Keizer, *Water*, 2001, **27**, 1.
170. M.-Y. Lee, K.-J. Hong, S.-Y. Yoshitsune and T. Kajiuchi, *J. Appl. Polym. Sci.*, 2005, **96**, 44.
171. N. Sankararamakrishnan, A. Dixit, L. Iyengar and R. Sanghi, *Bioresour. Technol.*, 2006, **97**, 2377.
172. M. L. Arrascue, H. M. Garcia, O. Horna and E. Guibal, *Hydrometallurgy*, 2003, **71**, 191.
173. H. Niu and B. Volesky, *Hydrometallurgy*, 2003, **71**, 209.
174. C. L. Lasko and M. P. Hurst, *Environ. Sci. Technol.*, 1999, **33**, 3622.
175. E. Guibal, C. Milot and J. Roussy, *Water Environ. Res.*, 1999, **71**, 10.
176. E. Guibal, M. Ruiz, T. Vincent, A. Sastre and R. Navarro-Mendoza, *Sep. Sci. Technol.*, 2001, **36**, 1017.
177. L. Wang, J. Tang and Y. J. Liu, *J. Tianjin Univ. Sci. Technol.*, 2011, **44**, 984.
178. V. S. Patil, C. D. Patil, B. K. Salunke, R. B. Salunkhe, G. A. Bathe and D. M. Patil, *Appl. Biochem. Biotechnol.*, 2011, **163**, 463.
179. V. S. Patil, R. B. Salunkhe, C. D. Patil, D. M. Patil and B. K. Salunke, *Appl. Biochem. Biotechnol.*, 2011, **162**, 1095.
180. Z. Zhang, S. Xia, J. Zhao and J. Zhang, *Colloids Surf. B*, 2010, **75**, 247.
181. L. R. Davidson, *Handbook of Water Soluble Gums and Resins*. McGraw Hill, New York, 1980.
182. W. Zhou, J. Wang, B. Shen, W. Hou and Y. Zhang, *Colloids Surf. B*, 2009, **72**, 295.
183. R. De Philippis, G. Colica and E. Micheletti, *Appl. Microbiol. Biotechnol.*, 2011, **92**, 697.
184. D. Inthorn, in *Photosynthetic Microorganisms in Environmental Biotechnology*, ed. H. Kojima and Y. K. Lee, Springer, Hong Kong, 2000, p. 111.
185. S. K. Mehta and J. P. Gaur, *Crit. Rev. Biotechnol.*, 2005, **25**, 113.
186. S. S. Ahluwalia and D. Goyal, *Bioresour. Technol.*, 2007, **98**, 2243.
187. G. M. Gadd, *J. Chem. Technol. Biotechnol.*, 2009, **84**, 13.
188. K. Chojnacka, *Environ. Int.*, 2010, **36**, 299.
189. S. Pereira, A. Zille, E. Micheletti, P. Moradas-Ferreira, R. De Philippis and P. Tamagnini, *FEMS Microbiol. Rev.*, 2009, **33**, 917.
190. R. De Philippis and E. Micheletti, in *Heavy Metals in the Environment*, ed. N. K. Shammas, Y. T. Hung, J. P. Chen and L. K. Wang, CRC, Boca Raton, FL, 2009, p. 89.
191. B. Volesky and Z. R. Holan, *Biotechnol. Prog.*, 1995, **11**, 235.
192. G. M. Gadd, *Experientia*, 1990, **46**, 834.
193. T. A. Davis, B. Volesky and A. Mucci, *Water Res.*, 2003, **37**, 4311.

CHAPTER 5

Zeolites in Wastewater Treatment

ABHA DUBEY,[a] DEEPTI GOYAL[b] AND
ANURADHA MISHRA*[b]

[a] Department of Chemistry, MMH College, Ghaziabad, India; [b] Department
of Applied Chemistry, School of Vocational Studies and Applied Sciences,
Gautam Buddha University, Greater Noida, Gautam Budh Nagar – 201310,
India
*Email: anuradha_mishra@rediffmail.com

5.1 Introduction

Water contains different types of suspended, dissolved, emulsified or colloidal
inorganic and organic pollutants. These pollutants have toxic effects on the
ecosystem. Several conditions are imposed by different government monitoring
agencies for the discharge of effluents into natural water resources. Cost
effective and environmentally friendly techniques are required for water
remediation. Worldwide, scientists are searching for newer and cheaper alternatives to treat wastewater. Adsorption is one such technology which has
gained popularity in recent years. Different natural materials have been used as
low-cost environmentally friendly adsorbents.

Zeolites are one such naturally occurring adsorbent which have obtained a
fair amount of attention in recent decades.[1,2] Natural zeolites are gaining research interest for environmental applications, mainly due to their unique
properties and worldwide occurrence. Zeolites are naturally occurring microporous crystalline solids with very well defined different cavity structures, large

RSC Green Chemistry No. 23
Green Materials for Sustainable Water Remediation and Treatment
Edited by Anuradha Mishra and James H. Clark

surface areas and physiochemical properties. These aluminosilicates of the alkali and alkaline earth metals have infinite three-dimensional structures. They are characterized by the ability to lose or gain water reversibly and to exchange certain constituent atoms also without major change of atomic structure. Zeolites have their framework made up of four connected networks of atoms (tetrahedra). A silicon or aluminum atom is in the middle and oxygen atoms at the corners.[3] These tetrahedral structures can be linked together by their corners. Oxygen atoms of each tetrahedron are shared with adjacent tetrahedra. The framework structure may contain linked cages, channels or cavities, which are of small size to allow small molecules to enter. The limiting pore sizes are roughly between 2 and 10 Å in diameter. These crystalline aluminosilicates contain silicon, aluminum and oxygen in their framework, and cations, water and/or other molecules within their pores. The aluminum ion is small enough to occupy the position in the center of the tetrahedron of four oxygen atoms; substitution of Si^{4+} by Al^{3+} defines the negative charge in the framework. The net negative charge is balanced by the monovalent or divalent exchangeable cations (sodium, potassium or calcium) located together with water; they are weakly held in the structure to compensate the charge imbalance. These cations are exchangeable with certain cations in solution. Many zeolites occur in nature in the form of minerals and are formed by interaction of glass-rich volcanic rocks with fresh water in lakes or sea water. The structural formation and adsorbent properties of zeolites make them work as chemical sieves, water softeners and adsorbents. Synthetic zeolites are also being produced in research laboratories as well as commercially. Natural and synthetic zeolites with or without modifications have been found to be effective for removal of ammonia, radioactive elements, heavy metals, organic and inorganic pollutants, as reported in the research literature.[2,4,5] These reports emphasize that natural zeolites exhibit excellent selectivity for a number of hazardous cations. Zeolites have become attractive adsorbents due to their high ion-exchange capacity, relatively high specific surface areas and, most importantly, their low cost. In this chapter the use of zeolites for water treatment, specifically for the removal of ammonia from aqueous solution, is discussed.

5.2 Synthesis and Properties of Zeolites

Interaction of volcanic rocks and ash with alkaline underground water (ponds, lakes and seawater) gives rise to the formation of zeolites. Many zeolites have been identified worldwide. The alkali metals Na and K and the alkaline earth metals Ca and Mg are present as cations in zeolites. An aluminosilicate framework, exchangeable cations and zeolitic water are three relatively independent components present in the zeolite structure. The general chemical formula of zeolites is $M_{x/n}[(AlO_2)_x(SiO_2)_y] \cdot zH_2O$, where M is (Na, K, Li) and/or (Ca, Mg, Ba, Sr), n is the charge on the cation, and the values of x, y, and z depend on the type of zeolite.[6] The structure type is defined by the most stable and conserved aluminosilicate framework (Figure 5.1). The water molecules can be present in voids of large cavities and bonded between framework ions

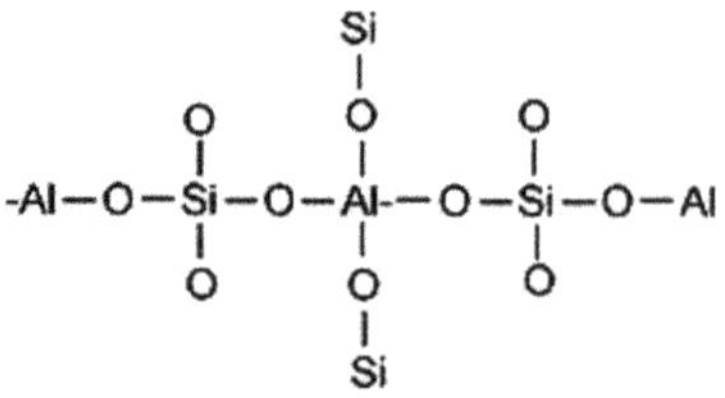

Figure 5.1 Basic structure of zeolites.

Table 5.1 Some important common natural zeolites.

Zeolite name	Chemical formula
Analcime	$(Na_{10})(Al_{16}Si_{32}O_{96}) \cdot 16H_2O$
Chabazite	$(Na_2Ca)_6(Al_{12}Si_{24}O_{72}) \cdot 40H_2O$
Clinoptilolite	$(Na_3K_3)(Al_6Si_{30}O_{72}) \cdot 24H_2O$
Erionite	$(NaCa_{0.5}K_9)(Al_9Si_{27}O_{72}) \cdot 27H_2O$
Faujasite	$(Na_{58})(Al_{58}Si_{134}O_{384}) \cdot 24H_2O$
Ferrierite	$(Na_2Mg_2)(Al_6Si_{30}O_{72}) \cdot 18H_2O$
Heulandite	$(Ca_4)(Al_8Si_{28}O_{72}) \cdot 24H_2O$
Laumontite	$(Ca_4)(Al_8Si_{16}O_{48}) \cdot 16H_2O$
Mordenite	$(Na_8)(Al_8Si_{40}O_{96}) \cdot 24H_2O$
Phillipsite	$(NaK)_5(Al_5Si_{11}O_{32}) \cdot 20H_2O$
Scolecite	$Ca_2Al_2Si_3O_{10} \cdot 3H_2O$
Stilbite	$(Na_2Ca_4)(Al_{10}Si_{26}O_{72}) \cdot 30H_2O$

and exchangeable ions *via* aqueous bridges. These can also serve as bridges between exchangeable cations. Some of the important zeolites with their chemical formulae are given in Table 5.1.

Atoms within the first set of parentheses are known as exchangeable ions as they can be replaced or exchanged more or less easily with other cations present in the aqueous solution without affecting the framework. Atoms or cations within the second set of parentheses are known as structural atoms, because with oxygen they make a rigid framework of the structure. This phenomenon is known as ion exchange or, commonly, cation exchange. The exchange process involves replacing one singly charged exchangeable atom in the zeolite by one singly charged atom from the solution or replacing two singly charged exchangeable atoms in the zeolite with one doubly charged atom from the solution. The magnitude of such cation exchange in a given zeolite is known as its cation exchange capacity (CEC). The CEC is commonly measured in terms of exchangeable cation per gram or 100 grams of zeolite or in terms of equivalents of exchangeable cation per gram or 100 grams of zeolite. As a rule, the greater the Al content (*i.e.* the more extra framework cations needed to balance the charge), the higher the CEC of the zeolite.[7] Some of the rare natural zeolites are offretite, paulingite, barrerite and mazzite. Natural zeolites vary in structure, crystal size, shape, pore diameter, porosity, chemical composition and purity, depending upon different types of deposits. The sorption capacity and selectivity for water and/or other molecules are defined by zeolite porosity, pore size distribution and specific surface. These properties affect the sorption

Table 5.2 Cation exchange capacities of clinoptilolite from different deposits.

Zeolite	Country	CEC value (meq g^{-1})	Ref.
Clinoptilolite	Australia	1.20	11
Clinoptilolite	Bulgaria	1.42	12
Clinoptilolite + mordenite	Chile	2.05	13
Clinoptilolite	China	1.03	14
Clinoptilolite	China	1.20	15
Clinoptilolite	China	1.20	16
Clinoptilolite	Croatia	1.45	17
Clinoptilolite + mordenite	Iran	1.20	18
Clinoptilolite	Romania	1.37	19
Clinoptilolite	Slovakia	1.17	20
Clinoptilolite	Turkey	1.6–1.8	21
Clinoptilolite	Turkey	1.84	19
Clinoptilolite + mordenite	Turkey	1.65	22
Clinoptilolite	Ukraine	0.64	23
Clinoptilolite	Ukraine	1.63	24

and ion exchange capacities of naturally occurring zeolites. The ion exchange behavior of natural zeolites is influenced by many factors, such as framework structure, ion size and shape, charge density of the anionic framework, ionic charge and concentration of the external electrolyte solution.[8,9] Clinoptilolite and heulandite are the most abundant natural zeolites.[1,2] These two zeolites can be differentiated on the basis of their Si/Al ratio and thermal stability.[10] The CECs of various clinoptilolites from different deposits are summarized in Table 5.2. The actual values are lower than the calculated ones.

Natural zeolites have been widely used for water remediation, and several research articles have been published in the last few decades. Clinoptilolite with a relatively low CEC is highly selective for the adsorption of some heavy metals,[25–27] radionuclides[28–30] and ammonium ions[10,25,31,32] and is the major zeolite used for water remediation. Some other naturally occurring zeolites such as mordenite, chabazite and phillipsite[25,26] also show high selectiveness for NH_4^+ and transition and radioactive elements.

5.3 Modification of Natural Zeolites

Adsorption is influenced by the zeolite structure, the Si/Al ratio and the type of the cation, number and location. Several chemical treatments have been used to improve these properties and hence the sorption efficiency of zeolites. Modifications are done mainly by surfactants to produce surfactant modified zeolites (SMZs), and by acid/base treatments. These modifications change the hydrophilic/hydrophobic properties for adsorption of different ions or organic compounds. To change the surface properties, one modification method widely employed is to use organic surfactants. In the past, many investigations have been conducted on modifications of natural zeolites with cationic surfactants and then use them to remove multiple types of contaminants from water.

5.3.1   Modification by Surfactants

Natural zeolites usually show little or no affinity for anions due to the net negative charge present on their framework. This results in low adsorption for organics in aqueous solution. Cationic (organic) surfactants are used to modify the surface properties of natural zeolites and make them good adsorbents for multiple contaminants. The common surfactants used are given in Table 5.3.

The degree of surfactant adsorption on zeolites is an important factor for the modification of natural zeolites. Surfactants form a monolayer or "hemimicelle" at the solid–aqueous interface *via* strong ionic bonds at surfactant concentrations at or below the critical micelle concentration (CMC) (Scheme 5.1). The hydrophobic tails of the surfactant molecules associate to form a bilayer or "admicelle" when the concentration of surfactant exceeds the CMC.[49]

The modification introduces complex functional groups in the zeolites for positively charged exchange sites formed by the positive groups of the surfactants, directed towards the surrounding solution in the bilayer. The organic-rich layer on the surface provides a partitioning medium for sorption of nonpolar organic compounds; the positively charged layer on the surface provides sites for sorption of anions.[10] Since the modifier surfactants are relatively large molecules they remain on the surface and do not enter the zeolite channels; as these channels remain negatively charged, they keep adsorbing inorganic cations.[50]

SMZs combine the cation sorption properties of unmodified zeolites with the additional ability to sorb anionic spices (arsenates, chromates, iodides, nitrates, phosphates, perchlorates, antimonates) and nonpolar hydrophobic organics, *e.g.* benzene, toluene, ethylbenzene, xylenes, phenols, pesticides, herbicides, dyes, *etc.*[10,38,51]

Polymer-modified zeolites also form a bilayer structure on zeolite surfaces and show similar anion sorption properties. Polyhexamethyleneguanidine is used for the modification.[10,52–54] The SMZs are stable in water and chemical solutions and are low-cost alternatives to other commercially available adsorbents.[38,52]

Table 5.3   Surfactants used to modify the surface properties of natural zeolites.

Surfactant	Ref.
Tetramethylammonium	33,34
Cetyltrimethylammonium (CTMA)	35,36
Hexadecyltrimethylammonium (HDTMA)	33,37–42
Octadecyldimethylbenzylammonium (ODMBA)	35,43–45
Cetylpyridinium (CPD)	46
Benzyldimethyltetradecylammonium (BDTDA)	42
Stearyldimethylbenzylammonium (SDBAC)	47,48
N,N,N,N,N,N-Hexamethyleneguanidine	55

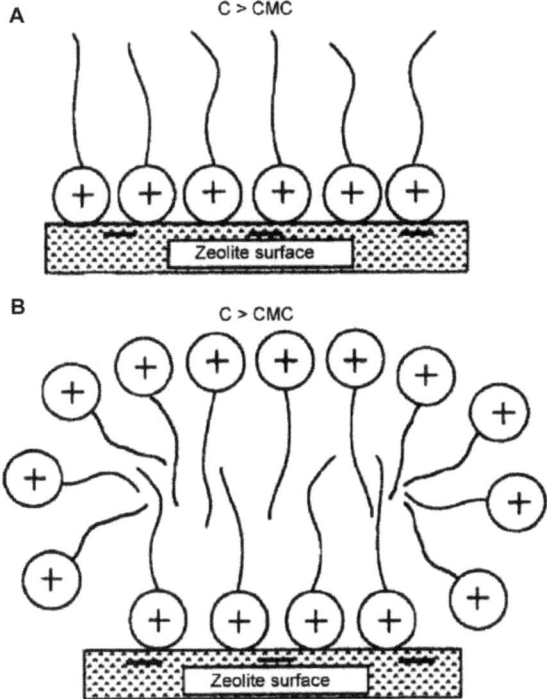

Scheme 5.1 Mechanism of modification of zeolites by CMC.

5.3.2 Modification by Acid/Base Treatment

An acid wash of natural zeolites removes the impurities blocking the pores and further eliminates cations to change into the H-form and finally dealuminate the structure. Proton exchanged zeolites are prepared by two methods: (i) direct ion exchange with dilute acid solution results in dealumination and reduction of thermal stability; (ii) ammonium ion exchange followed by calcination maintains the stable structure. Kurama *et al.*[55] converted the natural Turkish clinoptilolite and reported that ion exchange with hydrogen ions has a great influence on the effective pore volume and surface area of the zeolite. Another study[56] revealed that HCl treatment with varying concentrations and different temperatures affected the surface area and ultra-micropore volume, which in turn was affected by the degree of removal of aluminum from the structure. Treatment by hot HCl on the natural zeolite stilbite from China resulted in variation in the framework, eventually affecting the thermal stability and adsorption properties.

In another study, three zeolites from Armenia, Georgia and Greece were treated with dilute KOH and subsequently treated with HCl or heated at 700 °C. The results showed that the specific area was slightly increased and the microporosity increased due to dissolution of amorphous material by the KOH treatment. Both the specific area and microporosity significantly increased by the acid treatment but the cation exchange capacity was decreased due to partial dissolution of both the Si tetrahedral structure and free linkages, which yielded secondary micropores and destroyed specific exchange sites of the zeolites. Heating decreased both the specific area and microporosity.[57] The study concluded that acid treatment reduced the cation exchange capacity due to dealumination but it improved the Si/Al ratio, facilitating the adsorption/ separation of nonpolar molecules from water.

5.4 Synthetic Zeolites

Alkali treatment of silica- and alumina-based raw materials produces synthetic zeolites. These raw materials can be natural, synthetic or obtained from waste materials. Zeolite synthesis involves the hydrothermal (HT) crystallization of aluminosilicate solutions or gels in the presence of cations at pH > 10. This process is carried out at high temperature in a closed system. The crystallization process takes place in a few hours to several days. The nature of the reactant, the composition of the reaction mixture (Si/Al ratio) and the pH of the system affect the type and purity of the zeolite formed. Pretreatments such as pre-reaction seeding, hydrodynamic conditions, pressure and preliminary or post-heating ageing also play an important role.[58–61] Some of the important synthetic zeolites are given in Table 5.4.

5.4.1 Synthesis of Zeolites from Natural Materials

Zeolite A is the most important synthetic zeolite, which is commercially used for water softening, radioactive waste treatment and industrial wastewater treatment. The most common natural materials used for the synthesis of zeolites are given in Table 5.5.

These materials have high contents of silicon and aluminum, which easily dissolve and form zeolites under alkaline conditions.[82] The zeolites NaP, NaX, NaP1 and NaA with high CECs are prepared by fusion of low-grade natural clinoptilolite,[82,83] halloysite[67] and illite[68] with NaOH prior to hydrothermal treatment. Zeolites may be produced by chemical treatments of rocks for the conversion of Si and/or Al ingredients into alkaline silicates and aluminates and

Table 5.4 Some important synthetic zeolites.

Zeolite	Chemical formula
Zeolite NaA	$Na_{12}(Al_{12}Si_{12}O_{48}) \cdot 27H_2O$
Zeolite NaX	$Na_{86}(Al_{86}Si_{106}O_{384}) \cdot 264H_2O$
Zeolite NaY	$Na_{20}(Al_{20}Si_{48}O_{136}) \cdot 89H_2O$
Zeolite NaP	$Na_2(Al_2Si_{2-5}O_{8-14}) \cdot 5H_2O$
Zeolite NaP1	$Na_8(Al_8Si_8O_{32}) \cdot 16H_2O$
ZSM-5 Nan	$Na_n(Al_nSi_{96-n}O_{192}) \cdot 16H_2O$ ($n<27$)
Cancrinite	$Na_6Ca_2(Al_6Si_6O_{24}(CO_3)_2) \cdot 2H_2O$
Hydroxysodalite	$Na_6(Al_6Si_6O_4)_6 \cdot 8H_2O$

Table 5.5 Natural materials used for the synthesis of zeolites.

Clay mineral	Ref.
Bentonite	62,63
Diatomite	64–66
Halloysite	67
Illite	68
Interstratified illite–smectite	69
Kaolin	70–74
Montmorillonite	75,76
Mordenite	77
Perlite	78,79
Smectite	80,81

subsequent hydrothermal synthesis.[83] Korean serpentine, a source of amorphous Si, has been used for the synthesis of zeolite ZSM-5 with a high specific area.[84] Alkaline dissolution of Tunisian sand and aluminum scrap produced metasilicate and aluminate solutions, and a mixture of these two was used to synthesize a NaZ zeolite.[85] The synthesis of zeolites from natural materials involves high energy-consumption processes like grinding, calcination and fusion, which makes the process disadvantageous. Sometimes raw material mining also destroys the natural landscape. These disadvantages can be overcome by using waste products as raw materials for the synthesis of zeolites.

5.4.2 Synthesis of Zeolites from Industrial Wastes

Waste materials such as ash produced by municipal waste incineration, oil shale, rice husk, coal and other waste have been successfully used for the synthesis of zeolites.

5.4.2.1 Ash from Municipal Solid Waste Incineration

Owing to the presence of Al_2O_3 and SiO_2 and their high specific areas, municipal waste incineration has been explored as a potential starting material for zeolite synthesis.[86–91] Gismondine was synthesized by hydrothermal alkaline processing of the ash; a small amount of NaX was found in the samples. The cation exchange capacities of these zeolites were far less. The zeolites NaA, NaP and sodalite were successfully synthesized by fusion of incinerated ash prior to hydrothermal treatment. The cation exchange capacities of these zeolites were again found to be poor.[90] It was reported later that preliminary fusion of the ash under optimized conditions could increase the CEC of the product.[86]

5.4.2.2 Ash from Oil Shale

Use of fly ash from oil shale processing has been reported for the synthesis of zeolites.[92–94] NaP1 was produced by HT treatment of this ash under optimum

conditions. Purified ash by alkaline fusion followed by refluxing and hydrothermal treatment under optimum conditions produced NaX zeolites.

5.4.2.3   Ash from Rice Husk

Silica-rich rice husk ash has been used for the synthesis of zeolites. Incomplete
roasting of rice husk at 500–700 °C yields carbonized ash which contains carbonaceous material and highly reactive amorphous SiO_2 easily dissolved in an
alkaline medium.[95] This carbonized ash is used as a raw material for the
preparation of alkaline silicate solution. This solution is mixed with standard
aluminate solution and processed *via* hydrothermal treatment to produce
zeolites. Important zeolites thus produced are NaA,[96–102] NaX,[97,103,104]
NaY,[97,105–107] ZSM-5,[108–113] ZSM-48[113] and zeolite-beta.

5.4.2.4   Ash from Coal Fly

Thermal power plants generate millions of tons of ash which contains crystalline and amorphous aluminosilicates and thus can be used as a raw material
for the synthesis of zeolites. Alkaline activation of the coal fly ash (with or
without preliminary fusion with alkali) for the dissolution of Al- and Si-bearing
phases and precipitation gives zeolites.[114–123] Different types of zeolites, such as
chabazite, Linde F, NaA, NaX, NaY, NaP1, *etc.*,[3,114,115,119] could be synthesized under different treatment conditions. The types and yields of zeolite
synthesized were strongly affected by the chemical composition of the ash
used.[124] Other wastes such as paper sludge ash, waste porcelain and cupola slag
exhausted fluid cracking crystals have also been used[125–128] for the preparation
of zeolites and hence disposing them appropriately. Nonrecyclable glass and
thin walled aluminum scrap (Al foils and cans) can also be utilized as sources of
Al and Si.

5.5   Wastewater Treatment by Zeolites

A brief overview on the use of zeolites for water softening and their recent
applications for removal of ammonia from wastewater are summarized in this
section.

5.5.1   Water Softening

Zeolite NaA is used as a commercial water softener in detergent manufacturing
industries.[129] Most of the commercial washing powders contain zeolites, instead of harmful toxic phosphates which are responsible for water eutrophication.[130] Some zeolites have been successfully applied for elimination of
phosphate contamination *via* precipitation of calcium phosphates.[131–133]
A high affinity for exchange of Na cations with the hard Ca ions over a
broad range of pH facilitates the use of zeolite A as a water softener.[134] The rate
of exchange increases at higher temperatures as the hydrated shell of the Ca ion

is gradually removed. The NaP zeolite was reported to show similar results at room temperature; at 60 °C the affinity for Mg also increased.[135] The large pore diameter of zeolite X (0.74 nm) gives it a higher Mg ion binding capacity when compared to zeolite A and zeolite P and is used in detergent manufacturing.[136] Detergent manufacturers also use clinoptilolite[137,138] and zeolite 13X.[139] Zeolite A and zeolite X were more effective in cleaning than clinoptilolite at low temperatures, while all these zeolites had the same effectiveness at high temperatures.[138] Zeolite AX is a relatively new and effective zeolite for water softening applications and is a mixture of zeolite X (80%) and zeolite A (20%). As the purity of the ingredients is important for detergent manufacture of synthetic zeolites, chemicals or high-grade natural materials are preferably used. Some waste-derived zeolites, obtained by appropriate technology, can compete with commercially available products. Zeolite 4A (NaA) samples in a pure form with high crystallinity, obtained using coal fly ash,[140] behaved similar to commercial ones.

Zeolites can also remove dyes from washing liquor by hetero-coagulation and adsorption. Low concentrations of Na ions in zeolites reduce the risk of dyes discoloring other items.[137] The use of phosphate has been totally replaced by zeolites in many parts of the world.[141] Industrial water softening systems and domestic "on the tap" filters use zeolites. Hard cation-loaded zeolites are easily regenerated by treatment with concentrated sodium solutions.

5.5.2 Ammonia Removal

Ammonia, including the ionized NH_4^+ ion species, may have harmful effects on animal and human health.[50] Municipal sewage, agricultural waste, fertilizers and industrial wastewater are the major sources of nitrogen-containing ammonium ion contamination. Nitrogen contributes to accelerated eutrophication of lakes and rivers and decreases dissolved oxygen in water sources.[142] Ammonia in water affects the rubber components of water plumbing fittings. The World Health Organization has not recommended any health-based guidelines for ammonia in drinking water, but its concentration above 30 mg L^{-1} imparts taste problems and concentrations above 1.5 mg L^{-1} cause odor problems.[143] Removal of this ammoniacal nitrogen is essential and various methods have been used for this.[144–146] A high affinity for NH_4^+ ions, safety and low cost make zeolites effective adsorbents for ammonia.

In the last few decades, zeolitic materials have been extensively researched and the results are summarized in various reviews.[2,142] Researchers concluded that clinoptilolite and mordenite were the most effective zeolites for ammonia removal.[147] The adsorption capacity varies with the source and composition of the zeolite, with natural zeolites showing lower adsorption in comparison with synthetic zeolites. Some of the important synthetic zeolites used are NaP,[148,149] NaY,[148,150] NaX[151] and NaA,[151,152] all of whose adsorption capacities are typically in the range 20–50 mg g^{-1}.

Suitable pretreatments such as acid washing, heating, grinding, sieving and pre-exchange with strong Na$^+$ solutions result in improvement in the ammonia adsorption capacity of clinoptilolite.[153] Pre-exchange with the strongly ionic Na form of clinoptilolite has higher ammonia exchange capacity than the K^+ and Ca^{2+} forms, according to the order of affinity of clinoptilolite for alkali and alkaline earth cations: $K^+ > NH_4^+ > Ca^{2+} > Na^+ > Mg^{2+}$.[17] The actual ammonia adsorption capacity and efficiency of the NH_4^+ removal process depend upon the type of zeolite, the initial ammonia concentration, the contact time, the temperature, the amount of zeolite, its particle size and the presence of competitive ions.[2,50] The effect of different parameters on the equilibrium and kinetics of sorption has been discussed in several studies on effluents and model solutions.[147,148,154–157] A list of zeolites used for the removal of ammonia from aqueous solutions[158–176] is summarized in Table 5.6.

Studies of NH_4^+ removal using Croatian clinoptilolite and natural and modified bentonite clay showed that natural zeolite had a much higher removal efficiency than that of the clay under the same conditions. The removal

Table 5.6 Zeolites used for removal of ammonia from aqueous solutions.

Zeolite	Ref.
Natural	
Mordenite	109,147,159
Clinoptilolite	14,16,17,23,31,32,147,150,159–172
Modified	
Clinoptilolite	22,173
Na clinoptilolite	68,171
Ca clinoptilolite	174
Clinoptilolite, NaOH	163
Clinoptilolite, NaHCO₃	163
Clinoptilolite, HCl	163
Clinoptilolite, NaCl	159
Zeolite, microwaves	175
Blends	
Heulandite + montmorillonite + illite	160
Clinoptilolite + heulandites + mordenite	154
Synthetic	
Faujasite, CFA derived	158
NaA, synthetic	160
NaA, CFA derived	158
NaA, from halloysite	160
NaA, commercial	160
NaA, CFA derived	160
NaX, commercial	160
NaX, CFA derived	160
NaP1, CFA derived	160
NaX, with small amounts of HS and NaP, CFA derived	176
NaP, CFA derived	155
NaP, CFA derived, acid washed	155
NaP, CFA derived	160
NaP, from clinoptilolite	148
NaP and calcium silicate hydrate, CFA derived	156
Blend of NaX, HS, NaA and NaP, CFA derived	156
NaY, from clinoptilolite	148
NaY, from clinoptilolite	157
NaY, RHA derived	157

efficiency rapidly decreased with an increase of the initial concentration of ammonia, but removal was achieved in 60 min.[32] The results of equilibrium and kinetic studies of ammonia exchange with Australian clinoptilolite under binary and multi-component conditions revealed that the highest ammonia removal efficiency was achieved with the sodium form of the zeolite. For a multi-component system typically present in sewage a competitive effect between ammonia and other cations such as Ca^{2+}, Mg^{2+} and K^+ was reported.[176] Iranian clinoptilolite in millimeter and nanometer particle sizes was used by Malekian *et al.* in aqueous solutions with different Na ion concentrations. The lowest sodium concentration in the solution showed the maximum NH_4^+ ion exchange capacity.[175] In another study, comparison was done between New Zealand clinoptilolite and mordenite. The mordenite exhibited higher uptake at equilibrium than clinoptilolite at higher solution concentrations. The influence of other cations was less. Contrary to other reports, the ammonium ion showed a higher uptake than the potassium ion.

Another study showed that the presence of other components in the solution, such as heavy metal ions and organics, may affect ammonia exchange due to competitive adsorption.[17,148] The studies used leakage waters from waste dumps and pure ammonia solution on clinoptilolite and showed that it was higher for pure ammonia solution. The results indicated that other components prevented the exchange of ammonium ions.[17] Natural and modified Chilean clinoptilolite has also been used for ammonia uptake.[160] Batch experiments were conducted for 2 h by varying the feed solution concentrations at different pH values. The optimum pH was reported close to neutral (pH 5–8), similar to other studies.[154,148] The Langmuir isotherm model showed the best correlation for the equilibrium data.

The decrease of ammonia uptake at alkaline conditions (pH > 9) could be partially explained by the presence of a significant amount of electrically neutral NH_3 in the solution.[160] Theoretically, ammonia removal should be greater at lower pH and smaller at higher pH values, as the cation exchange mechanism occurs only by means of the ammonium ion.

Volatilization of ammonia contributes to the elimination of ammonia at alkaline pH values. Zhang *et al.*[155] indicated that about 5% of the ammonia is lost by volatilization at pH 11.4. A lower uptake of ammonia at pH > 8 might also be due to partial dissolution of the zeolite.[177] In acidic conditions (pH < 5), competition occurs between NH_4^+ and H^+ ions for the exchangeable sites on the zeolite, which results in the lower uptake.[155,160] Ammonia removal is very fast and rapid equilibrium is attained in 15 min, as shown by sorption kinetic studies.[160] Zeolites might undergo structural degradation in strong acidic solutions (pH < 2).[178] Similar results were reported in other studies also, using clinoptilolite from Turkey and NaA zeolites.[67,179]

Natural Turkish zeolites, containing clinoptilolite, heulandite and mordenite, were used in a batch mode experiment at room temperature and pH 8 to study the removal efficiency of NH_4^+ ions from aqueous solution.[154] The sorption kinetic study results concluded that the removal efficiency of the ammonium ion increased with increased shaking time and that major NH_4^+ ion removal

(75%) occurred within 15 min. It reached 80% within 30 min and after that became very slow. These results clearly showed that the rate of removal was very high in the beginning but reduced significantly after some time. This pattern can be explained by the presence of more vacant adsorption sites initially when the solute concentration gradient was high, but later it decreased and resulted in the slow uptake.[68] The contact time is an important parameter for continuous process operation.[160] Column experiments at different flow rates were carried out for ammonia uptake from drinking water using natural zeolites. The maximum adsorption occurred in 40 min at pH 4. The removal efficiency was reported to be 100% at lower ammonium ion concentrations.[23]

Natural zeolitic material containing 46% heulandite, 24% montmorillonite and 30% illite was used in another study.[177] The results showed that the optimum pH was 8, which indicated that adsorption followed a similar trend and the results obtained were similar to the experiments using clinoptilolite. The presence of other cations affected the removal of the ammonia. The order of preference of adsorption at identical mass concentrations was $Na^+ > K^+ > Ca^{2+} > Mg^{2+}$, and the effect of the presence of individual anions followed the order of preference $CO_3^{2-} > Cl^- > SO_4^{2-} > HPO_4^{2-}$. Owing to an increase of specific area on decreasing the particle size, the adsorption increased. Equilibrium was attained in 180 min and was not affected by the particle size of the adsorbent. An increase in zeolite size resulted in increased adsorption percent. The Freundlich model fitted well to the equilibrium isotherm data.

The equilibrium uptake behavior of ammonium ions on clinoptilolite and zeolite NaY obtained from it have been studied.[157] The results proved that owing to its lower Si/Al molar ratio and bigger aperture, zeolite NaY had a much (almost twice) higher ammonium exchange capacity than the natural zeolite. In both cases the adsorption isotherm data were similar and found to be in accordance with the Freundlich model. The influence of other cations (K, Ca, Mg) was studied and the results of both cases indicated a reduction in the uptake of ammonia. The natural zeolite showed a reduction in the order $K^+ > Ca^{2+} > Mg^{2+}$ (opposite to the order of the equivalent concentrations used); the results obtained were similar to earlier studies.[17] The order of preference was $Mg^{2+} > Ca^{2+} > K^+$, corresponding to their equivalent concentrations reported for zeolite NaZ. These results indicated that zeolite NaY has lower selectivity for these cations than the natural clinoptilolite. This can be explained by the bigger aperture present in zeolite NaY. The reason for this is that NaY has a much larger aperture and the ionic strength in the solution becomes the principal factor in ion exchange. Other studies using the natural clinoptilolite and synthetic zeolites NaY and NaP for ammonium ion exchange showed that the synthetic zeolites have much higher cation exchange capacity than the natural zeolites. Natural zeolites showed the highest selectivity for ammonium ions among all three materials.

Thermodynamic studies indicated that ammonia adsorption on zeolites at ambient conditions was an exothermic and spontaneous process, which was confirmed by several other investigations. Zeolite NaA, synthesized from natural halloysite, was used for studying the adsorption behavior of ammonium ions from aqueous solution. The effect of different parameters such as equilibrium time, pH, initial ammonium ion concentration and temperature were studied in the presence of other competitive cations.[151] Equilibrium was attained within 15 min and the Langmuir isotherm was successfully applied to the adsorption data.

The adsorption capacities of ammonium ions on the NaA zeolite were significantly reduced in the presence of competitive cations; the selectivity order was reported as $Na^+ > K^+ > Ca^{2+} > Mg^{2+}$, which was consistent with the results reported for natural zeolite (heulandite + montmorillonite),[177] but the order does not match with the selectivity order reported for clinoptilolite $(Ca^{2+} > K^+ > Mg^{2+})$ and coal fly ash derived NaX $(K^+ > Ca^{2+} > Na^+ > Mg^{2+})$.[164] These results suggest that different types of zeolite, and even different zeolites of the same type, may exhibit different cation selectivities. NaX obtained from coal fly ash showed that the increase of Ca^{2+}, Mg^{2+}, Na^+ and K^+ concentrations decreased the removal efficiency of ammonium ions. The order of anion effect on ammonia uptake was $CO_3^{2-} > Cl^- > SO_4^{2-}$, which was similar to the results obtained by other authors.[177]

The application of several synthetic zeolites for the decontamination of leachate produced in a municipal solid waste treatment plant and to liquid waste from a pig farm was done by Otal *et al.*[151] The results suggested that the behavior of any given zeolite with regard to the elimination of nitrogen was always better in the case of the pig slurry samples, which contains almost 10 times greater concentration of ammoniacal nitrogen. It was found that the performance of coal fly ash derived zeolites was similar to those of the commercial zeolites of the same type.

Kinetic and equilibrium studies on the removal of ammonium ions from aqueous solution were performed using zeolite NaY synthesized by ash from rice husk and powdered and granulated mordenite. The results showed that NaY had the superior adsorption capacity, three times higher than that of mordenite. The maximum monolayer adsorption capacities were obtained from the Langmuir plots for NaY, powdered mordenite and granulated mordenite. The initial uptake of ammonia occurred rather fast for NaY and took 30 min, whereas powdered mordenite attained equilibrium in 2 h; granular mordenite showed slower ammonia uptake and reached equilibrium in about 24 h. A pseudo-second-order model was followed in kinetic studies. The study concluded that zeolite Y synthesized from rice husk ash can be successfully utilized as an alternative sorbent to remove ammonia from water owing to its low production cost, fast adsorption rate and high adsorption capacitiy.[106]

Owing to the depletion of the free ion exchange sites, the ammonia removal efficiency of zeolites decreases after use for a long duration. This makes the studies of regeneration of zeolites of great practical interest. A number of methods was used for the regeneration of ammonium-loaded zeolites. The modified zeo-SBR was recommended for a new nitrogen removal process that

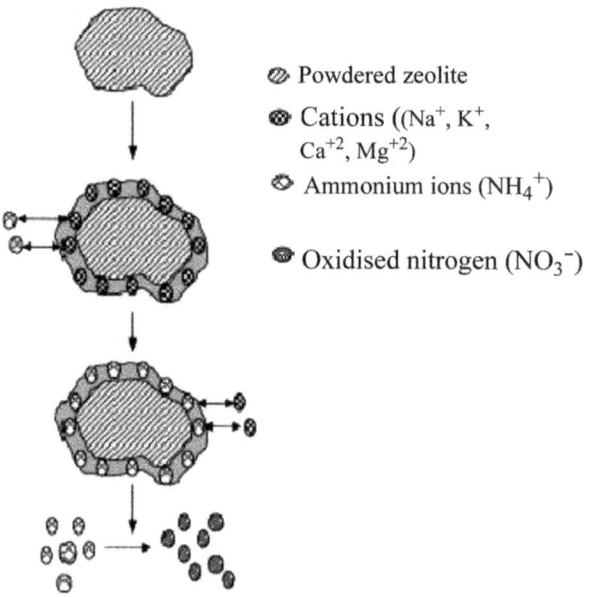

@ Powdered zeolite

Cations ((Na$^+$, K$^+$,
Ca^{+2}, Mg^{+2})

Ammonium ions (NH$_4$$^+$)

Oxidised nitrogen (NO$_3$$^-$)

Scheme 5.2   Adsorption of ammonium ions as well as regeneration of zeolites.

has a special function of consistent ammonium exchange as well as regeneration of zeolites (Scheme 5.2).

Different methods used for the regeneration of zeolites include heating (at 200–500 °C in an air stream), resulting in ammonia evaporation;[16] acid washing, leading to the ion exchange of NH$_4$$^+$ with H$^+$;[16,67] treatment with a sodium salt to obtain a Na zeolite;[16,23,68] electrochemical regeneration (using a Ti/IrO$_2$–Pt anode and Cu/Zn cathode) in the presence of chloride ions to convert ammonia into nitrogen gas;[180] and regeneration by nitrifying bacteria which convert ammonium ions to NO$_3$$^-$ on the surface of the zeolite in oxygen-enriched air.[2] The most effective and feasible method for regeneration of ammonium ion loaded zeolite was reported to be treatment with NaCl solution for half an hour to several hours. Regeneration of natural and modified Chinese clinoptilolite used for ammonia removal from drinking water was studied by heating, acid washing and NaCl treatment. The study concluded that the adsorption capacity of the modified clinoptilolite was lowered only by 4% after three regeneration cycles.[68] A general review of the above studies brings forth some salient factors, conditions and chemical mechanisms, as summarized below.

The ion exchange of ammonium ions by zeolites is a feasible, spontaneous and exothermic process. A high temperature is not favorable for such adsorption. The process is diffusion controlled and the rate is limited by heterogeneous diffusion on the zeolite/solution interface. A smaller particle size and rigorous stirring result in higher adsorption. The complete removal of ammonia is a time consuming process, whereas equilibrium is attained in 2–3 hours at higher zeolite dosage and lower ammonium ion concentration. The adsorption capacity of zeolites is higher at higher ammonium ion concentration, but the percentage of removal is lower; however, the removal efficiency increases with the increase of zeolite dosage. The optimum pH is reported near neutral values (5–8), and an increase in pH results in a decrease of the adsorption of ammonium ions due to transformation of the ions into neutral ammonia. The decrease of pH below the optimum range results in a decrease in adsorption, mainly due to competitive adsorption of hydrogen ions. Partial dissolution of the zeolites is also possible in acidic conditions. Although the removal is influenced by the presence of competitive alkali metal and alkaline earth metal ions, it is mainly dependent on the selectivities of the different types of zeolite used. Natural clinoptilolite has a higher selectivity for ammonium ions than synthetic zeolites, but has a higher affinity for potassium ions, so potassium ions show a strong negative effect on ammonia uptake.

The sodium form of natural and synthetic zeolites is the most suitable for ammonia removal. The zeolites can be effectively and easily regenerated so that they can be reutilized. Synthetic zeolites, namely types A, X and Y, despite their low selectivity to ammonium ions, have much higher ammonia exchange capacity than naturally occurring zeolites. Synthetic zeolites prepared from waste materials were successfully used for ammonia removal and can be considered as a good alternative to the natural zeolites and other commercial adsorbents.

Taking stock of available studies, and scanning them with the purpose of assessing the efficacy of different zeolites for water purification by ammonia removal, gives an insight into the mechanism and structural differentiation of the different components. The broad parameters affecting the efficiency of the process and the structural formation of zeolites are found to be of interest and worth further investigation.

5.6 Conclusion

The unique ion exchange and adsorption properties, high porosity and excellent thermal stability of zeolites make them very suitable for many applications, especially in water treatment processes. Many different studies have demonstrated their effectiveness in reducing the concentrations of contaminants (heavy metals, anions and organic matter) in water. Natural zeolites have proved their applicability in water remediation, although monitoring the changes in pH remain very important for their use in real environments. Zeolites can interact with hydrogen or hydroxyl ions present in solutions and, as a consequence, certain physicochemical phenomena such as hydrolysis of solids, degradation, dissolution and even phase transformations can occur. All these phenomena again depend on the structural characteristics and the chemical composition of the zeolite used. Nowadays, modified natural zeolites are increasingly used also for biological treatment of water, particularly for surface binding of biological agents from water. Suitable pretreatments such as acid washing, heating, grinding, sieving and pre-exchange with strong

Na$^+$ solutions result in an improvement of the ammonia adsorption capacity of natural zeolites. Further research should be focused on the optimization of the surface modification procedures to raise their efficiency and to enhance the capability of regeneration. Furthermore, detailed characterization of natural and modified zeolites is needed to better understand the structure–property relationship. To open up new areas for their application, possible further uses of zeolites as well as the behavior of zeolites at extreme conditions, including low temperatures, should be examined. The effectiveness of zeolites depends upon the starting material and the synthesis procedures used in the preparation of zeolites.

References

1. S. Babel and T. A. Kurniawan, *J. Hazard. Mater.*, 2003, **97**, 219.
2. A. Hedstrom, *J. Environ. Eng. (Reston, VA, U. S.)*, 2001, **127**, 673.
3. X. Querol, N. Moreno, J. C. Umaña, A. Alastuey, E. Hernández, A. L. Soler and F. Plana, *Int. J. Coal Geol.*, 2002, **50**, 413.
4. S. Kesraouiouki, C. R. Cheeseman and R. Perry, *J. Chem. Technol. Biotechnol.*, 1994, **59**, 121.
5. D. Caputo and F. Pepe, *Microporous Mesoporous Mater.*, 2007, **105**, 222.
6. R. Anuwattana and P. Khummongkol, *J. Hazard. Mater.*, 2009, **166**, 227.
7. F. A. Mumpton, *Proc. Natl. Acad. Sci. U. S. A.*, 1999, **96**, 3463.
8. *Natural Zeolites: Occurrence, Properties, Applications*, ed. D. L. Bish and D. W. Ming, Mineralogical Society of America/Geochemical Society, Washington, 2001.
9. D. Kallo, in *Natural Zeolites: Occurrence, Properties, Applications*, ed. D. L. Bish and D. W. Ming, Mineralogical Society of America/Geochemical Society, Washington, 2001, p. 519.
10. P. Misaelides, *Microporous Mesoporous Mater.*, 2011, **144**, 15.
11. S. Wang and H. Zhu, *J. Hazard. Mater.*, 2006, **B136**, 946.
12. N. L. Dimova, O. Petrov and Y. Tzvetanova, *Microporous Mesoporous Mater.*, 2010, **130**, 32.
13. A. H. Englert and J. Rubio, *Int. J. Miner. Process.*, 2005, **75**, 21.
14. Q. Du, S. Liu, Z. Cao and Y. Wang, *Sep. Purif. Technol.*, 2005, **44**, 229.
15. D. A. White and R. L. Bussey, *Sep. Purif. Technol.*, 1997, **11**, 137.
16. M. Li, X. Zhu, F. Zhu, G. Ren, G. Cao and L. Song, *Desalination*, 2011, **271**, 295.
17. A. Farkas, M. Rozic and Z. Barbaric-Mikocevic, *J. Hazard. Mater.*, 2005, **117**, 25.
18. M. Ghiaci, R. Kia, A. Abbaspur and F. Seyedeyn-Azad, *Sep. Purif. Technol.*, 2004, **40**, 285.
19. R. Petrus and J. Warchol, *Microporous Mesoporous Mater.*, 2003, **61**, 137.
20. S. K. Alpat, O. Ozbayrak, S. Alpat and H. Akcay, *J. Hazard. Mater.*, 2008, **151**, 213.
21. S. Capasso, E. Coppola, P. Iovino, S. Salvestrini and C. Colella, *Microporous Mesoporous Mater.*, 2007, **105**, 324.

22. M. Sarioglu, *Sep. Purif. Technol.*, 2005, **41**, 1.
23. M. Sprynskyy, M. Lebedynets, A. P. Terzyk, P. Kowalczyk, J. Namiesnik and B. Buszewski, *J. Colloid Interface Sci.*, 2005, **284**, 408.
24. R. M. Barrer, *J. Chem. Soc.*, 1948, 127.
25. D. Kalló, *Rev. Miner. Geochem.*, 2001, **45**, 519.
26. S. Ouki and M. Kavannagh, *Water Sci. Technol.*, 1999, **39**, 115.
27. T. Motsi, N. A. Rowson and M. J. H. Simmons, *Int. J. Miner. Process.*, 2009, **92**, 42.
28. L. L. Ames, *U.S. Pat.*, 30 017 242, 1962.
29. E. H. Borai, R. Harjula, L. Malinen and A. Paajanen, *J. Hazard. Mater.*, 2009, **172**, 416.
30. A. E. Osmanlioglu, *J. Hazard. Mater.*, 2006, **B137**, 332.
31. N. A. Booker, E. L. Cooney and A. J. Priestley, *Water Sci. Technol.*, 1996, **34**, 17.
32. M. Rozic, S. Cerjan-Stefanovic, S. Kurajica, V. Vancina and E. Hodzic, *Water Res.*, 2000, **34**, 3675.
33. S. C. Bouffard and S. J. B. Duff, *Water Res.*, 2000, **34**, 2469.
34. E. J. Sullivan, J. W. Carey and R. S. Bowman, *J. Colloid Interface Sci.*, 1998, **206**, 369.
35. D. Karadag, E. Akgul, S. Tok, F. Erturk, M. A. Kaya and M. Turan, *J. Chem. Eng. Data*, 2007, **52**, 2436.
36. D. Karadag, M. Turan, E. Akgul, S. Tok and A. Faki, *J. Chem. Eng. Data*, 2007, **52**, 1615.
37. Y. E. Benkli, M. F. Can, M. Turan and M. S. Celik, *Water Res.*, 2005, **39**, 487.
38. R. S. Bowman, *Microporous Mesoporous Mater.*, 2003, **61**, 43.
39. R. Cortes-Martinez, V. Martinez-Miranda, M. Solache-Rios and I. Garcia-Sosa, *Sep. Sci. Technol.*, 2004, **39**, 2711.
40. R. Cortes-Martinez, M. Solache-Rios, V. Martinez-Miranda and R. Alfaro-Cuevas, *Water, Air, Soil Pollut.*, 2007, **183**, 85.
41. G. M. Haggerty and R. S. Bowman, *Environ. Sci. Technol.*, 1994, **28**, 452.
42. A. Kuleyin, *J. Hazard. Mater.*, 2007, **144**, 307.
43. A. Dakovic, M. Tomasevic-Canovic, G. Rottinghaus, V. Dondur and Z. Masic, Colloids Surf, *B*, 2003, **30**, 157.
44. A. Dakovic, M. Tomasevic-Canovic, G. E. Rottinghaus, S. Matijasevic and Z. Sekulic, *Microporous Mesoporous Mater.*, 2007, **105**, 285.
45. A. Dakovic, S. Matijasevic, G. E. Rottinghaus, V. Dondur, T. Pietrass and C. F. M. Clewett, *J. Colloid Interface Sci.*, 2007, **311**, 8.
46. M. Ghiaci, A. Abbaspur, R. Kia and F. Seyedeyn-Azad, *Sep. Purif. Technol.*, 2004, **40**, 217.
47. J. Lemic, M. Tomasevic-Canovic, M. Adamovic, D. Kovacevic and S. Milicevic, *Microporous Mesoporous Mater.*, 2007, **105**, 317.
48. J. Lemic, D. Kovacevic, M. Tomasevic-Canovic, D. Kovacevic, T. Stanic and R. Pfend, *Water Res.*, 2006, **40**, 1079.
49. M. Noroozifar, M. Khorasani-Motlagh, M. N. Gorgij and H. R. Naderpour, *J. Hazard. Mater.*, 2008, **155**, 566.

50. K. A. Northcott, J. Bacus, N. Taya, Y. Komatsu, J. M. Perera and G. W. Stevens, *J. Hazard. Mater.*, 2010, **183**, 434.
51. R. E. Apreutesei, C. Catrinescu and C. Teodosiu, *J. Environ. Eng. Manage.*, 2008, **7**, 149.
52. N. Widiastuti, H. Wu, M. Ang and D. Zhang, *Desalination*, 2008, **218**, 271.
53. M. D. Richards and C. G. Pope, *J. Chem. Soc., Faraday Trans.*, 1996, **92**, 317.
54. E. I. Basaldella, J. C. Paladino, M. Solari and G. M. Valle, *Appl. Catal. B*, 2006, **66**, 186.
55. H. Kurama, A. Zimmer and T. Reschetilowski, *Chem. Eng. Technol.*, 2002, **25**, 301.
56. F. Cakicioglu-Ozkan and S. Ulku, *Microporous Mesoporous Mater.*, 2005, **77**, 47.
57. G. E. Christidis, D. Moraetis, E. Keheyan, L. Akhalbedashvili, N. Kekelidze, R. Gevorkyan, H. Yeritsyan and H. Sargsyan, *Appl. Clay Sci.*, 2003, **24**, 79.
58. C. S. Cundy and P. A. Cox, *Microporous Mesoporous Mater.*, 2005, **82**, 1.
59. Y. Marui, R. Irie, H. Takiyama, H. Uchida and M. Matsuoka, *J. Cryst. Growth*, 2002, **237**, 2148.
60. H. J. Koroglu, A. Sarioglan, A. M. Tatlier, A. Erdem-Senatalar and O. T. Savasci, *J. Cryst. Growth*, 2002, **241**, 481.
61. C. Kosanovic, T. A. Jelic, J. Bronic, D. Kralj and B. Subotic, *Microporous Mesoporous Mater.*, 2011, **137**, 72.
62. R. Ruiz, C. Blanco, C. Pesquera, F. Gonzalez, I. Benito and J. L. Lopez, *Appl. Clay Sci.*, 1997, **12**, 73.
63. H. Faghihian and N. Godazandeha, *J. Porous Mater.*, 2009, **16**, 331.
64. Y. Du, S. Shi and H. Dai, *Particuology*, 2011, **9**, 174.
65. A. Chaisena and K. Rangsriwatananon, *Mater. Lett.*, 2005, **59**, 1474.
66. B. Ghosh, D. C. Agrawal and S. Bhatia, *Ind. Eng. Chem. Res.*, 1994, **33**, 2107.
67. Y. Zhao, B. Zhang, X. Zhang, J. Wang, J. Liu and R. Chen, *J. Hazard. Mater.*, 2010, **178**, 658.
68. M. Mezni, A. Hamzaoui, N. Hamdi and E. Srasra, *Appl. Clay Sci.*, 2011, **52**, 209.
69. A. Baccouche, E. Srasra and M. E. Maaoui, *Appl. Clay Sci.*, 1998, **13**, 255.
70. E. I. Basaldella and J. C. Tara, *Zeolites*, 1995, **11**, 243.
71. C. A. Ríos, C. D. Williams and M. A. Fullen, *Appl. Clay Sci.*, 2009, **42**, 446.
72. D. Akolekar, A. Chaffee and R. F. Howe, *Zeolites*, 1997, **19**, 359.
73. M. Murat, A. Amokrane, J. P. Bastide and L. Montanaro, *Clay Miner.*, 1992, **27**, 119.
74. M. Meftah, B. Oueslati and A. Ben Haj Amara, *Phys. Procedia*, 2009, **2**, 1081.
75. P. Cañizares, A. Durán, A. F. Dorado and M. Carmona, *Appl. Clay Sci.*, 2000, **16**, 273.

76. I. D. R. Mackinnon, G. J. Millar and W. Stolz, *Appl. Clay Sci.*, 2010, **48**, 622.
77. C. Covarrubias, R. Garcia, R. Arriagada, J. Yanez and M. T. Garland, *Microporous Mesoporous Mater.*, 2006, **88**, 220.
78. U. Barth-Wirshing, H. Holler, D. Klammer and B. Konrad, *Miner. Petrol.*, 1993, **48**, 275.
79. G. E. Christidis, L. Paspaliaris and A. Kontopoulos, *Appl. Clay Sci.*, 1999, **15**, 305.
80. K. Abdmeziem and B. Siffert, *Appl. Clay Sci.*, 1994, **8**, 437.
81. K. Abdmeziem and B. Siffert, *Appl. Clay Sci.*, 1989, **4**, 1.
82. S. J. Kang, K. Egashira and K. A. Yoshida, *Appl. Clay Sci.*, 1998, **13**, 117.
83. D. Novembre, B. Di Sabatino, D. Gimeno, M. Garcia-Valles and S. Martinez-Manent, *Microporous Mesoporous Mater.*, 2004, **75**, 1.
84. D. J. Kim and H. S. Chung, *Appl. Clay Sci.*, 2003, **24**, 69.
85. H. Tounsi, S. Mseddi and S. Djemel, *Phys. Procedia*, 2008, **2**, 1065.
86. M. Sallam, *PhD thesis*, University of South Florida, 2006.
87. Y. Fan, F. S. Zhang, J. Zhua and Z. Liu, *J. Hazard. Mater.*, 2008, **153**, 382.
88. Z. Yao, C. Tamura, M. Matsuda and M. Miyake, *J. Mater. Res.*, 1999, **14**, 4437.
89. M. Miyake, C. Tamura and M. Matsuda, *J. Am. Ceram. Soc.*, 2002, **85**, 1873.
90. G. C. C. Yang and T. Y. Yang, *J. Hazard. Mater.*, 1998, **62**, 75.
91. R. P. Penilla, G. Bustos and G. Elizalde, *J. Am. Ceram. Soc.*, 2003, **86**, 1527.
92. R. Shawabkeh, R. A. Al-Harahsheh, M. Hami and A. Khlaifat, *Fuel*, 2004, **83**, 981.
93. N. R. C. F. Machado and D. M. M. Miotto, *Fuel*, 2005, **84**, 2289.
94. R. A. Shawabkeh, *Microporous Mesoporous Mater.*, 2004, **75**, 107.
95. H. Nur, *Indones. J. Agric. Sci.*, 2001, **1**, 40.
96. S. N. Azizi and M. Yousefpour, *J. Mater. Sci.*, 2010, **45**, 5692.
97. A. M. Yusof, N. A. Nizam and N. A. A. Rashid, *J. Porous Mater.*, 2010, **17**, 39.
98. C. Bhavornthayod and P. Rungrojchaipon, *Mater. Miner.*, 2009, **19**, 79.
99. E. L. Folletto, M. M. Castoldi, L. H. Oliveira, R. Hoffmann and S. L Jahn, *Lat. Am. Appl. Res.*, 2009, **39**, 75.
100. Z. Ghasemi and H. Younesi, J. Nanomater., *article ID*, 2011, 858961; doi: 10.1155/2011/858961.
101. D. I. Petkowicz, R. T. Rigo, C. Radtke, S. B. Pergher and J. H. Z. dos Santos, *Microporous Mesoporous Mater.*, 2008, **116**, 548.
102. T. Wajima, O. Kiguchi, K. Sugawara and T. Sugawara, *J. Chem. Eng. Jpn.*, 2009, **42**, 61.
103. H. Katsuki and S. Komarneni, *J. Solid State Chem.*, 2009, **182**, 1749.
104. P. Khemthong, S. Prayoonpokarach and J. Jatuporn Wittayakun, *Suranaree J. Sci. Technol.*, 2007, **14**, 367.

105. M. M. Rahman, N. Hasnida and W. B. Wan Nik, *J. Sci. Res.*, 2009, **1**, 285.

106. A. M. Yusof, L. K. Keat, Z. Ibrahim, Z. A. Majid and N. A. Nizam, *J. Hazard. Mater.*, 2010, **174**, 380.

107. J. Wittayakun, P. Khemthong and S. Prayoonpokarach, *Korean J. Chem. Eng.*, 2008, **25**, 861.

108. M. Chareonpanich, T. Namto and P. Kongkachuichay, *Fuel Process. Technol.*, 2004, **85**, 1623.

109. H. Katsuki, S. Furuta, T. Watari and S. Komarneni, *Microporous Mesoporous Mater.*, 2005, **86**, 145.

110. K. Kordatos, S. Gavela, A. Ntziouni, K. N. Pistiolas, A. Kyritsi and V. Kasselouri-Rigopoulou, *Microporous Mesoporous Mater.*, 2008, **115**, 189.

111. M. M. Mohamed, F. I. Zidan and M. Thabet, *Microporous Mesoporous Mater.*, 2008, **108**, 193.

112. W. Panpa and S. Jinawath, *Appl. Catal. B*, 2009, **90**, 389.

113. P. H. Wang, K. S. Lin, Y. J. Huang, M. C. Li and L. K. Tsaur, *J. Hazard. Mater.*, 1998, **58**, 147.

114. C. A. R. Ríos, C. D. Williams and C. L. Roberts, *Fuel*, 2009, **88**, 1403.

115. J. Garcia-Martiez, D. Cazorla-Amoros and A. Linares-Solano, *J. Chem. Technol. Biotechnol.*, 2002, **77**, 287.

116. N. Murayama, H. Yamamoto and J. Shibata, *Int. J. Miner. Process.*, 2002, **64**, 1.

117. H. Mimura, K. Yokota, K. Akiba and Y. Onodera, *J. Nucl. Sci. Technol.*, 2001, **38**, 766.

118. X. Querol, A. Alastuey, J. L Fernandez-Turiel and A. Lopez-Soler, *Fuel*, 1995, **74**, 1226.

119. X. Querol, J. C. Umana, F. Plana, A. Alastuey, A. Lopez-Soler, A. Medinaceli, A. Valero, M. J. Domingo and E. Garcia-Rojo, *Fuel*, 2001, **80**, 857.

120. N. Moreno, X. Querol and C. Ayora, *Environ. Sci. Technol.*, 2001, **35**, 3526.

121. G. Belardi, S. Massimilla and L. Piga, *Resour., Conserv. Recycl.*, 1998, **24**, 167.

122. S. S. Nam, M. W. Lee, S. B. Kim, K. W. Lee and M. K. Ko, *Environ. Eng. Res.*, 2000, **5**, 35.

123. C. F. Wang, J. S. Li, L. J. Wang and X. Y. Sun, *J. Hazard. Mater.*, 2008, **155**, 58.

124. A. Shoumkova and V. Stoyanova, *Proc. Bulg. Acad. Sci.*, 2011, **64**, 937.

125. T. Wajima, M. Haga, K. Kuzawa, H. Ishimoto, O. Tamada, K. Ito, T. Nishiyama, R. T. Downs and J. F. Rakovan, *J. Hazard. Mater.*, 2006, **B132**, 244.

126. R. Anuwattana, K. J. Bulkus Jr., S. Asavapisit and P. Khummongkol, *Microporous Mesoporous Mater.*, 2008, **111**, 260.

127. P. Misaelides, V. A. Nikashina, A. Godelitsas, P. A. Gembitskii and E. M. Kats, *J. Radioanal. Nucl. Chem.*, 1998, **227**, 183.

128. E. L Basaldella, R. M. Torres Sánchez and M. S. Conconi, *Appl. Clay Sci.*, 2009, **42**, 611.

129. Anon, *J. Am. Oil Chem. Soc.*, 1980, **228A**, 57.

130. Y. Yu, J. Zhao and A. E. Bayly, *Chin. J. Chem. Eng.*, 2008, **16**, 517.

131. Q. Guan, X. Hu, D. Wu, X. Shang, C. Ye and H. Kong, *Fuel*, 2009, **88**, 1643.

132. J. Chen, H. Kong, D. Wu, Z. Hu, Z. Wang and Y. Wang, *J. Colloid Interface Sci.*, 2006, **300**, 491.

133. N. Karapinar, *J. Hazard. Mater.*, 2009, **170**, 1186.

134. A. R. Loiola, J. C. R. A. Andrade, J. M. Sasaki and L. R. D. da Silva, *J. Colloid Interface Sci.*, 2010, **367**, 34.

135. I. Arrigo, P. Catalfamo, L. Cavallari and S. Di Pasquale, *J. Hazard. Mater.*, 2007, **147**, 513.

136. H. G. Hauthal, *SOFW J.*, 1996, **122**, 899.

137. M. Culfaz and N. E. Saracoglu, *J. Environ. Sci. Health*, 1999, **A34**, 1619.

138. M. Culfaz, N. Saracoglu and O. Ozdilek, *J. Chem. Technol. Biotechnol.*, 1996, **65**, 265.

139. A. de Lucas, M. A. Uguina, I. Covian and L. Rodriguez, *Ind. Eng. Chem. Res.*, 1992, **32**, 2134.

140. K. S. Hui and C. Y. H. Chao, *J. Hazard. Mater.*, 2006, **B137**, 401.

141. P. Krings and E. J. Smulders, *Journal of the 45th SEPAWA Congress*, Bad Durkheim, Germany, 1998, p. 70.

142. S. Wang and Y. Peng, *Chem. Eng. J.*, 2010, **156**, 11.

143. *Guidelines for Drinking-Water Quality: Incorporating First Addendum, vol. 1, Recommendations*, World Health Organization, Geneva, 3rd. edn., 2006.

144. C. Fux, M. Boehler, P. Huber, I. Brunner and H. Siegrist, *J. Biotechnol.*, 2002, **99**, 295.

145. T. A. Kurniawan, W. H. Lo and G. Y. S. Chan, *J. Hazard. Mater.*, 2006, **129**, 80.

146. H. Kurama, C. Karaguzel, T. Mergan and M. S. Celik, *Desalination*, 2010, **253**, 147.

147. L. R. Weatherley and N. D. Miladinovic, *Water Res.*, 2004, **38**, 4305.

148. Y. Wang, F. Lin and W. Pang, *J. Hazard. Mater.*, 2008, **160**, 371.

149. Y. Luna, E. Otal, L. F. Vilches, J. Vale, X. Querol and C. Fernandez-Pereira, *Waste Manage.*, 2007, **27**, 1877.

150. Y. F. Wang, F. Lin and W. Q. Pang, *J. Hazard. Mater.*, 2007, **142**, 160.

151. E. Otal, L. F. Vilches, N. Moreno, X. Querol, J. Vale and C. Fernandez-Pereira, *Fuel*, 2005, **84**, 1440.

152. W. H. Shih and H. L. Chang, *Mater. Lett.*, 1996, **28**, 263.

153. Z. Liang and J. Ni, *J. Hazard. Mater.*, 2009, **166**, 52.

154. K. Saltali, A. Ahmet Sari and M. Aydın, *J. Hazard. Mater.*, 2007, **141**, 258.

155. B. H. Zhang, D. Y. Wu, C. Wang, S. B. He, Z. J. Zhang and H. N. Kong, *J. Environ. Sci.*, 2007, **19**, 540.

156. M. Zhang, H. Zhang, D. Xu, L. Han, D. Niu, L. Zhang, W. Wu and B. Tian, *Desalination*, 2011, **277**, 46.
157. J. Gallant and A. Prakash, presented at the Canadian Nuclear Society 30th Annual CNS Conference and 33rd CNS/CAN Student Conference, University of Western Ontario, London, Ontario, Canada, 2009.
158. A. A. Halim, H. A. Aziz, M. A. M. Johari and K. S. Ariffin, *Desalination*, 2010, **262**, 31.
159. A. H. Englert and J. Rubio, *Int. J. Miner. Process.*, 2005, **75**, 21.
160. H. M. Huang, X. Xiao, B. Yan and L. Yang, *J. Hazard. Mater.*, 2010, **175**, 247.
161. R. Malekian, J. Abedi-Koupai, S. S. Eslamian, S. F. Mousavi, K. C. Abbaspour and M. Afyuni, *Appl. Clay Sci.*, 2011, **51**, 323.
162. X. Guo, L. Zeng, X. Li and H. S. Park, *J. Hazard. Mater.*, 2008, **151**, 125.
163. N. S. Bolan, C. Mowatt, D. C. Adriano and J. D. Blennerhassett, *Commun. Soil Sci. Plant Anal.*, 2003, **34**, 1861.
164. Y. Q. Wang, S. J. Liu, Z. Xu, T. W. Han, S. Chuan and T. Zhu, *J. Hazard. Mater.*, 2006, **136**, 735.
165. D. H. Wen, Y. S. Ho and X. Y. Tang, *J. Hazard. Mater.*, 2006, **133**, 252.
166. D. Karadag, Y. Koc, M. Turan and B. Armagan, *J. Hazard. Mater.*, 2006, **136**, 604.
167. D. Karadag, Y. Koc, M. Turan and M. Ozturk, *J. Hazard. Mater.*, 2007, **144**, 432.
168. A. Gunay, *J. Hazard. Mater.*, 2007, **148**, 708.
169. C. H. Liu and K. V. Lo, *J. Environ. Sci. Health, Part A*, 2001, **36**, 1671.
170. X. J. Guo, L. Zeng, X. M. Li and H. S. Park, *Sep. Sci. Technol.*, 2007, **42**, 3169.
171. S. N. Ashrafizadeh, Z. Khorasani and M. Gorjiara, *Sep. Sci. Technol.*, 2008, **43**, 960.
172. A. A. Halim, H. A. Aziz, M. A. M. Johari and K. S. Ariffin, *Desalination*, 2010, **262**, 31.
173. Z. Y. Ji, J. S. Yuan and X. G. Li, *J. Hazard. Mater.*, 2007, **141**, 483.
174. L. C. Lei, X. J. Li and X. W. Zhang, *Sep. Purif. Technol.*, 2008, **58**, 359.
175. M. Zhang, H. Zhang, D. Xu, L. Han, D. Niu, B. Tian, J. Zhang, L. Zhang and W. Wu, *Desalination*, 2011, **271**, 111.
176. E. L. Cooney, N. A. Booker, D. C. Shallcross and G. W. Stevens, *Sep. Sci. Technol.*, 1999, **43**, 2307.
177. H. M. Huang, X. Xiao, B. Yan and L. Yang, *J. Hazard. Mater.*, 2010, **175**, 247.
178. D. E. W. Vaughan, in *Natural Zeolites, Occurrence, Properties, Use*, ed. L. B. Sand and F. A. Mumpton, Pergamon, New York, 1978, p. 353.
179. A. Demir, A. Gunay and E. Debik, *Water S. Afr.*, 2002, **28**, 329.
180. X. Lei, M. Li, Z. Zhang, C. Feng, W. Baid and N. Sugiura, *J. Hazard. Mater.*, 2009, **169**, 746.

CHAPTER 6

Functionalized Silica Gel as Green Material for Metal Remediation

R. K. SHARMA,* GARIMA GABA, ANIL KUMAR AND ADITI PURI

Green Chemistry Network Centre, Department of Chemistry, University of Delhi, Delhi – 110007, India
*Email: rksharmagreenchem@hotmail.com

6.1 Introduction

Over the past few decades there has been a change in focus from scientific expansion and innovation to the widespread chronic effects of metals. This is due to concern towards worldwide environmental safety. Global growth and industrial development are the major factors that cause immense atmospheric depositions of metals. Actually, industrial operations like electroplating, mining and metal processing are the key contributors to unnatural metal contamination in the surroundings. The voluminous discharge of these metals has dramatically altered the biogeochemical cycles and there is no compartment of the atmosphere that has not been deteriorated by their toxic effects. This uncontrolled release of metals creates toxic or inhibitory effects and other degenerative diseases in living systems. For instance, Cd and Zn can lead to severe gastrointestinal and respiratory damage and acute heart, brain and kidney damage. Being persistent in natural ecosystems for an extended period, these metals accumulate in successive levels of the biological chain and thereby

RSC Green Chemistry No. 23
Green Materials for Sustainable Water Remediation and Treatment
Edited by Anuradha Mishra and James H. Clark

cause various damaging effects. The problem is even worse in developing countries owing to accelerated urban growth and lack of efficient control measures. Unfortunately, water bodies have been subjected to maximum exploitation and they are severely degraded or polluted because of these activities. This is due to the fact that the treatment of wastewater generated by industries is not given the necessary priority it deserves and this wastewater is discharged into water resources without treatment. This water is then used in various agricultural and day-to-day practices, posing several threats to mankind. Various rivers, lakes and estuaries have been found to be severely polluted with poisonous elements such as Cr, Cu, Cd and Ni to different degrees. Therefore, extraction and removal of toxic metal ions from various matrices at trace levels are of paramount importance.

As a solution to this, chelating agents immobilized on a solid support, known as chelating resins, are attracting great interest in environmental applications to remedy the polluted water resources. Solid supports can be mainly categorized as organic ones, which include synthetic as well as other polymeric compounds of natural origin, while inorganic solid supports comprise silica gel, alumina, magnesia, zirconia and other oxide species. Immobilization of organic complexing agents on the surface of an inorganic or organic solid support is usually aimed at modifying the surface with certain target functional groups that can be exploited for specific metal extraction.

Analysis of metal ions at trace levels poses a unique problem to analysts, because it involves the rigorous requirements of versatility, specificity, sensitivity and accuracy in the analysis. A wide variety of analytical techniques has been developed to determine concentrations of trace metals in various samples. However, the results may be erroneous because the metal ions may be present in these samples together with other elements at low levels. Thus, to quantify the desired analytes, the best way is to separate and preconcentrate the metal ions from the matrix constituents and determine them in an isolated state. Various techniques have been employed for this purpose. The most widely used method for the separation and preconcentration of metal ions with suitable complexing agent is solid phase extraction (SPE).

SPE is considered to be a powerful tool for the separation and enrichment of various inorganic as well as organic analytes.[1–5] The basic principle of SPE is the transfer of the analyte from the aqueous phase to the active sites of the solid phase. It has several advantages,[6,7] including:

1. Stability and reusability of the solid phase.
2. High preconcentration factors.
3. Ease of separation and enrichment under dynamic conditions.
4. No need for toxic and costly organic solvents.
5. Minimum costs due to low consumption of reagents.

SPE is mainly based on the utilization of inorganic and organic solid sorbents such as XAD resins,[8–10] ion exchange resins,[11] silica gel,[12,13] cellulosic derivatives,[14] polyurethane foam,[15] active carbon,[16] nanometer silica[17] and rice husks.[18] Extraction and removal of metal ions by these sorbents is well known and mainly based on the possible surface reactivity and adsorptive characters incorporated into these solid phases.[19] However, the basic disadvantage of these solid sorbents is the lack of metal selectivity, which leads to high interference of other existing species with the target metal ion(s).[20] To overcome this problem, a chemical or physical modification of the sorbent surface with a metal-selective complexing agent is required.

6.2 Benefits of Chelating Sorbents

Extraction of metal ions using chelating sorbents has the following advantages over more conventional methods:[21]

- Selective determination of metal ions is possible using a chelating sorbent having a ligand possessing high selectivity to the targeted metal ion.
- It is free from difficult phase separations, which are caused by the mutual solubility between water and organic solvent layers.
- The chelating sorbent method is economical since it uses only a small amount of ligand and can be reused a number of times.
- Trace metal ions at concentrations as low as parts per billion (ppb) can be determined.
- The adsorption of metal ions can be visibly estimated from the color intensity of the solid phase if the metal complex formed possesses adsorption in the visible wavelength region.
- Use of carcinogenic organic solvents is avoided and thus the technique is ecofriendly.

The chelators based on functionalized silica gel are preferred over conventional ion-exchange-based resins because of their high selectivity and capability of binding metals through multiple coordinating groups attached to the support.[22] The multiple coordination of metal ions on the chelating resin makes it particularly suitable for the collection of metal ions.

6.3 Silica Gel: An Ideal Support Material

Silica gel is a granular, porous form of silica and is made synthetically from sodium silicate or silicon tetrachloride or substituted chlorosilane/orthosilicate solution. Stober *et al.*[23] have reported the synthesis of spherical silica particles from tetraethoxysilane using NH_3 as catalyst.

Silica gel is commonly used as a rigid matrix for ligand immobilization. The chemical modification of silica gel surfaces with donor atoms such as N, S, O and P is primarily aimed at improving the adsorption and exchange properties of the silica gel along with incorporation of the particular selective characteristics into the modified silica gel phases towards certain metal ions. The immobilization of chelating materials containing donor atoms on the silica gel surface can occur *via* chemical bond formation between organic modifiers like amino- or chloro-modified silica gel phases or through simple physical adsorption processes.[24] Such a process often incorporates selectivity in the synthesized materials. Several new chemically modified silica gels have been synthesized and applied as normal or selective solid phase extractors for metal ions. The advantages of silica gel over other support materials are:[25,26]

- The polymeric support in the form of silica gel offers high chemical stability and a compact structure, so that during the synthesis of functionalized silica the functional structure of the chelating agents is not changed.
- Unlike organic polymers, which are flexible and can swell up to varying degrees depending on the solvent, temperature and pressure, silica gel has a rigid structure and hence during the synthesis of functionalized silica the functional structure of the chelating agent is not changed.
- Silica is less susceptible than organic polymers to chemical or thermal degradation, so functionalized silica gels can be used over a wide range of temperatures and under strongly acidic or alkaline conditions.

6.4 Functionalization of Silica Gel

A brief description of the various functional groups present on the silica surface followed by the methods employed in the synthesis of functionalized silica gels now follows.

6.4.1 Surface Chemistry of Silica Gel

The silica surface consists of two types of functional groups: siloxane (Si–O–Si) and silanol (Si–OH). Thus, silica gel modification can occur *via* the reaction of a particular molecule with either the siloxane (nucleophilic substitution at the Si) or silanol (direct reaction with the hydroxyl group) group, although it is generally accepted that the reaction with the silanol group constitutes the main modification pathway.

The silanols can be divided into isolated groups (or free silanols), where the surface silicon atom has three bonds into the bulk structure and the fourth bond attached to a single OH group, and vicinal silanols (or bridged silanols), where two isolated silanol groups are bridged by a hydrogen bond. A third type of silanol, called geminal silanols, consists of two hydroxyl groups attached to one silicon atom.

6.4.2 Chemical Modification of the Silica Surface

Silica gel can be used as a very successful adsorbing agent, as it does not swell or strain, has good mechanical strength and can undergo heat treatment. In addition, chelating agents can be easily loaded on silica gel with high stability, or be bound chemically to the support, affording a higher stability. The surface of silica gel is characterized by the presence of silanol groups, which are known to be weak ion-exchangers, causing low interaction, binding and extraction of various metal ions such as Cu, Ni, Co, Zn and Fe.[27,28] In addition, silica gel as a sorbent has a very low selectivity to extract a particular metal ion over background matrices. Consequently, modification of the silica gel surface has been performed to obtain solid sorbents with greater selectivity. In most of the methods for the preparation of immobilized silica gel, a two-step procedure has been used for loading the surface with specific organic compounds: physical adsorption and chemical functionalization. In the first case, the organic compound is directly adsorbed on the silanol group of the silica gel surface (impregnated or loaded sorbent), either by passing the reagent solution through a column packed with the adsorbent or by shaking the adsorbent in the reagent solution. In the second approach, a covalent bond is formed between the silica gel surface groups and those of the organic compound (functionalized sorbent). Chemisorption of chelating molecules on the silica surface provides immobility, mechanical stability and water insolubility, thereby increasing the efficiency, sensitivity and selectivity of the analytical applications.[29] Chemical modification of the silica surface by organic chelating groups acts as an ion exchanger, which provides greater selectivity for the analyte than that offered by a traditional ion exchanger. The most convenient way to develop a chemically modified surface is by simple immobilization (or fixing) of the group on the surface by adsorption, electrostatic interaction, hydrogen bond formation or other types of interaction.[30,31] Simple impregnation of the modifier solution[32] or covalent binding, so-called covalent grafting of the chelating molecule to the silica matrix *via* silanization, is the common practice for synthesizing a metal-specific functionalized silica gel.

6.5 Analytical Applications of Modified Silica Gels as Chelating Sorbents

Many functionalized silica gels (Table 6.1) have been synthesized for scavenging metal ions.[33–92] Their metal extraction ability is not just limited to wastewater treatment but they also help to remediate various environmental compartments. A brief discussion of various functionalized silica gels follows the table.

8-Hydroxyquinoline (HQN) functionalized silica gel has been used by many workers for the preconcentration of trace elements. The silica gel modified with (3-aminopropyl)triethoxysilane was reacted with 5-formyl-8-hydroxyquinoline to anchor the 8-quinolinol ligand on the silica gel. It was used for the preconcentration of Cu^{2+}, Pb^{2+}, Ni^{2+}, Fe^{3+}, Cd^{2+}, Zn^{2+} and Co^{2+} prior to their determination by flame atomic absorption spectrometry (FAAS). The optimum pH ranges for quantitative sorption are 4.0–7.0, 4.5–7.0, 3.0–6.0, 5.0–8.0, 5.0–8.0, 5.0–8.0 and 4.0–7.0 for Cu, Pb, Fe, Zn, Co, Ni and Cd, respectively. All the metals can be desorbed with 2.5 M HCl or HNO_3. The sorption capacity for these metal ions is in range of 92–448.0 $\mu mol\ g^{-1}$ and follows the order Cd < Pb < Zn < Co < Ni < Fe < Cu. Good tolerance limits of

Table 6.1 Analytical application of functionalized silica gels for scavenging trace amounts of metal ions.

Functional group anchored	Metal ions separated and/or preconcentrated	Ref.
8-Hydroxyquinoline	Cu^{2+}, Pb^{2+}, Fe^{3+}, Zn^{2+}, Co^{2+}, Ni^{2+}, Cd^{2+}	33,34
Dimethylglyoxime	Ni^{2+}	35
Dithizone and zinc dithizone	Ag^{+}, Pb^{2+}, Hg^{2+}	36,37
Imidazole	Cr^{3+}, Mn^{2+}, Fe^{3+}, Co^{2+}, Cu^{2+}, Zn^{2+}, Hg^{2+}, Cd^{2+}, Al^{3+}	38
Diaza-18-crown-6	Cr^{3+}, Mn^{2+}, Fe^{3+}, Co^{2+}, Cu^{2+}, Zn^{2+}, Hg^{2+}, Cd^{2+}, Al^{3+}	38
Dibenzo-18-crown-6	Cr^{3+}, Mn^{2+}, Fe^{3+}, Co^{2+}, Cu^{2+}, Zn^{2+}, Hg^{2+}, Cd^{2+}, Al^{3+}	38
Dithioacetal	Hg^{2+}	39
Ethylene sulfide	Co^{2+}, Ni^{2+}, Cu^{2+}, Hg^{2+}, Pb^{2+}	40
1,8-Dihydroxyanthraquinone	Zn^{2+}, Pb^{2+}, Cd^{2+}	41
1-Aminoanthraquinone	Cr^{3+}, Cu^{2+}	42
Resacetophenone	Fe^{3+}, Co^{2+}, Ni^{2+}, Cu^{2+}, Zn^{2+}, Pb^{2+}, Cd^{2+}	43
Salicylaldoxime	Cu^{2+}, Co^{2+}, Ni^{2+}, Zn^{2+}	44
Diphenylisothiourea	Ag^{+}, Au^{3+}, Pd^{2+}, Pt^{4+}	45
2-Hydroxynaphthaldehyde	Cu^{2+}, Zn^{2+}, Cd^{2+}, Hg^{2+}, Pb^{2+}	46
β-Diketone	Mn^{2+}, Co^{2+}, Ni^{2+}, Cu^{2+}, Zn^{2+}	47
4,4,4-Trifluoro-1-(2-thienyl)butane-1,3-dione	Co^{2+}, Ni^{2+}	48
5-Formyl-3-(1′-carboxyphenylazo)salicylic acid	Cr^{3+}, Mn^{2+}, Fe^{3+}, Co^{2+}, Cu^{2+}, Zn^{2+}, Cd^{2+}, Pb^{2+}	49
o-Dihydroxybenzene	Fe^{3+}, Ni^{2+}, Cu^{2+}, Zn^{2+}, Ca^{2+}, Pb^{2+}, Cd^{2+}	50
Thioglycolic acid	Cu^{2+}	51
N-(3-Propyl)-o-phenylenediamine	Cr^{3+}, Mn^{2+}, Fe^{3+}, Co^{2+}, Ni^{2+}, Cu^{2+}, Zn^{2+}, Pb^{2+}, Cd^{2+}	52
3-(2-Aminoethylamino)propyl group	Cu^{2+}, Au^{3+}, Pd^{2+}	53
2-Amino-1,3,4-thiadiazole	Hg^{2+}	54
Polyamidoamine	Cu^{2+}, Zn^{2+}, Cd^{2+}, Hg^{2+}, Ag^{+}, Au^{3+}, Pd^{2+}, Pt^{4+}	55
Polyethyleneimine	Cu^{2+}, Zn^{2+}, Cd^{2+}	56
2-Aminothiazole	Hg^{2+}	57
5-Benzylidene-2-thiobarbituric acid	Cu^{2+}, Pb^{2+}, Cd^{2+}, Hg^{2+}	58
4,4′-Diaminodiphenyl ether and 4,4′-Diaminodiphenyl sulfone salicylaldehyde Schiff bases	Cr^{3+}, Mn^{2+}, Zn^{2+}	59
Thioetheric sites	Au^{3+}, Pd^{2+}, Pt^{4+}	60
Mercaptothiazoline	Hg^{2+}	61
Mercaptopyridine	Hg^{2+}	61
Mercaptobenzothiazole	Hg^{2+}	61
(E)-N-(1-Thien-2′-ylethylidene)phenylene-1,2-diamine	Hg^{2+}	62
Di-ionizable calix(4) arene	Pb^{2+}	63
2-Aminothiophenol	Co^{2+}, Ni^{2+}, Cu^{2+}, Zn^{2+}, Cd^{2+}, Tl^{+}, Tl^{3+}	64
Benzophenone 4-aminobenzoylhydrazone	Co^{2+}, Ni^{2+}, Cu^{2+}, Zn^{2+}	65
4-Phenylacetophenone 4-aminobenzoylhydrazone	Co^{2+}, Ni^{2+}, Cu^{2+}	66

Table 6.1 (*Continued*)

Functional group anchored	Metal ions separated and/or preconcentrated	Ref.
Niobium(V) oxide	V^{3+}, Co^{2+}, Ni^{2+}, Cu^{2+}, Zn^{2+}, Pb^{2+}, Cd^{2+}	67
p-Toluenesulfonylamide	Cr^{3+}, Cu^{2+}, Zn^{2+}, Pb^{2+}	68
4,4′-Oxy-bis(chlorophenylglyoxime)	Co^{2+}, Ni^{2+}, Cu^{2+}	69
Diethylenetriamine	Ag^+, Cu^{2+}, Ni^{2+}, Hg^{2+}, Zn^{2+}, Pb^{2+}	70
Ethylenediaminetetraacetic acid	Co^{2+}, Ni^{2+}	71
5-(4-Dimethylaminobenzylidene)-rhodanine	Ag^{2+}	72
N,N-Bis(carboxymethyl)dithiocarbamate	Cu^{2+}, Pb^{2+} and Ni^{2+}	73
Poly[4-(4-vinylbenzyloxy)-2-hydrobenzaldehyde]	Cd^{2+}	74
4-Chloroisonitroacetophenone 4-aminobenzoylhydrazone	Cu^{2+}	75
2-Mercaptoimidazole	Cr^{3+}	76
Diphenyl diketone monothiosemicarbazone	Pd^{2+}	77
3-Hydroxy-2-methyl-1,4-naphthoquinone	Fe^{2+}, Cu^{2+}, Co^{2+}, Zn^{2+}	78
o-Vanillin	Fe^{2+}, Cu^{2+}, Co^{2+}, Zn^{2+}	79
Aurintricarboxylic acid	Cu^{2+}	80
Poly(amidoxime)	Pd^{2+}, Cu^{2+}, Ni^{2+}, Cd^{2+}	81
Poly(methacrylic acid)	Cd^{2+}	82
(3-Glycidoxypropyl)trimethoxysilane	Cu^{2+}	83
Bayberry tannin	Au^{3+}	84
Dithiocarbamate	Pb^{2+}, Cd^{2+}, Cu^{2+}, Hg^{2+}	85
(3-Aminopropyl)triethoxysilane and (3-Glycidoxypropyl)trimethoxysilane	Hg^{2+}	86
Zircon mineral	U^{6+}	87
Bayberry tannin	Cr^{3+}	88
1-Hexylpyridinium hexafluorophosphate	Fe^{3+}	89
Glycerol	Al^{3+}	90
2-Hydroxyacetophenone 3-thiosemicarbazone	Cd^{2+}	91
N-Methyl-D-glucamine	B^{3+}	92

this method for various electrolytes such as $NaNO_3$, $NaCl$, $NaBr$, Na_2SO_4, Na_3PO_4, glycine, sodium citrate, EDTA and humic acid are also reported. The preconcentration factors are 150, 250, 200, 300, 250, 300 and 200 for Cd, Co, Zn, Cu, Pb, Fe and Ni, respectively, and with $t_{1/2}$ values <1 min except for Ni. The 95% extraction by the batch method takes ≤25 min. The simultaneous enrichment and determination of all the metals are possible if the total load of the metal ions is less than the sorption capacity. In river water samples, all these metal ions were enriched with the present ligand anchored silica gel and determined with FAAS (relative standard deviation, RSD ≤ 6.4%). The Co content in pharmaceutical samples (vitamin tablets) were also preconcentrated with this modified silica gel and estimated by FAAS, with RSD ≈ 1.4%. The results are in good agreement with the certified value (1.99 μg g^{-1} of the tablets). Iron and copper in the certified reference materials (synthetic) SLRS-4 and SLEW-3 have been enriched with the modified silica gel and estimated with RSD ≤ 5%.[33]

Bernal *et al.* have also synthesized 8-hydroxyquinoline functionalized silica and studied the adsorption characteristics of Cu^{2+}, Cd^{2+}, Zn^{2+}, Fe^{3+} and Pb^{2+}. Experiments were carried out in both single and multiple cation solutions. The influence of the aqueous media composition on the adsorption process, the possible adsorbed species, and the adsorption kinetics were also discussed.[34]

Selective solid-phase extraction of Ni^{2+} was accomplished with dimethylglyoxime (DMG) functionalized silica, but the pore width influenced the sorption efficiency of the material. With microporous silica as the host, the stoichiometry of the Ni^{2+}–DMG complex was $1:1$ rather than $1:2$. The capacity of the DMG functionalized mesoporous silica was only $9 \, \mu mol \, Ni \, g^{-1}$ because of leaching of the complexing agent. However, the microporous material showed no loss of DMG, but low permeability lowered the capacity.[35]

Zaporozhets *et al.* have used silica gel functionalized with dithizone and zinc dithiazone for the determination of Ag^+, Hg^{2+} and Pb^{2+} present in buttermilk and natural, mineral and wastewater samples.[36] Mahmoud *et al.* have employed dithizone functionalized silica gel for extraction of Hg^{2+} from natural tap water with a preconcentration factor of 200 using an off-line method based on cold vapor atomic absorption analysis.[37]

Hanzel *et al.* have prepared various silica gel functionalized materials with imidazole (see Figure 6.1), diaza-18-crown-6 (DA18C6) and dibenzo-18-crown-6 (DB18C6) for the adsorption of Co^{2+}. The sorption kinetics, the influence of cobalt concentration and a suitable pH for metal complexation in static conditions with various kinds of silica gel matrix were analyzed. The influence of the presence of other heavy or toxic metals (Hg^{2+}, Cd^{2+}, Mn^{2+}, Zn^{2+}, Cu^{2+}, Fe^{3+}, Cr^{3+}, Al^{3+}) and electrolytes (Na^+, K^+) on the sorption of Co(II) from aqueous solutions was also investigated. The sorption of cobalt decreased in order $Hg > Cu > Cd > Zn \approx Fe > Mn > Al \approx Cr$. The presence of Na^+ and K^+ at 0.05 M concentration significantly influenced the adsorption of Co^{2+} in the case of silica gel functionalized with DB18C6.[38]

Mahmoud *et al.* have prepared silica gel functionalized with a dithioacetal (Figure 6.2). Different *para* substituents (on the dithioacetal) have been used ($X = H$, CH_3, OCH_3, Cl, NO_2). This functionalized silica gel has been employed in the scavenging of Hg^{2+} in the presence of other interfering metal ions. The range of adsorption of Hg^{2+} was found to be 94–100%. The potential application for selective extraction of Hg(II) from two different natural water samples, namely sea and drinking tap water, spiked with 1.0 and $10.0 \, ng \, mL^{-1}$

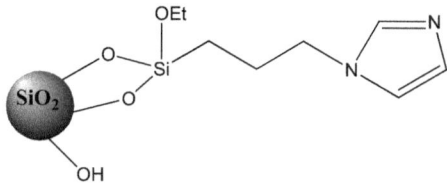

Figure 6.1   Silica gel functionalized materials with imidazole.

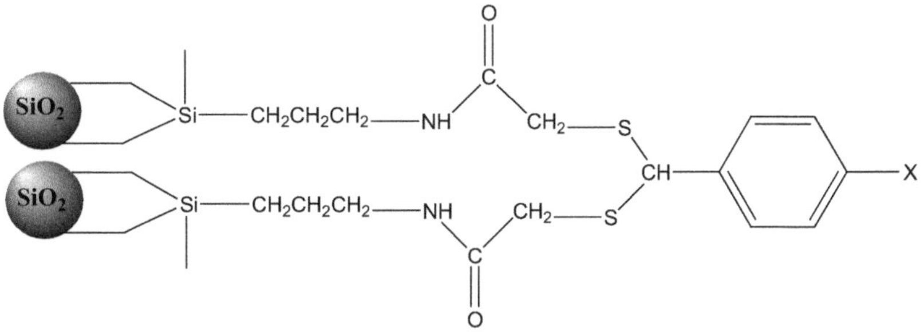

Figure 6.2 Silica gel functionalized with *para*-substituted dithioacetals.

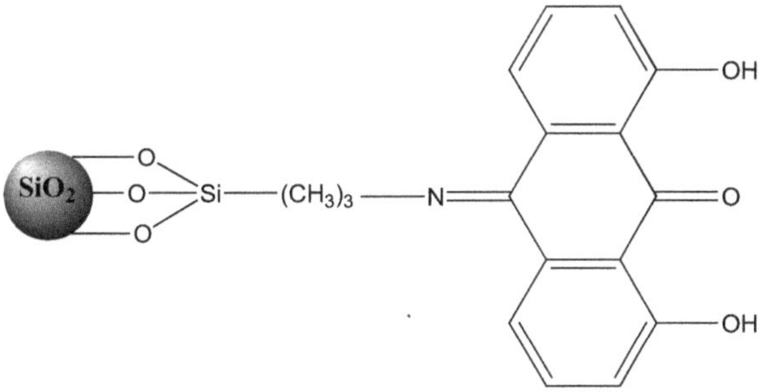

Figure 6.3 Chelating matrix prepared by immobilizing 1,8-dihydroxyanthraquinone on silica gel modified with (3-aminopropyl)triethoxysilane.

Hg(II) were also studied by a column technique followed by cold vapor atomic absorption analysis of the unretained Hg(II). The results showed a good extraction efficiency of modified silica gel (90–100 ± 3%) for the spiked Hg(II) ions.[39]

Airoldi *et al.* have synthesized ethylene sulfide functionalized silica gel for scavenging Hg^{2+}, Cu^{2+}, Ni^{2+}, Co^{2+} and Pb^{2+}.[40] A new chelating matrix was prepared by immobilizing 1,8-dihydroxyanthraquinone (DHAQ) on silica gel modified with (3-aminopropyl)triethoxysilane (Figure 6.3), which was used to preconcentrate Pb(II), Cd(II) and Zn(II). The optimum pH ranges for quantitative sorption are 6.0–7.5, 7.0–8.0 and 6.0–8.0 for Pb, Zn and Cd, respectively. All the metal ions can be desorbed with 2 M HCl/HNO$_3$. The sorption capacity of the matrix was found to be 76.0, 180.0 and 70.2 μmol g^{-1} for Pb, Zn and Cd, respectively, with a preconcentration factor of $\sim$200. The lowest concentration of metal ions for quantitative recovery was 5.0 ng mL^{-1}.[41]

Mahmoud *et al.* have synthesized 1-aminoanthraquinone functionalized silica gel for scavenging Cu^{2+} and Cr^{3+}, with good extraction values of 98.5–96.3 ± 2.0–3.0%. Also, a very high preconcentration factor of 500 was achieved with this adsorbent.[42]

Silica gel modified with (3-aminopropyl)triethoxysilane was further functionalized with resacetophenone (Figure 6.4) and used for the scavenging of Fe^{3+}, Co^{2+}, Ni^{2+}, Cu^{2+}, Zn^{2+}, Pb^{2+} and Cd^{2+}. The optimum pH range for

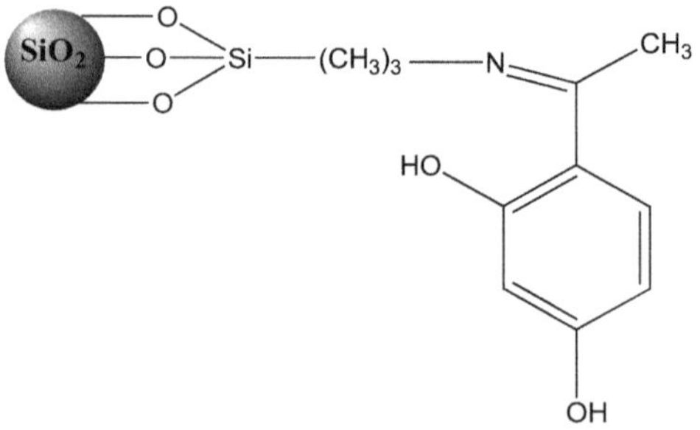

Figure 6.4   Silica gel functionalized with resacetophenone.

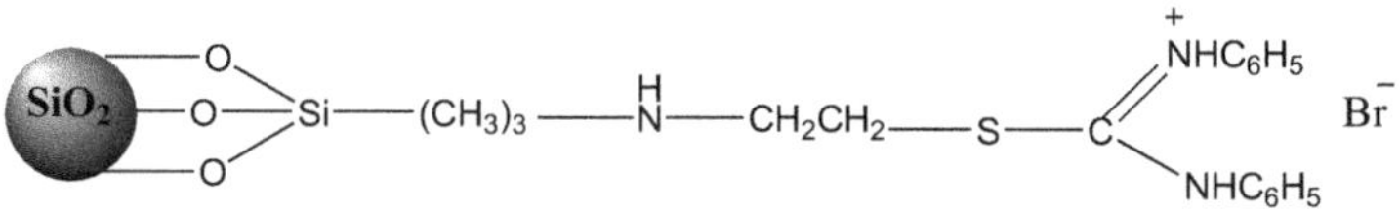

Figure 6.5   Diphenylisothiourea functionalized silica.

quantitative adsorption was 6.0–7.5, 5.5–7.5, 5.0–7.0, 6.5–7.5, 6.0–7.5, 6.0–7.0 and 6.0–7.5 for Cu^{2+}, Pb^{2+}, Fe^{3+}, Zn^{2+}, Co^{2+}, Ni^{2+} and Cd^{2+}, respectively. All the metals can be desorbed with 3 M HCl or HNO_3. The sorption capacity for these metal ions is in range of 57.8–365.0 $\mu mol\ g^{-1}$. The preconcentration factor was found to be 200, 300, 150, 250, 250, 200 and 200 for Cd^{2+}, Co^{2+}, Zn^{2+}, Cu^{2+}, Pb^{2+}, Fe^{3+} and Ni^{2+}, and the time required for 50% adsorption ($t_{1/2}$) was <1 min. The copper content in a pharmaceutical sample (vitamin tablet) was found to be in good agreement with the certified value of 0.4 mg g^{-1} of sample.[43]

Sarkar *et al.* have synthesized salicylaldoxime functionalized silica gel for scavenging Cu^{2+}, Ni^{2+}, Co^{2+} and Zn^{2+}. The experimental conditions were optimized both in batch and column processes to achieve the maximum efficiency. Kinetic and thermodynamic parameters as well as isotherm constants were evaluated to test the feasibility of the process. The role of various metal ions and different anions were tested in order to monitor the process in the case of real samples.[44]

Diphenylisothiourea functionalized silica (Figure 6.5) has been employed in the scavenging of Ag^+, Au^{3+}, Pd^{2+} and Pt^{2+}. The noble metals retained on the microcolumn were effectively eluted with thiourea solution. The analytical results obtained by the proposed method for geological and metallurgical samples were in good agreement with the certified values.[45]

Osman *et al.* have synthesized 2-hydroxynaphthaldehyde functionalized silica gel (Figure 6.6) for scavenging Cu^{2+}, Zn^{2+}, Cd^{2+}, Hg^{2+} and Pb^{2+}. The potential applications of this functionalized silica gel as a scavenger for the five metal ions spiked in drinking tap water (1 $\mu g\ mL^{-1}$) were found to give percentage recovery values in the range 90.2–96.3 ± 4.1–6.3%, while preconcentration of the same five metal ions spiked in drinking tap water (50.0 $ng\ mL^{-1}$) was successfully accomplished with a percentage recovery range of 92.6–95.8 ± 4.8–5.7%.[46]

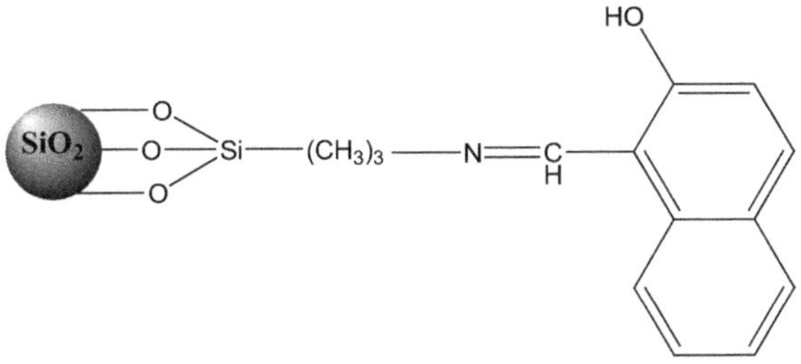

Figure 6.6 2-Hydroxynaphthaldehyde functionalized silica gel.

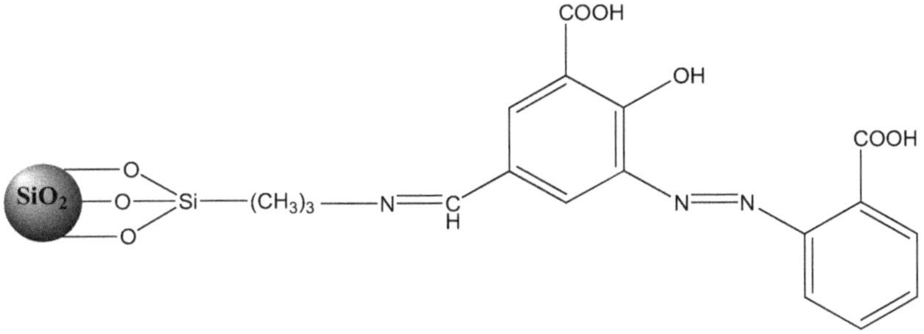

Figure 6.7 5-Formyl-3-(1'-carboxyphenylazo)salicylic acid functionalized silica gel.

Kokusen *et al.* have synthesized a β-diketone functionalized on an octadecyl group bonded to silica gel and used it in the scavenging of Mn^{2+}, Co^{2+}, Ni^{2+}, Cu^{2+} and Zn^{2+} in the pH region 2.0–6.0. The adsorption followed the order $Cu^{2+} > Ni^{2+} > Co^{2+} > Zn^{2+} > Mn^{2+}$.[47] Also, 4,4,4-trifluoro-1-(2-thienyl)butane-1,3-dione functionalized silica gel has been employed in the extraction of Co^{2+} and Ni^{2+}.[48] Akl *et al.* have developed 5-formyl-3-(1'-carboxyphenylazo)salicylic acid functionalized silica gel (Figure 6.7) for scavenging Cd^{2+}, Zn^{2+}, Fe^{3+}, Cu^{2+}, Pb^{2+}, Mn^{2+}, Cr^{3+} Co^{2+} and Mn^{2+}. These metal ions can be quantitatively adsorbed by the adsorbent from aqueous systems at pH 7.0–8.0. The adsorbed metal ions can be readily desorbed using 1.0 M HNO_3 or 0.05 M Na_2EDTA. The adsorption capacity was 0.32–0.43 meq g^{-1}. Nanogram concentrations (0.07–0.14 ng mL^{-1}) of these metal ions could be determined reliably with a preconcentration factor of 100.[49]

Figure 6.7 5-Formyl-3-(1'-carboxyphenylazo)salicylic acid functionalized silica gel.

Singh *et al.* have synthesized *o*-dihydroxybenzene (DHB) functionalized silica gel (Figure 6.8) for scavenging Cu^{2+}, Pb^{2+}, Fe^{3+}, Zn^{2+}, Ca^{2+}, Ni^{2+} and Cd^{2+}. The sorption capacity varies from 32 to 348 μmol g^{-1} and is highest for copper. Desorption was found to be quantitative with 1.0–3.0 M HCl/HNO_3. The preconcentration factors are between 100 and 300 and $t_{1/2}$ values ≤17 min. The present matrix coupled with FAAS has been used to enrich and determine

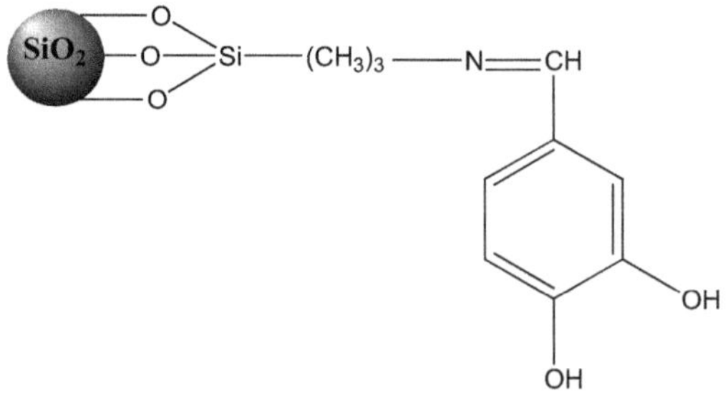

Figure 6.8   *o*-Dihydroxybenzene functionalized silica gel.

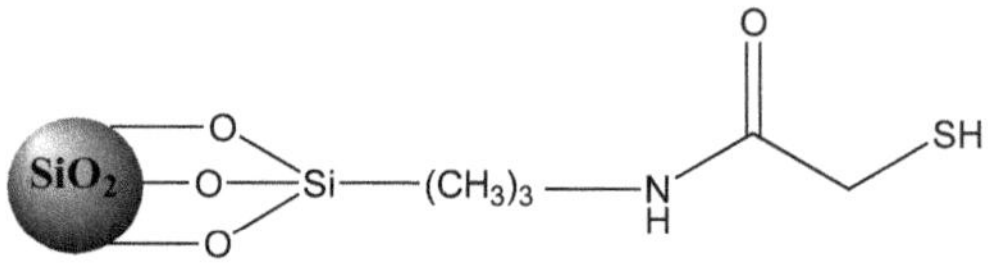

Figure 6.9   Thioglycolic acid functionalized silica gel.

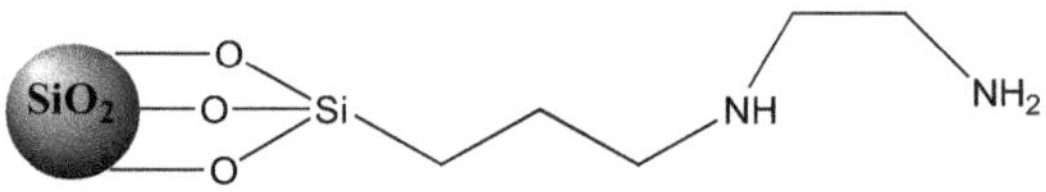

Figure 6.10   3-(2-Aminoethylamino)propyl bonded silica gel.

the seven metal ions in river and tap water samples (RSD 1.4–7.0%) and in synthetic certified water samples SLRS-4 (NRC, Canada) with an RSD of 2.73–2.83%. The cobalt present in pharmaceutical vitamin tablets and Zn in milk powder was also preconcentrated on DHB-anchored silica gel and determined by FAAS (RSD 2.00–2.72%).[50]

Fonseca *et al.* have synthesized thioglycolic acid functionalized silica gel (Figure 6.9) as an efficient scavenger for Cu^{2+} from aqueous/ethanolic solutions. This new modified silica showed good sorption ability at lower temperatures and its reusability was demonstrated.[51]

N-(3-Propyl)-*o*-phenylenediamine (NPPDA) functionalized silica gel has been employed in the scavenging of trace amounts of Cr^{3+}, Mn^{2+}, Fe^{3+}, Co^{2+}, Ni^{2+}, Cu^{2+}, Zn^{2+}, Pb^{2+} and Cd^{2+}. The optimum pH range for 90–100% extraction was found to be 7.0–8.0; the adsorption capacity was 350–450 $\mu mol\,g^{-1}$. Elution was found to be quantitative with 1–2 M HNO_3 or 0.05 M Na_2EDTA. Cu^{2+}, Fe^{3+}, Mn^{2+} and Zn^{2+} present in pharmaceutical vitamin tablets were also preconcentrated. Nanogram concentrations (0.07–0.14 ng mL^{-1}) of these metal ions were determined quantitatively with a preconcentration factor of 100.[52]

A preconcentration method for gold, palladium and copper based on the sorption of Au^{3+}, Pd^{2+} and Cu^{2+} ions on a column packed with 3-(2-aminoethylamino)propyl bonded silica gel (Figure 6.10) has been described. The recoveries of Au^{3+}, Pd^{2+} and Cu^{2+} were 98.93 $\pm$ 0.51,

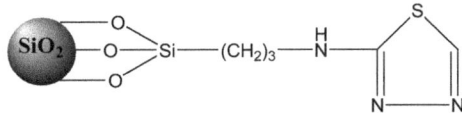

Figure 6.11 Silica gel functionalized with 2-amino-1,3,4-thiadiazole.

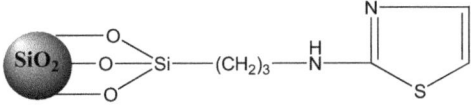

Figure 6.12 2-Aminothiazole functionalized silica gel.

98.81 ± 0.36 and 99.21 ± 0.42% at 95% confidence level, respectively. The detection limits of the elements were 0.032, 0.016 and 0.012 $\mu g\,mL^{-1}$, respectively.[53]

Silica gel functionalized with 2-amino-1,3,4-thiadiazole (Figure 6.11) has been used for the scavenging of Hg^{2+}.[54]

Qu *et al.* have prepared silica gels functionalized with ester- and amino-terminated dendrimer-like polyamidoamine polymers. They were used for the extraction of Cu^{2+}, Zn^{2+}, Cd^{2+}, Hg^{2+}, Ag^+, Au^{3+}, Pd^{2+} and Pt^{4+}. Both these ester- and amino-terminated polyamidoamine functionalized silica gels exhibited better adsorption capabilities for Au^{3+} and Pd^{2+} than for the base metal ions Cu^{2+}, Zn^{2+}, Cd^{2+} and Hg^{2+} and the noble metals Ag^+ and Pt^{4+}.[55]

Polyethyleneimine (PEI) has been functionalized on the surface of silica gel (PEI/SiO$_2$) *via* the coupling effect of (3-chloropropyl)trimethoxysilane. PEI/SiO$_2$ has been employed in the scavenging of Cu^{2+}, Zn^{2+} and Cd^{2+}. Adsorption follows the order $Cu^{2+} > Cd^{2+} > Zn^{2+}$. The isothermal adsorption data fit the Langmuir equation, and the adsorption is typical chemical adsorption with a monomolecular layer. The pH value has a great influence on the sorption and complexation, and at pH 6–7 the adsorption efficiency is a maximum. As dilute HCl is used as an eluent, the adsorbed metal ions are eluted easily from PEI/SiO$_2$ and regeneration and reuse without decreasing sorption for PEI/SiO$_2$ was demonstrated.[56]

Filho *et al.* have developed an electroanalytical method using 2-aminothiazole functionalized silica gel (Figure 6.12) for efficient and selective scavenging of Hg^{2+}. The precision for six determinations of 0.02 and 0.20 mg L^{-1} Hg^{2+} was 4.1% and 3.5% (RSD), respectively, and the detection limit was estimated as 0.10 μg L^{-1} for mercury(II).[57]

Silica gel functionalized with 5-benzylidene-2-thiobarbituric acid (BZTBA) (Figure 6.13) has been employed in the scavenging of Cu^{2+}, Pb^{2+}, Cd^{2+} and Hg^{2+}. The metal sorption properties of Si-BZTBA were also studied and the evaluated results refer to the high metal sorption of Si-BZTBA for Cu^{2+}, Hg^{2+}, Cd^{2+} and Pb^{2+} with the same order. These four Si-BZTBA–metal complexes were also synthesized and the stoichiometric ratios were identified as 1 : 1, except for the lead complex which was found to give a 1 : 2 ratio.[58]

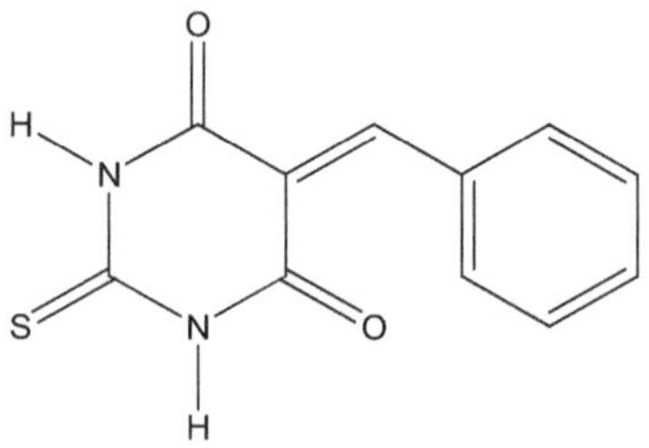

Figure 6.13   5-Benzylidene-2-thiobarbituric acid.

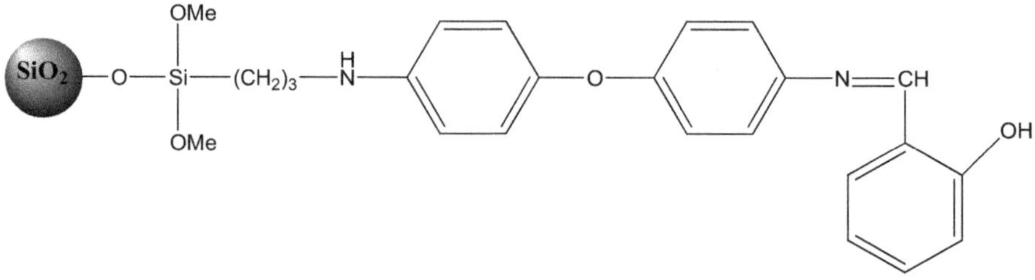

Figure 6.14   Chelating materials synthesized using Schiff bases of 4,4'-diaminodiphenyl ether and *o*-hydroxybenzaldehyde.

Dey *et al.* have synthesized two new chelating polymers (Si-DDE-*o*-HB and Si-DDS-*o*-HB) (Figure 6.14) by modifying the activated silica gel phase with Schiff bases of 4,4'-diaminodiphenyl ether (DDE)/4,4'-diaminodiphenyl sulfone (DDS) and *o*-hydroxybenzaldehyde (*o*-HB). Extraction of Cr^{3+}, Mn^{2+} and Zn^{2+} using Si-DDE-*o*-HB was found to be higher than that by Si-DDS-*o*-HB. The order of metal adsorption was $Zn^{2+} > Mn^{2+} > Cr^{3+}$. The optimum pH for adsorption was 7.5. Dilute HNO_3 was used as an eluent and the pre-concentration factor was 66.2 for Zn^{2+}.[59]

Gentscheva *et al.* have synthesized silica gel functionalized with thioether (sulfide) sites for scavenging Au^{3+}, Pt^{4+} and Pd^{2+}. Au^{3+} is quantitatively (>95%) sorbed in the pH region 1–9. The sorption of Pt^{4+} starts at pH 1 and does not exceed 25% in the entire pH region examined. The sorption of Pd^{2+} starts at pH 7 and reaches 80% at pH 9. Au^{3+} was quantitatively eluted with a 5% aqueous solution of thiourea and the adsorption capacity was 195 mg g^{-1}.[60]

Perez-Quintanilla *et al.* have synthesized mercaptothiazoline (MTZ), mercaptopyridine (MP) and mercaptobenzothiazole (MBT) (Figure 6.15) functionalized silica gels and employed them for scavenging Hg^{2+}. SBA-15 and MCM-41 functionalized with MTZ by the homogeneous method present good mercury adsorption values [1.10 and 0.7 mmol Hg(II)/g of silica, respectively]. This fact suggests a better applicability of such mesoporous silica supports to extract Hg^{2+} from aqueous solutions. In earlier reports it had been seen that mercury can be removed by different modified amorphous silicas in the pH range 2.0–6.0. Therefore this pH range was selected in order to evaluate the

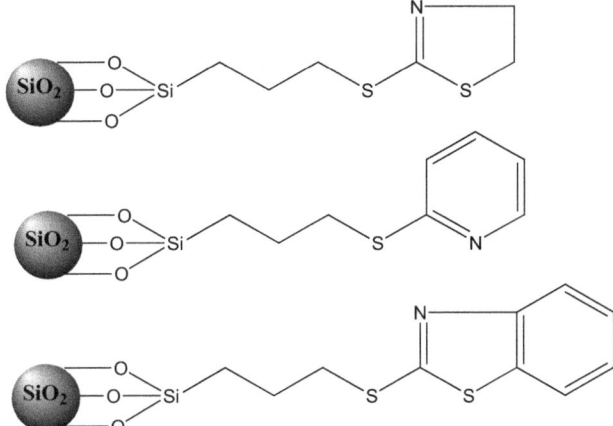

Figure 6.15   Silica gels functionalized with mercaptothiazoline (*top*), mercaptopyridine (*middle*) and mercaptobenzothiazole (*bottom*).

influence of this parameter on the adsorption process. The saturation of the monolayer covalently bound to silica by mercury was achieved at pH 4 for silica modified with MP. However, lower mercury adsorption values were obtained at low pH values. This can be attributed to protonation of the nitrogen donor atom in the heterocycle, which inhibits binding of the organic moiety with the metal ion. Basically, at low pH values, the strong sulfur atoms are the only available adsorption sites. However, when the pH is raised, both nitrogen and sulfur donor atoms are available and this increases mercury complexation. Also, it was found that mercury adsorption by MBT-modified silica was independent of pH.[61]

(*E*)-*N*-(1-Thien-2'-ylethylidene)phenylene-1,2-diamine (TEPDA) functionalized silica gel has been employed for scavenging low concentration of Hg^{2+} (30 pg mL^{-1}) from tap water with a preconcentration factor of 200.[62] Di-ionizable calix(4) arenes with two lower-rim *N*-sulfonylcarboxamides linked to silica gel *via* a single attachment on the upper rim have been synthesized and employed for scavenging Pb^{2+} from aqueous solutions. Pb^{2+} sample solutions were prepared with pH values of 2.1, 3.2, 4.0, 4.6 and 5.3 and it was found that uptake of Pb^{2+} by all of these ion exchange resins decreased markedly as the sample solution pH decreased. The concentration factor value of 55 at pH 5.3 drops to 45, 35, 10 and 0.3 at pH values of 4.6, 4.0, 3.2 and 2.1, respectively. It was found that the Pb^{2+} desorption process is quite rapid. Using 0.01 M HNO_3 as an eluting agent, most of the bound Pb^{2+} ions on the resins were eluted in the first 4.0 mL of stripping solution at a sample pH of 5.3.[63] Khosravan *et al.* have synthesized 2-aminothiophenol functionalized silica gel (Figure 6.16) using (3-chloropropyl)trimethoxysilane as a linking agent. It has been used in the scavenging of Fe^{3+}, Co^{2+}, Ni^{2+}, Cu^{2+}, Zn^{2+}, Cd^{2+}, Tl^+ and Tl^{3+}. The order of increasing metal ion uptake coincided with increasing hardness of these metal ions.[64]

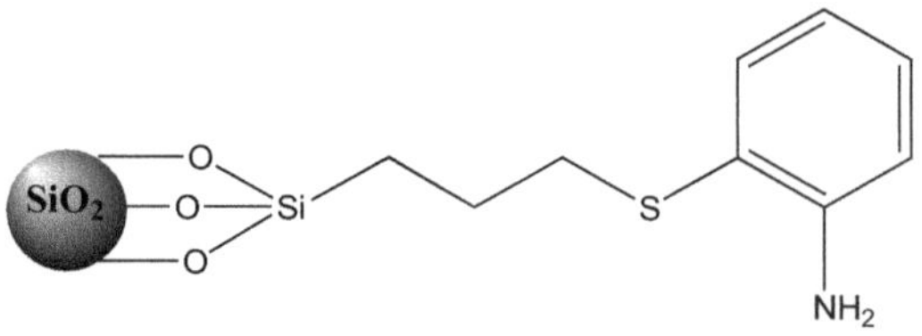

Figure 6.16   2-Aminothiophenol functionalized on silica gel.

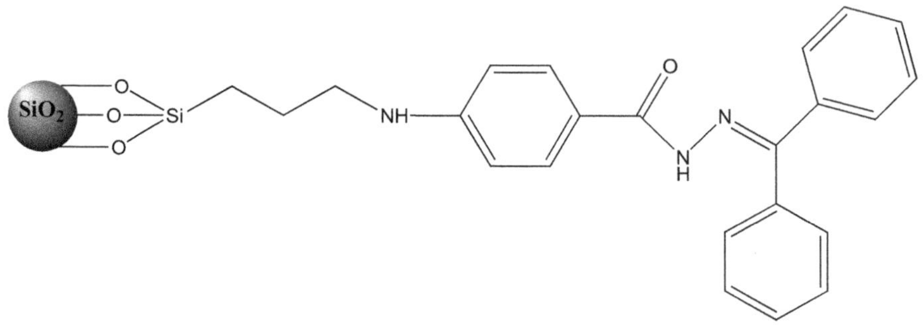

Figure 6.17   Benzophenone 4-aminobenzoylhydrazone modified silica gel.

Similarly, benzophenone 4-aminobenzoylhydrazone (BAH) has been functionalized on silica gel after surface modification with (3-chloropropyl)-trimethoxysilane (Figure 6.17). It has been employed in the scavenging of Co^{2+}, Ni^{2+}, Cu^{2+} and Zn^{2+} from aqueous samples. The mean sorption energy of metal adsorption using BAH-functionalized silica gel was calculated from Dubinin–Raduskevich isotherms, indicating a chemical sorption mode for the four cations. Thermodynamic parameters, *i.e.* $\triangle G$, $\triangle S$ and $\triangle H$, were also calculated for the system. From these parameters, $\triangle H$ values were found to be endothermic: 27.0, 22.7, 32.6 and 34.6 kJ mol^{-1} for Cu^{2+}, Ni^{2+}, Co^{2+} and Zn^{2+} metal ions, respectively; $\triangle S$ values were calculated to be positive for the sorption of the same sequence of divalent cations onto the sorbent. Negative $\triangle G$ values indicated that the sorption process for these metal ions onto immobilized silica gel is spontaneous.[65]

Hatay *et al.* have synthesized 4-phenylacetophenone 4-aminobenzoylhydrazone functionalized silica gel (Figure 6.18) using (3-chloropropyl)-trimethoxysilane for scavenging Co^{2+}, Ni^{2+} and Cu^{2+}. The sorption of these metal ions onto the modified silica gel correlated well with the Langmuir-type adsorption isotherm and adsorption capacities were found to be 0.012, 0.014 and 0.018 mmol g^{-1} for Cu^{2+}, Ni^{2+} and Co^{2+}, respectively. Thermodynamic parameters such as the standard free energy change ($\triangle G^{\circ}$), enthalpy change ($\triangle H^{\circ}$) and entropy change ($\triangle S^{\circ}$) were calculated to determine the nature of the sorption process. From these parameters, $\triangle H^{\circ}$ values were found to be endothermic at 38.39, 18.0 and 14.7 kJ mol^{-1} for the same sequence of divalent cations, and $\triangle S^{\circ}$ values were calculated to be positive for the sorption of each metal ion onto the modified silica gel. Negative $\triangle G^{\circ}$ values indicated that the

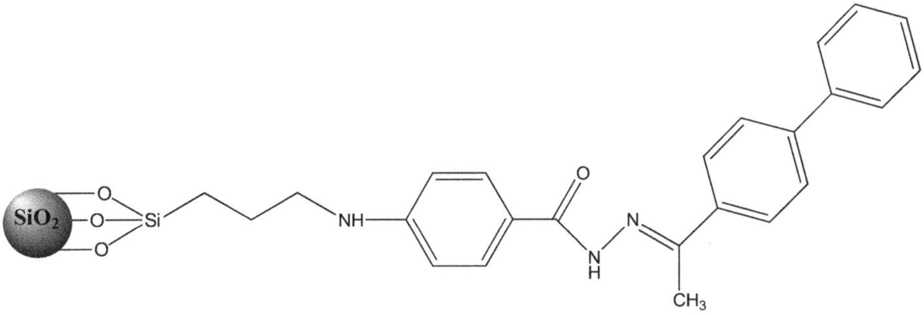

Figure 6.18 4-Phenylacetophenone 4-aminobenzoylhydrazone functionalized silica gel.

sorption process for all metal ions was spontaneous in nature, although they presented an endothermic enthalpy for the interaction, resulting in an entropically favored process.[66]

Maltez *et al.* have synthesized niobium(V) oxide functionalized silica gel and developed an online preconcentration system to simultaneously determine V^{3+}, Co^{2+}, Ni^{2+}, Cu^{2+}, Zn^{2+}, Pb^{2+} and Cd^{2+} in water by inductively coupled plasma mass spectrometry detection. The system is based on cationic retention of the analytes onto a mini-column filled with modified silica gel. The optimized operating conditions were pH 7.0, a sample flow rate of $6.0\,mL\,min^{-1}$, an eluent flow rate of $2.0\,mL\,min^{-1}$ and an eluent (HNO_3) concentration of 2.5 M. The relative standard deviation ($n=7$), enrichment factor and linear working range were 0.8–4.5%, 23.3–37.2 and 0.05–$25.0\,\mu g\,L^{-1}$, respectively. The limits of detection were between 0.01 and $0.03\,\mu g\,L^{-1}$.[67] *p*-Toluenesulfonylamide was immobilized on silica gel and on nanometer-sized SiO_2. The adsorption efficiency toward metal ions was investigated by the batch equilibrium technique. At pH 4.0 the adsorption capacity of the functionalized silica gel was found to be 4.9, 5.0, 33.2 and $12.6\,mg\,g^{-1}$ for Cr^{3+}, Cu^{2+}, Pb^{2+} and Zn^{2+}, respectively. However, the adsorption capacity of the nm-SiO_2 adsorbent towards Cr^{3+} was $26.7\,mg\,g^{-1}$ under ultrasonic dispersing. The potential application of *p*-toluenesulfonylamide-modified silica gel for simultaneous preconcentration of trace Cr, Cu, Pb and Zn from standard reference materials and food samples was performed, with satisfactory results.[68]

Silica gel modified with 4,4′-oxybis(chlorophenylglyoxime) (Figure 6.19) was synthesized and the sorption capacity towards Co^{2+}, Ni^{2+} and Cu^{2+} from aqueous solutions was studied. The optimum pH values for the separation of these divalent cations on the sorbent were 5.0, 6.0 and 6.0 for Cu, Co and Ni, respectively. The process of metal separation was followed by the batch method, and fitted to Langmuir and Freundlich sorption isotherms. The maximum sorption capacities (0.055, 0.042 and $0.034\,mmol\,g^{-1}$) were found from the Langmuir equation and the enthalpies of binding were 44.96, 71.63 and $68.14\,kJ\,mol^{-1}$ for Cu, Co and Ni, respectively. The other thermodynamic parameters calculated from the adsorption results were used to explain the nature of the adsorption.[69]

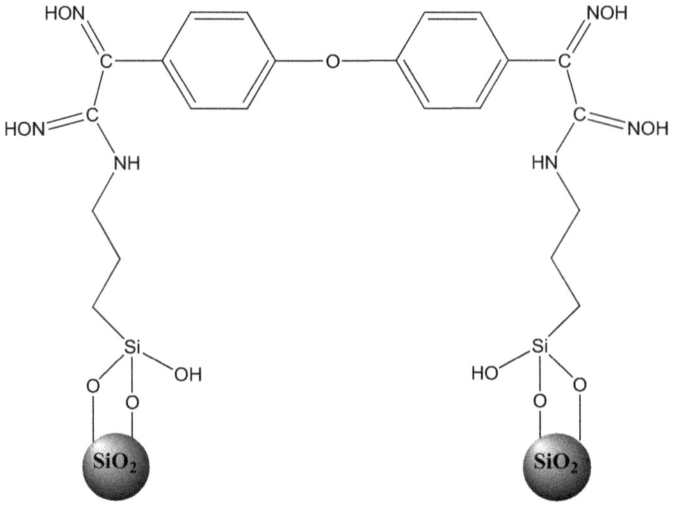

Figure 6.19   Silica gel modified with 4,4′-oxybis(chlorophenylglyoxime).

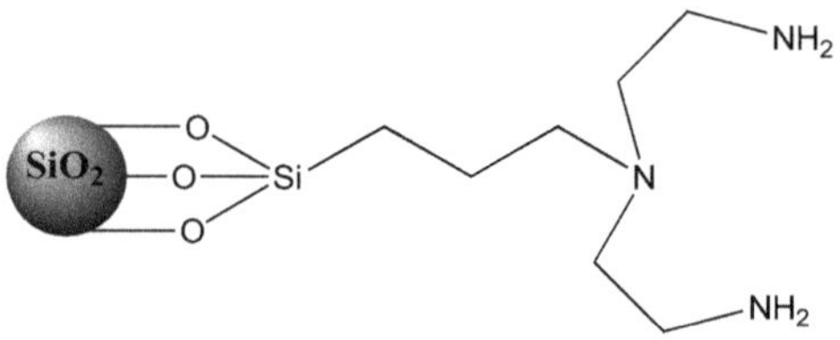

Figure 6.20   Silica gel-supported diethylenetriamine chelating resins.

Silica gel-supported diethylenetriamine chelating resins were prepared by functionalization of silica gel (Figure 6.20). Elemental analysis revealed that the direct amination routes and homogeneous conditions were more beneficial than the corresponding end-group protection routes and heterogeneous conditions to the syntheses of chelating resins with high N content. Several metal ions, such as Ag^+, Cu^{2+}, Ni^{2+}, Hg^{2+}, Zn^{2+} and Pb^{2+}, were chosen as representatives to investigate the relationship between adsorption capacities and N content of the ligands onto the surface of the silica gel. Saturated adsorption experiments for all metal ions at pH 5.0 were studied at 25 °C. In order to avoid the decrease of coordination capacity of adsorbent due to the protonation of N atoms at lower pH values and formation of precipitates of the metal ions at high pH, a value of pH 5.0 was selected for quantitative sorption of Ag^+, Cu^{2+}, Ni^{2+}, Hg^{2+}, Zn^{2+} and Pb^{2+}. Twenty-four hours of contact time was selected in this study to ensure that the metal ions could be completely adsorbed. The kinetic experiment also found that the equilibrium adsorption times of the resins for Hg^{2+} were about 6–8 h.[70]

The removal of Co^{2+} and Ni^{2+} ions from contaminated water was carried out using silica gel functionalized with both ethylenediaminetetraacetic acid (EDTA) (Figure 6.21) and diethylenetriaminepentaacetic acid (DTPA).

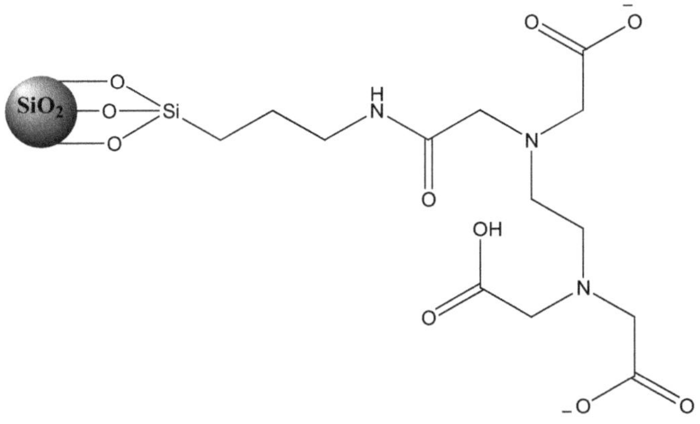

Figure 6.21 Silica gel functionalized with ethylenediaminetetraacetic acid.

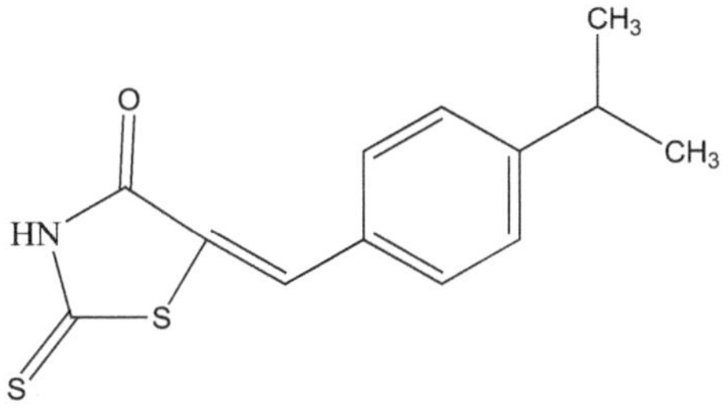

Figure 6.22 5-(4-Dimethylaminobenzylidene)rhodanine.

The adsorption and regeneration studies were conducted in batch mode. The optimum conditions for the removal of both metals at an initial concentration of 10 mg L^{-1} were 2 g L^{-1} for the adsorbent dose, pH 3, 50 rpm agitation speed and 4 h of contact time. The removal of Co^{2+} and Ni^{2+} by EDTA- and/or DTPA-modified silica gels was substantially higher than that by the unmodified form. The maximum Co^{2+} and Ni^{2+} uptakes by the EDTA-modified silica gel were 20.0 and 21.6 mg g^{-1}, comparable to their adsorption capacities by DTPA-modified silica gel (Co^{2+}: 16.1 mg g^{-1}; Ni^{2+}: 16.7 mg g^{-1}). At the same concentration of 10 mg L^{-1}, the removal of both metals by the modified adsorbents ranged from 96% to 99%. The two-site Langmuir model was representative to simulate the adsorption isotherms.[71]

Silica gel doped with 5-(4-dimethylaminobenzylidene)rhodanine (Figure 6.22) was prepared for separation, preconcentration and determination of silver ions by AAS. This new modified silica gel was used as an effective adsorbent for the solid phase extraction of Ag^+ from aqueous solutions. The variables that influence the adsorption/desorption of trace levels of Ag^+ were optimized in the column process. The preconcentration factor and capacity of the adsorbent at optimum conditions were found as 220 and 420 mg Ag^+ per gram of adsorbent, respectively. The relative standard deviation and the detection limit for measurement of Ag^+ in the experiments were less than 1.5%

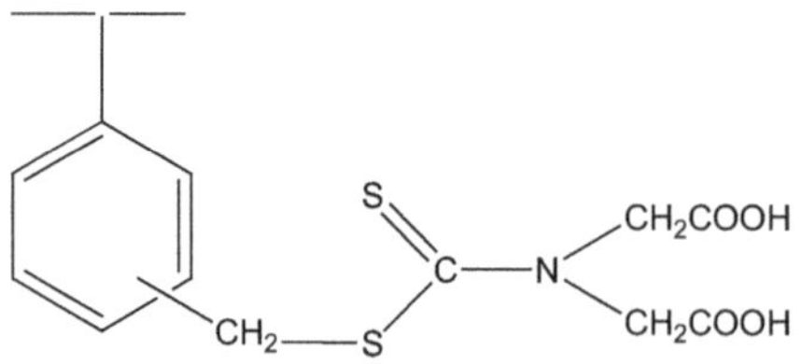

Figure 6.23   *N,N'*-Bis(carboxymethyl)dithiocarbamate.

($n = 10$) and $0.02\,\text{mg}\,\text{L}^{-1}$, respectively. This influenced the optimized adsorption/desorption of trace levels of Ag^+ in the column process. The preconcentration factor and capacity of the adsorbent at optimum conditions were found as 220 and $420\,\mu\text{g}\,Ag^+$ per gram of adsorbent, respectively. The detection limit for measurement of Ag^+ was $0.02\,\mu\text{g}\,\text{L}^{-1}$. Common coexisting ions did not interfere with the separation and determination of the silver ions. The proposed method was successfully applied for preconcentration and determination of silver ions in spiked water samples.[72]

N,N'-Bis(carboxymethyl)dithiocarbamate chelating resin (PSDC) (Figure 6.23) was synthesized by anchoring the chelating agent *N,N'*-bis-(carboxymethyl)dithiocarbamate to the chloromethylated PS-DVB matrix as a new adsorbent for removing divalent metal ions from waste streams. The adsorption performances of PSDC towards metals such as Cu^{2+}, Pb^{2+} and Ni^{2+} were systematically investigated, based upon which the adsorption mechanisms were deeply exploited. The adsorption behavior of Pb^{2+}, Cu^{2+} and Ni^{2+} by PSDC was pH dependent, with the most effective uptake at pH 5–6. Adsorption isotherms correlated well with the Langmuir equation and the maximum adsorption capacities were in the order $Pb^{2+} > Cu^{2+} > Ni^{2+}$. Enthalpy change values were respectively 16.80, 17.49 and $12.54\,\text{kJ}\,\text{mol}^{-1}$ for Cu^{2+}, Pb^{2+} and Ni^{2+}.[73]

Poly[4-(4-vinylbenzyloxy)-2-hydroxybenzaldehyde] (Figure 6.24) has been found to be capable of the removal of low concentrations of cadmium metal ions at ppb levels from aqueous media. The adsorption is best described with the Langmuir model, indicating a monolayer coverage of the metal ion on the surface of the chelating polymer. The R_L values indicate that the adsorption in the range of all the concentrations studied is favorable. The kinetic study has revealed a very fast adsorption process, which takes less than 40 s to reach the equilibrium capacity and follows the pseudo-second-order kinetic model.[74]

4-Chloroisonitroacetophenone 4-aminobenzoylhydrazone (CAAH), chemically anchored on a silica gel surface (Figure 6.25), has been used for Cu(II) sorption from aqueous solution. The sorption behavior of Cu(II) was evaluated by the use of batch and column methods. In acidic media at pH ≤ 3, no sorption was detected due to protonation. The low pH value avoids any interaction of metal ions with the Si–CAAH surface. However, a significant increase in the maximum value of the acid–base interaction was detected at pH 5–5.5. The obtained dynamic data were fitted to Freundlich, Langmuir and Dubinin–Raduskevich (DR) isotherms. The mean sorption energy of copper sorption

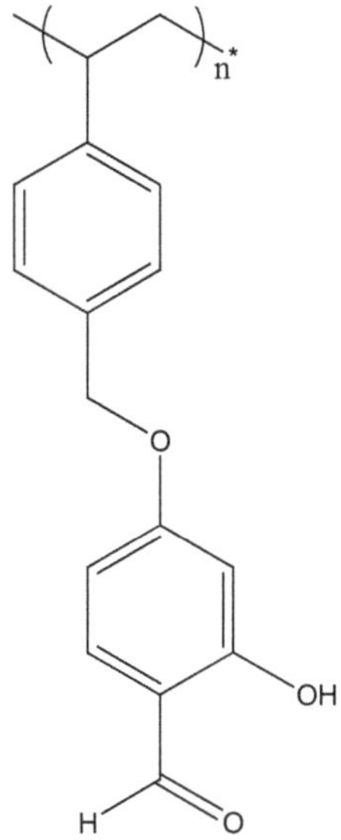

Figure 6.24 Poly[4-(4-vinylbenzyloxy)-2-hydroxybenzaldehyde].

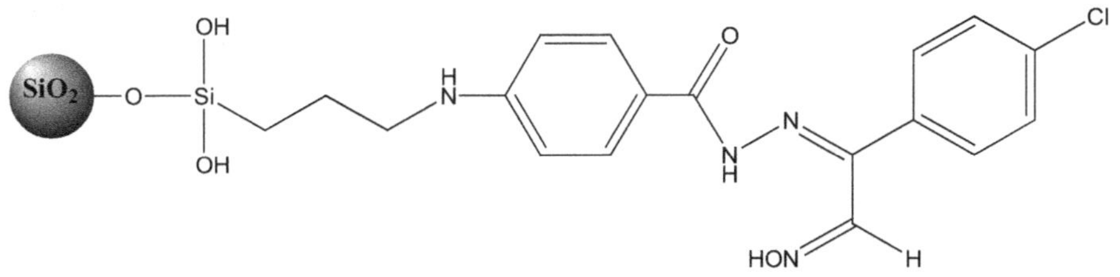

Figure 6.25 A4-Chloroisonitroacetophenone 4-aminobenzoylhydrazone functionalized silica gel.

onto silica gel was calculated from the DR isotherm, indicating a chemical ion exchange. Thermodynamic parameters such as Gibbs free energy change, enthalpy change and entropy change were calculated from the adsorption isotherms, which were used to explain the mechanism of the adsorption.[75]

The organic ligand 2-mercaptoimidazole was successfully immobilized onto silica gel (Figure 6.26) through reaction with (3-chloropropyl)trimethoxysilane as intermediate reagent. This modified silica (Si-CTS-MCP) can adsorb Cr^{3+} ions in solutions having pH 5 with an interaction time of 80 min. The optimum condition of these parameters was applied to determine the removal percentage of Cr^{3+} in water samples using this modified silica gel. When the modified silica was applied to remove Cr^{3+} from water samples, 16.067% of Cr^{3+} could be adsorbed by Si-CTS-MCP under optimum conditions.[76]

A novel, highly selective, efficient and reusable chelating resin, diphenyl diketone monothiosemicarbazone modified silica gel (Figure 6.27), was prepared and applied for the on-line separation and preconcentration of Pd(II)

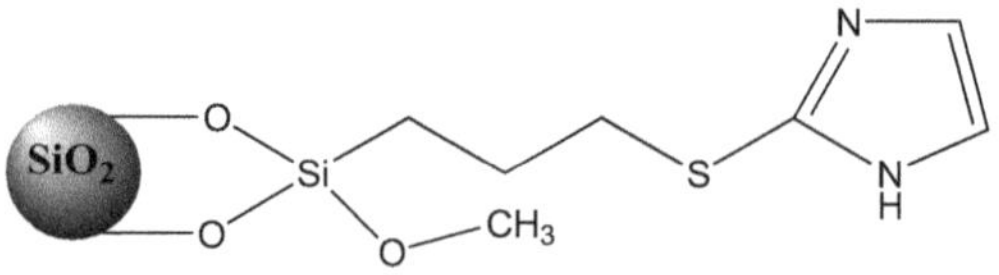

Figure 6.26   2-Mercaptoimidazole-immobilized silica gel.

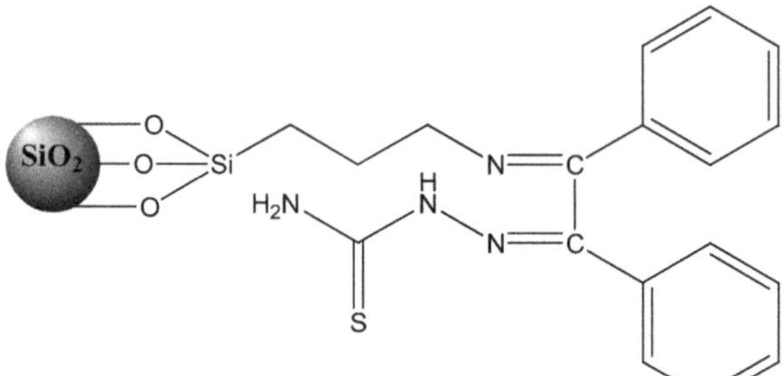

Figure 6.27   Diphenyl diketone monothiosemicarbazone-modified silica gel.

ions in a catalytic converter with spiked tap water samples. The proposed solid-phase extraction system for the first time enabled an effective on-line palladium preconcentration and final determination of trace amounts of Pd(II) ions in various samples with complex and variable matrices as a result of its excellent analytical characteristics, such as low detection limit ($5\,ng\,mL^{-1}$), high enrichment factor (335), high adsorption capacity ($0.73\,mmol\,g^{-1}$) and good selectivity and precision. The resin could continuously be used for a long period of time without any appreciable change in its sorption properties towards Pd^{2+} ions. The method was not only practicable for palladium trace analysis in a variety of matrices in order to reduce its hazardous impact on the ecosystem, but also a valuable addition because of the recovery of costly palladium metal.[77]

3-Hydroxy-2-methylnaphtho-1,4-quinone-immobilized silica gel has been used for the adsorption and estimation of copper, cobalt, iron and zinc by both batch and column techniques. The distribution coefficient D determined for each metal ion was as follows ($mL\,g^{-1}$): Fe, 3.6×10^2; Cu, 3.9×10^2; Co, 3.8×10^2; Zn, 4.1×10^2. The method was successfully implemented to estimate zinc, copper and cobalt in milk, steel and vitamin samples.[78] *o*-Vanillin-immobilized silica gel was synthesized for the adsorption of copper, cobalt, iron and zinc by both batch and column techniques. The metal ions were quantitatively retained on the column packed with the immobilized silica gel in the pH range 4.0–6.0 for Cu, 5.0–6.0 for Co, 4.5–6.0 for Fe and 6.0–8.0 for Zn.[79]

A green analytical method was developed for separation and preconcentration of trace amounts of copper from aqueous samples using aurintricarboxylic acid-immobilized silica gel (ATA-SG). This was done by determining the chemical speciation and stability constant of the Cu-ATA complex. The pH–metric studies indicated strong complexation of Cu with ATA (log $\beta_{Cu2ATA} = 19.56$). The species distribution curve for the Cu–ATA complex indicated almost 100% complexation of copper, with ATA forming Cu_2ATA as the predominant species. Considering the selectivity and strong interaction of ATA toward copper, ATA was immobilized on a polymeric matrix of silica gel and the conditions were optimized for the uptake of copper from the aqueous solutions. The uptake of Cu ions by ATA-SG was studied both by batch and column methods. The method was developed for the estimation of trace amounts of copper in various samples. Molecular modeling studies were also performed on ATA-immobilized silica gel to gain an insight into the active structure of ATA on the silica surface.[80]

Gao *et al.* have synthesized the composite chelating particles poly-(amidoxime)/SiO_2 (Figure 6.28) *via* an amidoximation transformation reaction; they were found to possess strong chelating adsorption abilities for Cu^{2+}, Pb^{2+}, Ni^{2+} and Cd^{2+}. The adsorption using the particles is selective and the adsorption capacities are in the order $Cu^{2+} > Ni^{2+} > Pb^{2+} > Cd^{2+}$.[81]

Poly(methacrylic acid) (PMAA) was grafted onto the surface of silica gel particles using (3-methacryloxypropyl)trimethoxysilane (MPS) as intermediate, and the grafted particle PMAA/SiO_2 (Figure 6.29) with strong adsorption ability for Cd^{2+} was prepared. The adsorption ability of PMAA/SiO_2 is largely dependent on the pH value and the temperature of the Cd^{2+} solution. The adsorption efficiency was a maximum in a solution of pH 7. The experimental results showed that PMAA/SiO_2 possesses very strong adsorption ability for Cd^{2+}, and the saturated adsorption amount could reach up to 42.6 mg g^{-1}. The equilibrium adsorption data were found to be consistent with the Langmuir isotherm model, indicating favorable monolayer adsorption. PMAA/SiO_2 can be easily regenerated using dilute HCl solution as eluent. In addition, PMAA/SiO_2 has excellent reusability: it can be used repeatedly over 10 times without significantly losing any adsorption capacity.[82]

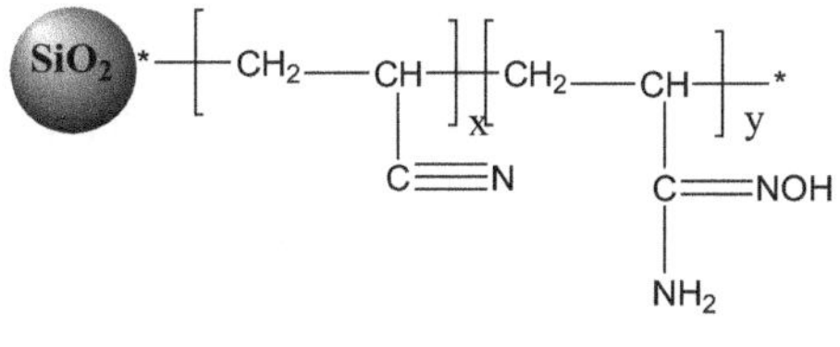

Figure 6.28 Poly(amidoxime)/silica gel immobilized chelating particles.

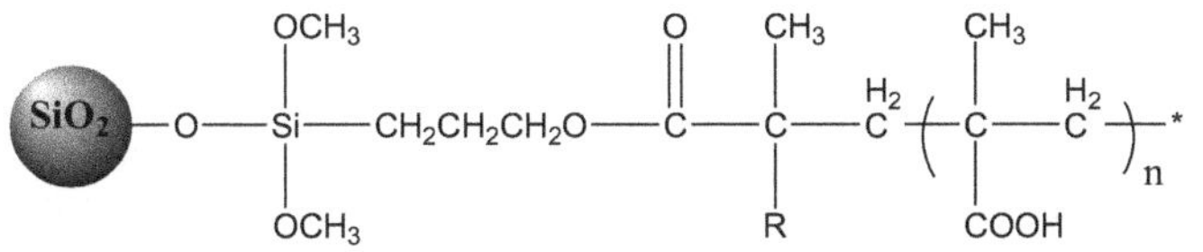

Figure 6.29 Poly(methacrylic acid)-immobilized silica gel.

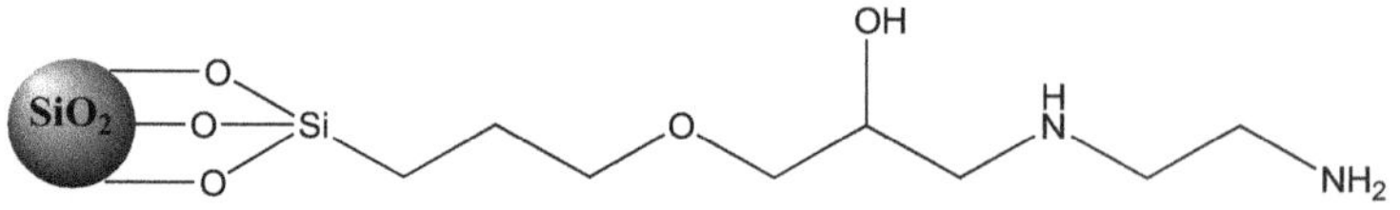

Figure 6.30   Mesoporous   silica   functionalized   with   (3-glycidoxypropyl)-
trimethoxysilane.

Functionalized hexagonal mesoporous silica gels were prepared by chemical
modification of a surfactant-free mesoporous silica (OSU-6-W) with (3-glyci-
doxypropyl)trimethoxysilane (Figure 6.30). Different degrees of derivatization
with 3-glycidoxypropyl were realized by using either a single silylation reaction
or two silylation reactions with an intermediate hydrolysis step. The basic
centers of the attached pendant groups provide the capacity to extract copper
from aqueous solution *via* an adsorption process that followed the Langmuir
model and had a remarkably high capacity of 6.75 mmol g^{-1} for adsorption of
copper.[83]

Bayberry tannin (BT) was immobilized on a mesoporous silica matrix to
prepare a novel adsorbent, which was subsequently used for the adsorptive
recovery of Au^{3+} from aqueous solutions. It was found that BT-immobilized
mesoporous silica (BT-SiO$_2$) was able to effectively recover Au^{3+} from acidic
solutions (pH 2.0). The equilibrium adsorption capacity of Au^{3+} on BT-SiO$_2$
was high, up to 642.0 mg g^{-1} at 323 K. Owing to its mesoporous structure,
BT-SiO$_2$ exhibited an extremely fast adsorption rate to Au^{3+} compared with
other tannin gel adsorbents. The presence of other coexisting metal ions, such
as Pb^{2+}, Ni^{2+}, Cu^{2+} and Zn^{2+}, did not decrease the adsorption capacity of
Au^{3+} on BT-SiO$_2$, and BT-SiO$_2$ had almost no adsorption capacity for the
coexisting metal ions, which suggested a high adsorption selectivity of BT-SiO$_2$
for Au^{3+}. Additionally, about 73% of the adsorbed Au^{3+} can be desorbed
using aqua regia, and the Au^{3+} solution was concentrated about 18 times
compared with the original solution. Consequently, the outstanding
characteristics of BT-SiO$_2$ provide the possibility of effective recovery and
concentration of Au^{3+} from dilute solutions.[84]

Silica-supported dithiocarbamate adsorbent (Si-DTC) was synthesized by
anchoring the chelating agent of macromolecular dithiocarbamate (MDTC) to
the chloro-functionalized silica matrix (SiCl) as a new adsorbent for adsorption
of Pb^{2+}, Cd^{2+}, Cu^{2+} and Hg^{2+} from aqueous solutions. The experimental data
were exploited for kinetic and thermodynamic evaluations related to the adsorption processes. The characteristics of the adsorption process were evaluated using the Langmuir, Freundlich and Dubinin–Radushkevich adsorption
isotherms, and the adsorption capacities were found to be 0.34, 0.36, 0.32 and
0.40 mmol g^{-1} for Pb^{2+}, Cd^{2+}, Cu^{2+} and Hg^{2+}, respectively.[85]

Monoamine- and pentamine-immobilized silica gels were prepared through
the treatment of silica with (3-aminopropyl)triethoxysilane and (3-glycidoxy-
propyl)trimethoxysilane (Figure 6.31). The uptake behavior of the modified
silica towards mercury ions at different experimental conditions of pH, time,
concentration and temperature, using batch and column methods, was studied.

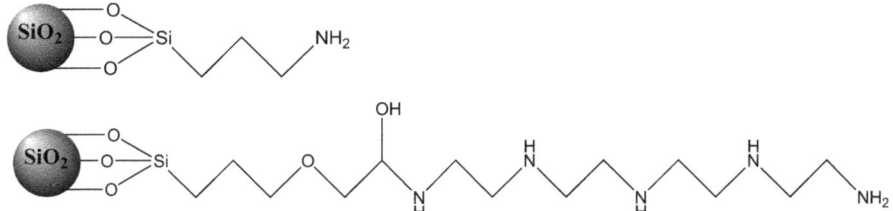

Figure 6.31 Silica gel immobilized with (3-aminopropyl)triethoxysilane (*top*) and (3-glycidoxypropyl)trimethoxysilane (*bottom*).

The maximum uptake values at 293 K were 1.70 and 3.50 mmol g^{-1}. Kinetic and thermodynamic data showed a pseudo-second-order adsorption process with an endothermic nature. Regeneration of the modified silica was performed using 0.50 mol dm^{-3} KI.[86]

A solution of sodium silicate produced from the alkali fusion of Egyptian zircon mineral as a waste was used to prepare silica gel in the pH range 6–7. The surface of the obtained silica was functionalized with diethylenetriamine and tetraethylenepentamine to give triamine-modified silica (TAMS) and pentamine-modified silica (PAMS), respectively. The uptake behavior of the modified silica towards U(VI) ions at different experimental conditions of pH, time, concentration and temperature was studied. The maximum uptake values at 25 °C were found to be 90.3 and 112 mg g^{-1} for TAMS and PAMS, respectively. Kinetic and thermodynamic studies showed an endothermic pseudo-second-order adsorption process. Regeneration of the loaded silica was performed using 1 M HNO$_3$. The modified silica gels have successfully been applied for the extraction of U(VI) obtained from alkaline leaching of Egyptian monazite sand.[87]

Huang *et al.* synthesized Bayberry tannin-immobilized mesoporous silica bead (BT-SiO$_2$) and this supporting matrix was used for selective adsorption of Cr^{3+}. It was found that the adsorption of Cr^{3+} onto BT-SiO$_2$ was a maximum in the pH range 5.0–5.5. The adsorption capacity was 1.30 mmol g^{-1} at 303 K and pH 5.5, when the initial concentration of Cr^{3+} was 2.0 mmol L^{-1}. The adsorption kinetic data can be well described using a pseudo-first-order model and the equilibrium data can be well fitted by the Langmuir isotherm model. Further investigations revealed that the coexisting metal ions in solution had no influence on the adsorption of Cr^{3+} onto BT-SiO$_2$, and no bayberry tannin was leached out during the adsorption process.[88]

1-Hexylpyridinium hexafluorophosphate ionic liquid was firstly used for chemical modification of silica utilizing acid-catalyzed sol–gel processing and was used as a sorbent for removal of trace levels of Fe^{3+} ions from aqueous samples. In the optimum experimental conditions, the limit of detection, limit of quantification and preconcentration factor were 0.7 µg L^{-1}, 2.5 µg L^{-1} and 200, respectively. The RSD for nine replicate determinations of 25 µg L^{-1} of Fe was 3.2%. The calibration graph using the preconcentration system was linear in the range of 2.5–50 µg L^{-1}, with a correlation coefficient of 0.9997.

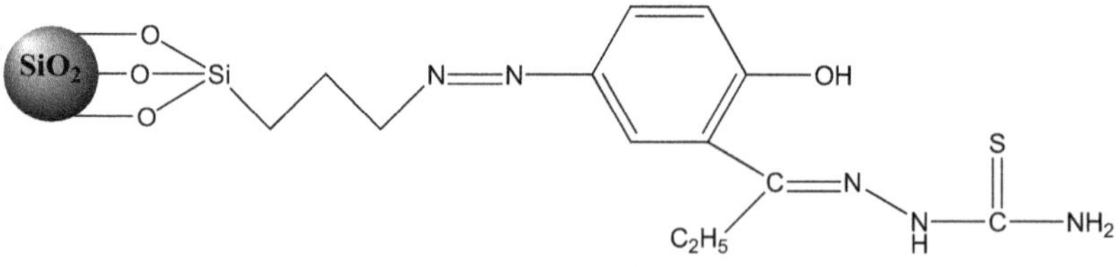

Figure 6.32   2-Hydroxyacetophenone 3-thiosemicarbazone-functionalized silica.

The developed method was validated by the analysis of certified reference material and applied successfully to the separation and determination of iron in several water samples.[89]

A solid-phase extraction system was proposed for the determination of aluminum after preconcentration with glycerol-bonded silica gel. The method is rapid and efficient for the enrichment of aluminum ions at trace levels. Optimal sorption conditions were found for sorption and desorption of aluminum ions. The effects of diverse ions on the sorption and recovery of aluminum have been studied and it was shown that the selectivity of the sorption process was very good. A very satisfactory preconcentration factor of 500 was achieved by this method. The lowest concentration of aluminum ions for quantitative recovery was $2\,\mathrm{ng\,mL^{-1}}$. The capacity of sorbent was $400\,\mu g$ per gram of sorbent. The method showed good reproducibility ($RSD = 2.2\%$ for $n = 7$) and was applied to the determination of aluminum in mineral water, hair and a green tea sample.[90]

Kumar *et al.* prepared 2-hydroxyacetophenone 3-thiosemicarbazone-functionalized silica gel (Figure 6.32) and investigated its removal behavior towards Cd^{2+} in aqueous solutions. The adsorption was described by the Langmuir and Freundlich adsorption isotherms. The maximum adsorption capacity was $313\,\mu g\,g^{-1}$ from the Langmuir isotherm at an initial concentration of $1000\,\mu g\,g^{-1}$. The kinetic results show that the adsorption is well described by the pseudo-second-order kinetic model. According to the results, the functionalized silica can be used as a good adsorbent for the removal of metal pollutants.[91]

As a novel adsorbent for boron removal from aqueous solutions, silica-supported *N*-methyl-D-glucamine adsorbent (Si-MG) was synthesized by anchoring the *N*-methyl-D-glucamine-modified (3-glycidoxypropyl)trimethoxy-silane to the silica surface. The characteristics of the adsorption process were also evaluated using the Langmuir and Freundlich adsorption isotherms, and the maximum adsorption capacity of Si-MG was found to be $1.54\,\mathrm{mmol\,g^{-1}}$. The adsorption equilibrium could be obtained within $30\,\mathrm{min}$ and the experimental data were well described by the pseudo-second-order kinetic model. It was therefore concluded that the adsorbent Si-MG could be promising for boron removal from salt lake brine.[92]

6.6   Conclusion

Accurate analysis and removal of metal ions from water bodies, especially at trace levels, is one of the most difficult and complicated analytical tasks, since it requires the rigorous requirements of versatility, specificity, sensitivity and accuracy in the analysis. However, solid phase extraction, using functionalized silica gel, offers many advantages over conventional methods. It is convenient, economical, offers high chemical stability and can be used over a wide temperature range. The chelating resins are synthesized as per the need, and selective preconcentration and determination of the metal ions is possible using metal-selective chelating ligands (immobilized on silica gel) having high affinity for the targeted metal ion. Moreover, this is a green analytical method because it does not involve the use of large amounts of toxic and chlorinated solvents, and helps in complete remediation of metals just by tuning the optimum conditions.

References

1. D. Bilba, D. Bejan and L. Tofan, *Croat. Chem. Acta*, 1998, **71**, 155.
2. M. E. Mahmoud, M. M. Osman and M. E. Amer, *Anal. Chim. Acta*, 2000, **415**, 33.
3. M. E. Mahmoud, *Anal. Chim. Acta*, 1999, **398**, 297.
4. K. S. Abou-El-Sherbini, I. M. M. Kenawy, M. A. Ahmed, R. M. Issa and R. Elmorsi, *Talanta*, 2002, **58**, 289.
5. A. R. Ghiasvand, R. Ghaderi and A. Kakanejadifard, *Talanta*, 2004, **62**, 287.
6. R. K. Sharma and P. Pant, *J. Hazard. Mater.*, 2009, **163**, 295.
7. R. K. Sharma and P. Pant, *Int. J. Environ. Anal. Chem.*, 2009, **87**, 503.
8. U. Pyell and G. Stork, *Fresenius J. Anal. Chem.*, 1992, **342**, 376.
9. Y. Guo, B. Din, Y. Liu, X. Chang, S. Meng and J. Liu, *Talanta*, 2004, **62**, 209.
10. Y. S. Kim, G. In, C. W. Han and J. M. Choi, *Microchem. J.*, 2005, **80**, 151.
11. V. A. Lemos, D. G. da Silva, A. L. de Carvalho, D. A. Santana, G. S. Novaes and A. S. Passos, *Microchem. J.*, 2006, **84**, 14.
12. O. Abollino, M. Aceto, C. Sarzanini and E. Mentasti, *Anal. Chim. Acta*, 2000, **411**, 223.
13. P. K. Jal, S. Patel and B. K. Mishra, *Talanta*, 2004, **62**, 1005.
14. A. R. Khorrami, T. Hashempur, A. Mahmoudi and A. R. Karimi, *Microchem. J.*, 2006, **84**, 75.
15. V. Gurnani, A. K. Singh and B. Venkataramani, *Talanta*, 2003, **61**, 889.
16. P. Pohl and B. Prusisz, *Anal. Chim. Acta*, 2004, **502**, 83.
17. I. Narin, M. Soylak, L. Elci and M. Dogan, *Talanta*, 2000, **52**, 1041.
18. Y. Cui, X. Chang, Y. Zhai, X. Zhu, H. Zheng and N. Lian, *Microchem. J.*, 2006, **83**, 35.
19. C. R. T. Tarley, S. L. C. Ferreira and M. A. Z. Arruda, *Microchem. J.*, 2004, **77**, 163.
20. P. Liu, Q. Pu, Z. Hu and Z. Su, *Analyst*, 2000, **125**, 1205.
21. R. K. Sharma, B. S. Garg, N. Bhojak and S. Mittal, *Microchem. J.*, 1999, **61**, 94.
22. G. Schmuckler, *Talanta*, 1965, **12**, 281.

23. M. Feldman and P. Desrochers, *Ind. Innovation*, 2003, **10**, 5.
24. R. K. Sharma, S. Mittal and M. Koel, *Crit. Rev. Anal. Chem.*, 2003, **33**, 183.
25. R. K. Sharma and S. Dhingra, *Designing and Synthesis of Functionalized Silica Gels and their Applications as Metal Scavengers, Sensors, and Catalysts: A Green Chemistry Approach*, LAP Lambert, Saarbrücken, Germany, 2011.
26. R. K. Sharma, S. Mittal, S. Azmi and A. Adholeya, *Surf. Eng.*, 2005, **21**, 232.
27. R. J. Kvitek, J. F. Evans and P. W. Carr, *Anal. Chim. Acta*, 1982, **144**, 93.
28. A. R. Sarkar, P. K. Datta and M. Sarkar, *Talanta*, 1996, **43**, 1857.
29. P. K. Jal, R. K. Dutta, M. Sudershan, A. Saha, S. N. Bhattacharya, S. N. Chintalapudi and B. K. Mishra, *Talanta*, 2001, **55**, 233.
30. S. B. Savvin and A. V. Mikhailova, *Zh. Anal. Khim.*, 1996, **51**, 49.
31. V. V. Sukhan, O. A. Zaporozhets, N. A. Lipkovskaya, L. B. Pogasi and A. A. Chuiko, *Zh. Anal. Khim.*, 1994, **49**, 700.
32. V. M. Ostrovskaya, *Zh. Anal. Khim.*, 1977, **32**, 1820.
33. A. Goswami, A. K. Singh and B. Venkataramani, *Talanta*, 2003, **60**, 1141.
34. J. P. Bernal, E. R. De San Miguel, J. C. Aguilar, J. C. Salazer and J. De Gyves, *Sep. Sci. Technol.*, 2005, **35**, 1661.
35. J. Seneviratne and J. A. Cox, *Talanta*, 2000, **52**, 801.
36. O. Zaporozhets, N. Petruniock and V. Sukhan, *Talanta*, 1999, **50**, 865.
37. M. E. Mahmoud, M. M. Osman and M. E. Amer, *Anal. Chim. Acta*, 2000, **415**, 33.
38. R. Hanzel and P. Rajec, *J. Radioanal. Nucl. Chem.*, 2000, **246**, 607.
39. M. E. Mahmoud and G. A. Gohar, *Talanta*, 2000, **51**, 77.
40. A. G. S. Prado, L. N. H. Arakaki and C. Airoldi, *J. Chem. Soc., Dalton Trans.*, 2001, 2206.
41. A. Goswami and A. K. Singh, *Talanta*, 2002, **58**, 669.
42. M. E. Mahmoud, *Anal. Lett.*, 2002, **35**, 1251.
43. A. Goswami and A. K. Singh, *Anal. Chim. Acta*, 2002, **454**, 229.
44. M. Sarkar, M. Das and P. K. Datta, *J. Colloid Interface Sci.*, 2002, **246**, 263.
45. P. Liu, Z. Su, X. Wu and Q. Pu, *J. Anal. At. Spectrom.*, 2002, **17**, 125.
46. M. M. Osman, S. A. Kholeif, N. A. Abou Al-Matay and M. E. Mahmoud, *Microchim. Acta*, 2003, **143**, 25.
47. H. Kokusen, *J. Ion Exchange*, 2005, **16**, 115.
48. M. Iiyama, S. Oshima, H. Kokusen, M. Sekita, S. Tsurubou and Y. Komatsu, *Anal. Sci.*, 2004, **20**, 1463.
49. M. A. A. Akl, I. M. M. Kenawy and R. R. Lasheen, *Microchem. J.*, 2004, **78**, 143.
50. G. Venkatesh, A. K. Singh and B. Venkataramani, *Microchim. Acta*, 2004, **144**, 233.
51. R. S. A. Machado, M. G. da Fonseca, L. N. H. Arakaki, J. G. P. Espinola and S. F. Oliveria, *Talanta*, 2004, **63**, 317.
52. M. A. A. Akl, I. M. Kenawy and R. R. Lasheen, *Anal. Sci.*, 2005, **21**, 923.

53. M. Imamoglu, A. O. Aydin and M. S. Dundar, *Cent. Eur. J. Chem.*, 2005, **3**, 252.
54. N. L. D. Filho, L. Caetano, D. R. do Carmo and A. H. Rosa, *J. Braz. Chem. Soc.*, 2006, **17**, 473.
55. R. Qu, Y. Niu, C. Sun, C. Ji, C. Wang and G. Chang, *Microporous Mesoporous Mater.*, 2006, **97**, 58.
56. B. Gao, F. An and K. Liu, *Appl. Surf. Sci.*, 2006, **253**, 1946.
57. N. L. D. Filho, D. R. do Cramo and A. H. Rosa, *Electrochim. Acta*, 2006, **52**, 965.
58. M. E. Mahmoud, S. S. Haggag and A. H. Hegazi, *J. Colloid Interface Sci.*, 2006, **300**, 94.
59. R. K. Dey, U. Jha, A. C. Singh, S. Samal and A. R. Ray, *Anal. Sci.*, 2006, **22**, 1105.
60. G. Gentscheva., P. Tzvetkova, P. Vassileva, L. Lakov, O. Peshev and E. Ivanova, *Microchim. Acta*, 2007, **156**, 303.
61. D. P. Quintanilla, I. Hierro, F. C. Hermosilla, M. Fajardo and I. Sierra, *Anal. Bioanal. Chem.*, 2006, **384**, 827.
62. A. Moghimi, *Chin. J. Chem.*, 2007, **25**, 1536.
63. D. Zang, J. Wang, T. R. Lawson and R. A. Bartsch, *Tetrahedron*, 2007, **63**, 5076.
64. M. Khosravan, S. J. Fatemi, A. S. Saljooghi and A. Badiei, *Asian J. Chem.*, 2007, **19**, 1131.
65. I. H. Gubbuck, R. Gup and M. Ersoz, *J. Colloid Interface Sci.*, 2008, **320**, 376.
66. I. Hatay, R. Gup and M. Ersoz, *J. Hazard. Mater.*, 2008, **150**, 546.
67. H. F. Maltez, M. A. Vieira, A. S. Ribeiro, A. J. Curtius and E. Carasek, *Talanta*, 2008, **74**, 586.
68. Q. He, X. Chang, X. Huang and Z. Hu, *Microchim. Acta*, 2008, **160**, 147.
69. I. H. Gubbuk, I. Hatay, A. Coskun and M. Ersöz, *J. Hazard. Mater.*, 2009, **172**, 1532.
70. Y. Zhang, R. Qu, C. Sun, H. Chen, C. Wang, C. Ji, P. Yin, Y. Sun, H. Zhang and Y. Niu, *J. Hazard. Mater.*, 2009, **163**, 127.
71. E. Repo, T. A. Kurniawan, J. K. Warchol and M. E. T. Sillanpaa, *J. Hazard. Mater.*, 2009, **171**, 1071.
72. G. Azimi, J. Zolgharnein, M. R. Sangi and S. Ebrahimi, *Anal. Sci.*, 2009, **25**, 711.
73. X. S. Jing, F. Q. Liu, X. Yang, P. P. Ling, L. J. Li, C. Long and A. M. Li, *J. Hazard. Mater.*, 2009, **167**, 589.
74. P. A. Amoyaw, M. Williams and X. R. Bu, *J. Hazard. Mater.*, 2009, **170**, 22.
75. I. H. Gubbuk, R. Gup, H. Kara and M. Ersoz, *Desalination*, 2009, **249**, 1243.
76. H. Budiman, F. Sri and A. H. Setiawan, *Eur. J. Chem.*, 2009, **6**, 141.
77. R. K. Sharma, A. Pandey, S. Gulati and A. Adholeya, *J. Hazard. Mater.*, 2012, **209–210**, 285.
78. R. K. Sharma, B. S. Garg, J. S. Bist and N. Bhojak, *Talanta*, 1996, **43**, 2093.

79. R. K. Sharma, B. S. Garg, N. Bhojak, J. S. Bist and S. Mittal, *Talanta*, 1999, **48**, 49.
80. R. K. Sharma, *Pure Appl. Chem.*, 2001, **73**, 181.
81. B. Gao, Y. Gao and Y. Li, *Chem. Eng. J.*, 2010, **158**, 542.
82. W. Wang, *Process Saf. Environ. Prot.*, 2011, **89**, 127.
83. Z. A. Alothman and A. W. Apblett, *Chem. Eng. J.*, 2009, **155**, 916.
84. X. Huang, Y. Wang, X. Liao and B. Shi, *J. Hazard. Mater.*, 2010, **183**, 793.
85. L. Bai, H. Hu, W. Fu, J. Wan, X. Cheng, L. Zhuge, L. Xiong and Q. Chen, *J. Hazard. Mater.*, 2011, **195**, 261.
86. A. A. Atia, A. M. Donia and W. A. Al-Amrani, *Desalination*, 2009, **246**, 257.
87. A. M. Donia, A. A. Atia, A. M. Daher, O. A. Desouky and E. A. Elshehy, *Int. J. Miner. Process.*, 2011, **101**, 81.
88. X. Huang, X. Liao and B. Shi, *J. Hazard. Mater.*, 2010, **173**, 33.
89. H. A. Zadeh, M. G. Assadi, S. Shabkhizan and H. Mousazadeh, *Arab. J. Chem.*, 2011; doi: 10.1016/j.arabjc.2011.07.006.
90. A. Safavi, S. Momeni and N. Saghir, *J. Hazard. Mater.*, 2009, **162**, 333.
91. R. Kumar, T. N. Abraham and S. K. Jain, *Int. Proc. Chem., Biol. Environ. Eng.*, 2011, **4**, 48.
92. L. Xu, Y. Liu, H. Hu, Z. Wu and Q. Chen, *Desalination*, 2012, **294**, 1.

CHAPTER 7

Nanomaterials for Water Remediation

DEEPTI GOYAL,[a] GEETA DURGA[b] AND
ANURADHA MISHRA*[a]

[a] Department of Applied Chemistry, School of Vocational Studies and
Applied Sciences, Gautam Buddha University, Greater Noida, Gautam Budh
Nagar – 201310, India; [b] Department of Applied Sciences, School
of Engineering and Technology, Sharda University, Greater Noida,
Gautam Budh Nagar – 201306, India
*Email: anuradha_mishra@rediffmail.com

7.1 Introduction

When water is adversely contaminated by organic pollutants, bacteria or microorganisms, industrial effluent containing heavy metals and various anions, and/or any compound that deteriorates its initial quality, it is called wastewater. It can be categorized into municipal wastewater and industrial wastewater. The municipal wastewater is mainly composed of proteins, carbohydrates, fats and oils, and trace amounts of priority pollutants, surfactants and emerging contaminants,[1] whereas industrial wastewater could be designated as the effluent produced from any industrial activity such as agriculture, food industry, iron and steel industry, mines and quarries, *etc.* The treatment process in wastewater treatment plants is directly related to the composition of wastewater. Conventional remediation and treatment technologies have shown limited effectiveness in reducing the levels of pollutants.[2] Nanotechnologies have the potential to overcome the drawbacks of existing

RSC Green Chemistry No. 23
Green Materials for Sustainable Water Remediation and Treatment
Edited by Anuradha Mishra and James H. Clark
© The Royal Society of Chemistry 2013
Published by the Royal Society of Chemistry, www.rsc.org

technologies and have come to significantly impact the remediation of environmental problems through development of new "green" technologies that minimize production of undesirable byproducts, as well as remediation of existing waste sites and polluted water sources. Removal of the finest contaminants from water can be envisaged using nanomaterials.

The use of nanomaterials is being researched to fabricate separation of contaminants from water.[3] Additionally, the use of nanomaterials to bioremediate and disinfect wastewater represents a new generation of environmental remediation technologies that could provide cost-effective solutions to some of the most challenging environmental cleanup problems.[4,5] The large surface areas and high surface reactivity of nanomaterials makes them important in maintaining fundamental issues such as health, energy and water.[6] Various nanomaterials, *viz.* nanoparticles, nanomembranes and nanopowders, are used for detection and removal of chemical and biological substances, including metals such as cadmium, copper, lead, mercury and zinc, nutrients, cyanide, organics including antibiotics, algae (*e.g.* cyanobacterial toxins), viruses, bacteria and parasites.[7] Herein we give a brief description of various contaminants present in wastewater, conventional water treatment processes and an overview of the use of nanomaterials in water remediation.

7.2 Contamination in Water and Remediation Techniques

Thousands of contaminants, such as heavy metals, plastics, lubricants, solvents, fuels, refrigerants and pesticides, are released into the environment through various industries. Other contributing factors to environmental pollution include our foods, water, soil, plants and even our bodies. The various pollutants present in our environment can be categorized as shown in Figure 7.1.

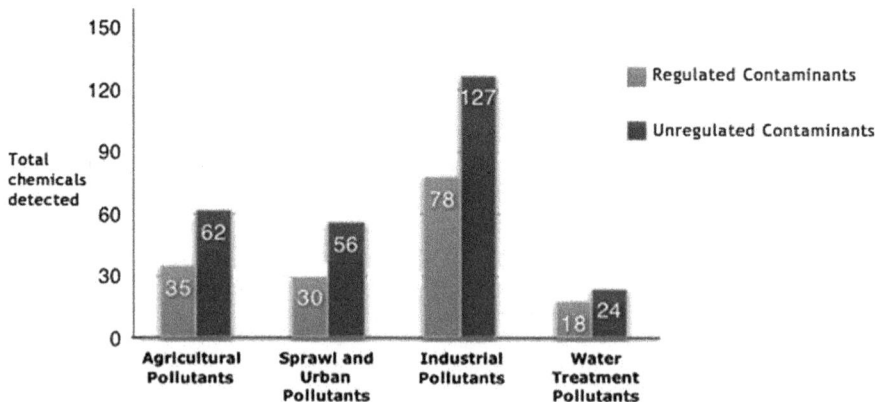

Figure 7.1 Contaminants present in water.

Some organic contaminants which are present in water are dioxins, polycyclic aromatic hydrocarbons (PAHs), DDT and polybrominated diphenyl ethers (PBDEs), *etc.* Dioxins are a group of chemically related compounds and produced in various industrial processes by burning waste, heating metals or bleaching paper. PAHs are a group of over 100 different chemicals produced by incomplete burning of coal, oil and gas, garbage or other organic substances like tobacco or charbroiled meat.[8] They enter into water through discharges from industrial and wastewater treatment plants and do not dissolve easily in water. They stick to solid particles and settle to the bottoms of lakes or rivers. Along with the above-mentioned organic pollutants, some heavy metal ions like mercury, arsenic and chromium are also present in water. Widely distributed substances, such as arsenic, heavy metals, halogenated aromatics, nitrosamines, nitrates, phosphates, and so on, are known to cause harm to humans and the environment.[9]

The technologies and processes commonly used for water purification are coagulation and flocculation; sedimentation; dissolved air flotation; filtration; steam distillation; ion exchange; deionization; reverse osmosis; and disinfection. Generally, conventional water treatment includes four stages,[10] as shown in Scheme 7.1.

Materials commonly used in these technologies are sediment filters, activated carbon, water softeners, ion exchangers, ceramics, activated alumina, organic polymers and many hybrid materials. With the advancement of science in the 20th and 21st centuries, scientists are now able to create lighter and stronger materials for remediation of contaminated water.[11] Nanomaterials have been extensively used for rapid or cost-effective cleanup of wastes when compared to

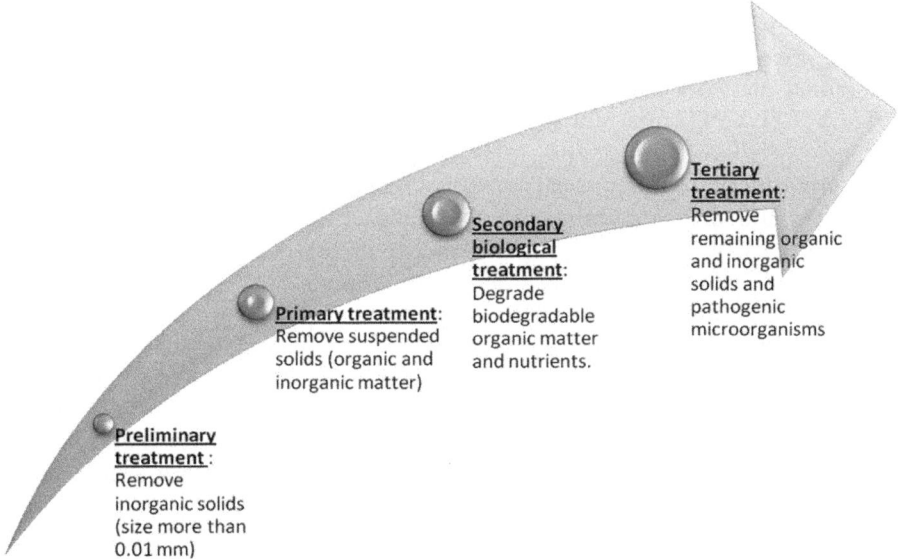

Scheme 7.1 Conventional sewage treatment process.

conventional approaches. The impacts of nanomaterials are increasingly evident in all areas of science and technology, including the field of environmental studies and treatment. Experts anticipate the development and implementation of environmentally beneficial nanotechnologies in the categories of sensing and detecting, pollution prevention, and treatment and remediation. Of the three, the category of treatment and remediation has seemingly experienced the most growth in recent years. In the following sections, we provide an overview of the application of nanomaterials in the water remediation process.

7.3   Nanotechnologies in Water Remediation

Nanotechnology has developed tremendously in the past decade and scientists have synthesized many novel materials with a vast range of potential applications.[12,13] Compared to other absorbents, nanoparticles have a greater advantage due to their larger surface area per unit mass. Nanoparticles also possess unique structural and electronic properties, which makes them powerful adsorbents. Various kinds of nanoscale materials are being evaluated as functional materials for water purification, *e.g.* metal-containing nanoparticles, carbonaceous nanoparticles, nanocrystalline zeolites, photocatalysts, magnetic nanoparticles and dendrimers, *etc.*, which are summarized in Table 7.1.[14-32]

7.3.1   Carbon Nanotubes

Carbon nanotubes (CNTs) are very thin, hollow cylinders made of carbon atoms having diameters in the range of 1 nm. They are about 10 000 times thinner than a human hair. CNTs are produced using various thermal processes to strip carbon atoms from carbon-bearing materials and use them to form a hexagonal network of carbon atoms that is rolled up into a cylinder (Figure 7.2). CNTs are 300 times stronger and six times lighter than steel, having a very large surface area, high tensile strength, can dissipate heat better than any other known material, possess excellent thermal and electrical conductivity, and are chemically and thermally stable. These properties make CNTs very useful in several areas. In water treatment, CNTs are very powerful absorbents for a wide variety of organic compounds.[33] Many water pollutants have very high affinity for carbon nanotubes and pollutants can be removed from contaminated water by filters made of this nanomaterial, for example water-soluble drugs which can hardly be separated from water by activated carbon. By using CNTs the problems due to the filters' saturation can be reduced because of their large surface area and consequently a very high capacity to retain pollutants.[34,35] The unique properties of CNTs allow water molecules to pass through the interior of the cylinders while preventing chemical and microbial contaminants from so doing. This can be done by applying a little pressure input to push the water through the nanotubes, giving CNTs a big advantage over current membrane technologies.[36,37]

Table 7.1 Nanomaterials used in water remediation.

Nanoparticles/nanomaterials	Pollutants	Ref.
1. Carbonaceous nanomaterials		
i. Single-wall carbon nanotubes (SWCNTs)	Heavy metal ions	14
ii. Multi-wall carbon nanotubes (MWCNTs)	Heavy metal ions	15
	Trihalomethanes (THMs)	16
	Chlorophenol	17
iii. Activated carbon fibers	Benzene, toluene, xylene	18
iv. Graphene	Dyes, heavy metal ions	19
2. Nanocrystalline zeolites	Heavy metal ions	20
3. Zero-valent iron nanoparticles (nZVIs)	Inorganic ions	21
	Heavy metal ions	22
	Chlorinated organic compounds	23
4. Silver nanoparticles	Bacteria	24
5. TiO_2 photocatalyst		
i. Nanocrystalline TiO_2	Heavy metal ions	25
ii. Fe(III)-doped TiO_2	Phenol	26
iii. TiO_2-based p-n junction nanotubes	Toluene	27
6. Bimetallic nanoparticles		
i. Pd/Fe nanoparticles	Brominated organic compounds, chloromethane	28,29
ii. Ni/Fe nanoparticles	Brominated organic compounds	28
iii. Pd/Au nanoparticles	Trichloroethene	30
7. Single enzyme nanoparticles	Not tested	31
8. Dendrimers	Metal ions, bacteria	32

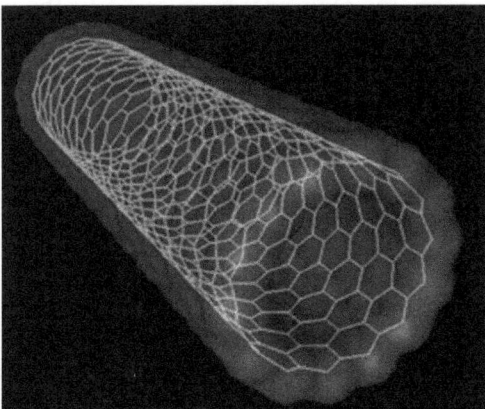

Figure 7.2 A carbon nanotube.

Along with these properties, CNTs were also found to be able to adsorb some toxic chemicals such as dioxins,[38] fluoride,[39] lead[40] and alcohols.[37] The adsorption of dioxins, which are very common and persistent carcinogenic

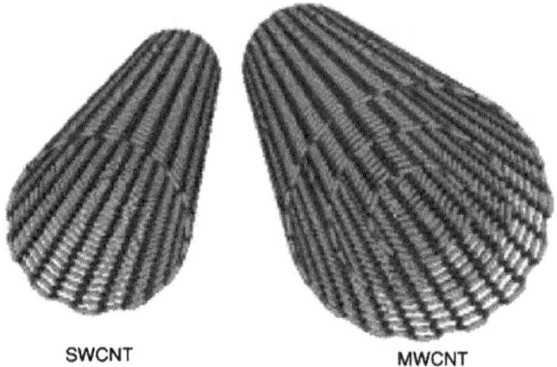

SWCNT   MWCNT

Figure 7.3   SWCNTs and MWCNTs.

byproducts of many industrial processes, is a good example of the potential of nanotubes in this field. It is well reported in the literature that nanotubes[39] can attract and trap more dioxins than activated carbon or other polycyclic aromatic materials that are currently used as filters. This improvement is probably due to the stronger interaction forces that exist between dioxin molecules and the curved surfaces of nanotubes compared to those for flat graphene sheets.

Single-walled carbon nanotubes (SWCNTs) and multi-walled carbon nanotubes (MWCNTs) (Figure 7.3) are well reported in the literature for their uses as efficient adsorbent materials. It has also been reported that the capacity of MWNTs to adsorb lead from water is higher than that for activated carbon.[40] MWNTs were found to be good adsorbents for the removal of dichlorobenzene from wastewaters over a wide pH range. It has also been shown that SWNTs act as molecular sponges for molecules such as CCl_4; the nanotubes were in contact with a support surface which also adsorbs molecules, although more weakly than the nanotubes.[41] The possibility of using graphite nanofibers to purify water from alcohols has also been explored.[42] These experimental results suggest that CNTs might be promising adsorbents for removing polluting agents from water.

7.3.2   Graphene

Graphene is an allotrope of carbon whose structure is a single planar sheet of sp^2-bonded carbon atoms, densely packed in a honeycomb crystal lattice. Graphene is the most easily visualized as atomic-scale chicken wire made of carbon atoms and their bonds. The crystalline or "flake" form of graphite consists of many graphene sheets stacked together (Figure 7.4).

Graphene is the subject of research into many different applications, one of which is water treatment processes. The pore size in the graphene membrane is about 1–12 nm, just wide enough to selectively let some small molecules through. It is a stronger material than those presently used for reverse osmosis. It is reported that graphene sand composite, synthesized from cane sugar,

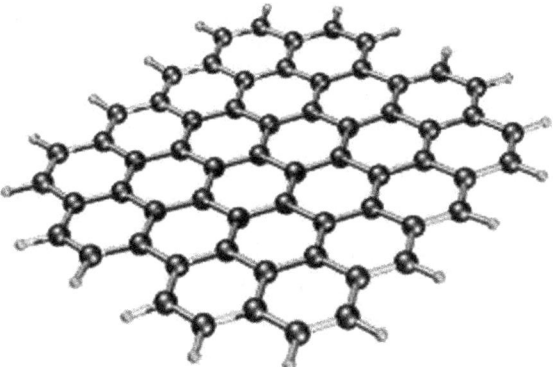

Figure 7.4   A graphene molecule.

effectively removed contaminants such as rhodamine 6G dye and chloropyrifos (a pesticide) from water.[43]

Graphene-like nanoflakes (GNFs), having a high specific surface area, have also been prepared and used as electrodes for capacitive deionization. The electrosorption performance of prepared GNFs was found much better than commercial activated carbon, and suggests a great potential in capacitive deionization applications. The specific electrosorptive capacity of GNFs for sodium ions (Na^+) was found higher than that of previously reported data using graphene and activated carbon under the similar experimental condition; thus the chemically synthesized GNFs can be used as effective electrode materials in water desalination processes.[44] Along with this, reduced graphene oxide (RGO)–metal/metal oxide composites can also be used in water purification processes. RGO composites can be synthesized by a redox-like reaction between RGO and the metal precursor. RGO–MnO_2 and RGO–Ag were found to be efficient adsorbent candidates for removal of mercury ions from water. RGO composites supported on river sand, using chitosan as the binder, were also found to be efficient adsorbent materials for heavy metal removal from water.[19] A novel composite based on graphene oxide (GO), crosslinked with iron(III) hydroxide, was also developed for effective removal of arsenate from contaminated drinking water. GO was first treated with iron(II) sulfate, then the iron(II) compound crosslinked with GO was *in situ* oxidized to the iron(III) compound by hydrogen peroxide, followed by treating with ammonium hydroxide. These composites were evaluated as absorbents for arsenate removal in a wide range of pH from 4 to 9 and it was found that arsenate removal increased with an increase in pH up to 8, but started to decrease with any increase in pH higher than 8.[45] Magnetite–graphene hybrids, synthesized by chemical reaction between graphene and magnetite particles, also showed high binding capacity for As(III) and As(V) compared to bare magnetite particles. Their high binding capacity is due to the increased adsorption sites in the prepared composite, which occurs by reducing the aggregation of bare magnetite. Since the composites show near complete (over 99.9%) arsenic removal within 1 ppb, they are practically usable for arsenic separation from water.[46]

7.3.3 Fullerenes

Fullerenes are spherical carbon molecules which can be made substantially stronger than diamond (Figure 7.5). Fullerenes are used for adsorption of organic compounds, but they are very efficient in the removal of organometallic compounds as they are weak sorbents for a wide variety of organic compounds (*e.g.* phenols, PAHs, amines). nC60 is a strong, broad-spectrum antibacterial agent, able to maintain its toxicity under varying environmental conditions with significant potential as an antibacterial agent for water treatment and biofouling control.[47]

Absorption of fullerenes depends greatly on the dispersion state of the C60, which is insoluble in water. However, because C60 clusters with water, there are interstitial spaces into which the compounds can diffuse, leading to a significant adsorption/desorption hysteresis. While this is a more alarming prospect in a release scenario, it is a desirable property for an effective antimicrobial agent. The low concentrations of nC60 needed to exert an antibacterial effect and its increasing availability and affordability make it a potentially attractive, viable agent for wastewater and drinking water treatment. The versatility and potency of nC60 indicate potential biological disruptions in the event of exposure to soil or water ecosystems. The ability of nC60 to kill different types of bacteria under oxic/anoxic, light/dark conditions foreshadows the effects of nC60 on soil and water microbial communities. However, there are several factors which may mitigate the antibacterial activity of nC60. Previous work has shown that salt concentrations increased the nC60 particle size[48,49] and an increasing particle size results in increasing MIC (minimum inhibiting concentration, *i.e.* decreased toxicity).[50] In marine settings, the ionic strength can reach 1 M, instigating nC60 coagulation and a loss of toxicity either due to precipitation or an unknown factor. In soil or water, nC60 may sorb to particulate matter and be immobilized or even neutralized, though the toxicity of sorbed nC60 has not been determined.

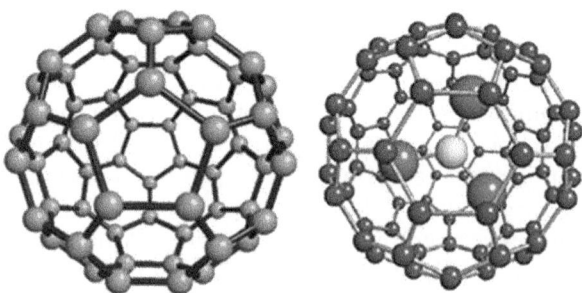

Figure 7.5   A fullerene molecule.

7.3.4 Nanocrystalline Zeolites

The term zeolite represents a very broad group of crystalline structures generally composed of silicon, aluminum and oxygen. The zeolite framework consists of four-connected networks of atoms. These tetrahedrally shaped molecules link together by their corners to form several structures. The framework contains cages, cavities or channels, which allows small molecules to enter (Figure 7.6). Zeolites have been widely studied in the past 10 years owing to their attractive properties such as molecular sieving, high cation exchange capacities, and their affinity for heavy metals. As catalysts and sorbents, they can remove atmospheric pollutants, including engine exhaust gases. Zeolites are also widely used as ion exchange beds in domestic and commercial water purification, softening and other applications.[20] Although conventional synthetic methods produce zeolites on the scale of 1–10 μm, nanoscale zeolites which comprise discrete uniform structures with dimensions ranging from 5 to 10 nm have been successfully synthesized.[51] Greater external surface areas, smaller diffusion path lengths and a greater erosion to coke formation make nanocrystalline zeolites superior to traditional micron-sized zeolites.[52] It is well reported in the literature that nanocrystalline NaY zeolites are capable of absorbing 10% more toluene compared to commercial NaY zeolites; similarly, ZSM-5 zeolites with a particle size 15 nm absorb 50% more toluene than ZSM-5 having a large particle size.[51] The fact that cations displaced from zeolites are relatively harmless makes zeolites attractive for the removal of undesirable and toxic heavy metal ions from acid marine disposal (AMD) effluents. Other factors that make natural zeolites attractive alternatives for the treatment of AMD are their lower cost since they are relatively abundant,[53] their high surface area due to their porous and rigid structure,[20] and the fact that they also act as molecular sieves and this property can easily be modified to

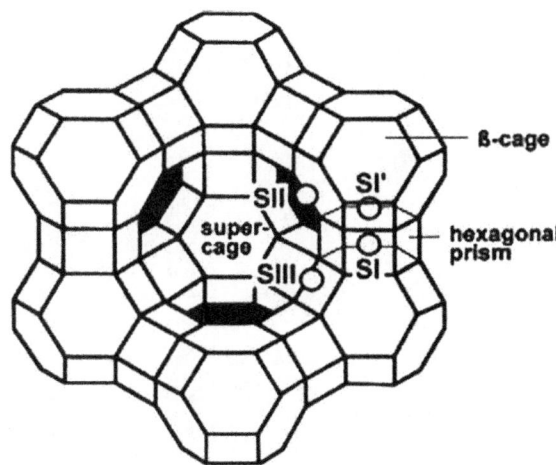

Figure 7.6 Zeolite molecules.

increase their performance. Because of these attractive characteristics, there has been a growing interest in adsorbing heavy metals from solution using natural zeolites. The ion exchange property of zeolites is also used in controlling soil pH and nutrient levels, and they are also used also as animal feed supplements.[20]

7.3.5   Magnetic Nanoparticles

Magnetic nanoparticles (MNPs) are nanoparticles which are called magnetic due to metal components such as iron, nickel or cobalt in them. MNPs are used for a wide range of applications, including data storage, magnetic fluids, catalysis, biotechnology, biomedicine and environmental remediation. Magnetic nanoparticles can be easily separated from water using a magnetic field. This property of magnetic nanoparticles distinguishes them from nonmagnetic nanoparticles as they can perform separation by magnetic gradients, or high magnetic gradient separation. This process allows the particles not only to remove compounds from water but also to be recycled and regenerated. Some magnetic nanoparticles used in water remediation are discussed below.

7.3.5.1   Magnetic Carbon Nanotubes

Magnetic carbon nanotubes have sufficient activity to remove contaminants from water. These nanotubes are made by capturing iron-based nanoparticles within the graphitic tube walls (Figure 7.7). These hybrid materials shows superior stability, with increased external and internal surface areas. Water-soluble CNTs have been enhanced with magnetic iron nanoparticles for the removal of aromatic compounds from water.

Figure 7.7   Multi-walled CNTs with iron nanoparticles.

7.3.5.2 Nanoscale Zero-Valent Iron

Nanoscale zero-valent iron (ZVI) having a large surface area and high surface reactivity is a newly developed nanotechnology for water purification, providing enormous flexibility for *in situ* applications.[54] Arsenic contamination in drinking water is a global problem. The existing ways for arsenic removal require an extensive hardware set-up of energy-consuming high-pressure pumps. ZVI provides a low-cost technology for cleaning arsenic from drinking water since iron oxides are known to bind arsenic. One effective property of iron oxide is that it possesses an exceptionally high surface area, which allows more binding spots for arsenic. Nanoscale ZVI with a 12 nm diameter can capture 100 times more arsenic than the larger iron oxide filters used today. Once this arsenic has been captured by nano-iron, these nanoparticles can be easily removed from water by simply using a hand magnet. By the application of nano-iron, it is possible to transform a wide variety of heavy metals and also highly toxic arsenic into an insoluble form, which remains tightly bonded in the rock environment.[55] A wide range of pollutants such as chlorinated hydrocarbons (*e.g.* carbon tetrachloride, tetrachloroethylene, trichloroethylene, dichloro-ethylene, vinyl chloride, chloroform, polychlorinated biphenyls) and other organic substances (*e.g.* nitrobenzene, dioxins) can be successfully removed by the use of Fe(0) nanoparticles and decomposed into simpler nontoxic phases (Scheme 7.2). Modified iron nanoparticles, such as catalyzed and supported nanoparticles, have been synthesized to further enhance the speed and efficiency of remediation.[56] Thus, these pollutants cannot migrate any further, which leads to a significant reduction in their ecological impacts.

The reducing potential of ZVI nanoparticles is very high, reducing other matter contained in water like dissolved oxygen, sulfates, nitrates, *etc*. As ZVI is oxidized to ferrous and/or ferric iron, the pH increases, hydrogen is evolved,

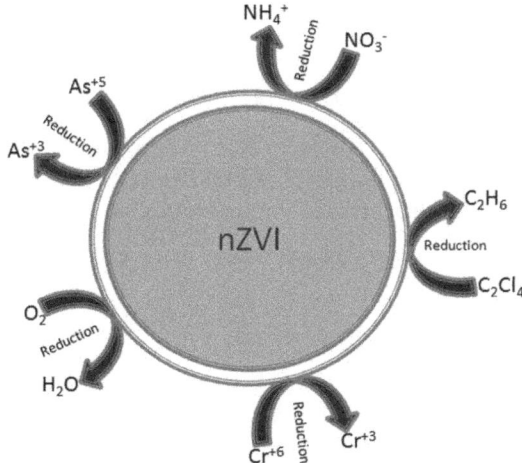

Scheme 7.2 Mechanism of degradation of contaminants using zero-valent iron nanoparticles.

oxidizable materials are consumed, and the strong reducing conditions created are favorable for pathways (oxide-mediated electron transfer from the metal to the chlorinated organic, reduction of the chlorinated organic by the ferrous iron and reduction by evolved hydrogen) leading to complete dechlorination. Eventually, ferric or ferrous iron may precipitate as a solid or remain in solution, depending on the pH and redox conditions. For example, oxygen present in the environment is reduced, which leads to a decrease of the redox potential and a reduction of monitored contaminants (eqn 7.1):

$$Fe^0 + \tfrac{1}{2}O_2 + 2H^+ \rightleftharpoons Fe^{2+} + H_2O \tag{7.1}$$

The mechanism of degradation of some contaminants using Fe^0 nanoparticles is summarized below:

1. Reduction of nitrates. The reactions between nano-Fe and nitrates produces nitrogen gas and eventually ammonia.[57] The mechanism of nitrate reduction is mediated by free electrons, which are released directly or indirectly by the corroding process of elemental Fe^0 (eqn 7.2):

$$4Fe^0 + NO_3^- + 10H^+ \rightleftharpoons 4Fe^{2+} + NH_4^+ + 3H_2O \tag{7.2}$$

2. Reduction of chromium. Hexavalent chromium is reduced to trivalent chromium, which is less soluble, less toxic and less mobile.[58] The reduction of hexavalent chromium is described by (eqn 7.3):

$$3Fe^0 + 2Cr^{6+} \rightleftharpoons 3Fe^{2+} + 2Cr^{3+} \tag{7.3}$$

3. Reduction of arsenic. In a groundwater environment, Fe^0 is oxidized into oxides and hydroxides of Fe^{2+} and Fe^{3+}. These oxides and hydroxides then adsorb and coprecipitate As^{3+} and As^{5+}. The reduction of arsenic with hydrated oxides is described as follows (eqns 7.4, 7.5):[59,60]

$$Fe(OH)^0 + AsO_4^{3-} + 3H^+ \rightleftharpoons FeH_2AsO_4(s) + H_2O \tag{7.4}$$

$$Fe(OH)^0 + AsO_4^{3-} + 2H^+ \rightleftharpoons FeHAsO_4^-(s) + H_2O \tag{7.5}$$

4. Reduction of chlorinated hydrocarbons. These are the most widely known and most common pollutants of groundwater.[61] ZVI nanoparticles react with chlorinated hydrocarbons, *e.g.* for tetrachloroethylene (eqn 7.6):

$$C_2Cl_4 + 5Fe^0 + 6H^+ \rightleftharpoons C_2H_6 + 5Fe^{2+} + 4Cl^- \tag{7.6}$$

7.3.6 Silver Nanoparticles

Silver nanoparticles are the most popular nanoparticles due to their relatively low manufacturing costs and many applications, *viz.* medical uses, water purification, antimicrobial uses, paints, coatings, food packaging. The size of these nanosilver particles can range between 10 and 200 nm with high surface reactivity and strong antimicrobial properties; thus they are used in many products as a bactericide.[31] Nanosilver is primarily produced as colloidal silver, spun silver, nanosilver powder and polymeric silver.[62] Various studies have been done to evaluate the antimicrobial effect of nanosilver; one of them was to study the effect of nanosilver on the growth of *E. coli*, which showed that bacterial growth was inhibited when nanosilver was introduced. The potent toxicity of nanosilver towards beneficial microorganisms is the major challenge that researchers face in its large-scale implementation in water treatment processes.[63]

7.3.7 TiO$_2$ Nanoparticles

The photocatalytic properties of TiO$_2$ nanomaterials have been exploited in water treatment applications. TiO$_2$ nanoparticles have been extensively studied for oxidative and reductive transformations of organic and inorganic species present as contaminants in air and water.[64] The high surface area of TiO$_2$ nanoparticles creates a larger catalytic surface for the production of hydroxyl radicals that are strong oxidizing agents.[65] When UV light is used to activate the nanoparticles, efficient removal of aromatic organic compounds can be achieved. The removal of heavy metals can also be done by the use of TiO$_2$. Other advantages of using TiO$_2$ for water treatment applications are its low cost, resistance to corrosion and overall stability. TiO$_2$ nanoparticles have been used as a solid-phase extraction packing material for the remediation of surface waters. TiO$_2$ nanoparticles when used as a packing material can effectively preconcentrate and extract heavy metals from river water and seawater. This has been effectively accomplished in batch and column experiments at the natural pH of coastal waters.[66] Along with TiO$_2$ nanoparticles, TiO$_2$ nanowire membranes have also been successfully fabricated with the capability of filtering organic contaminants from water with simultaneous photocatalytic oxidation.[67] Nanowire membranes with nanowires of 20–100 nm in diameter have uniform thickness and flexibility. Composite photocatalytic membranes, *viz.* TiO$_2$/Al$_2$O$_3$, were effectively used to remove dye when photocatalysis was coupled with membrane separation.[68] Similarly, inside-out tubular TiO$_2$/Al$_2$O$_3$ composite membranes have also been prepared, which degraded a great amount of the water pollutant of concern from the target wastewater.[69]

7.3.8 Bimetallic Nanoparticles

Bimetallic nanoparticles have been proven to be effective in removal of organic contaminants from wastewater. These bimetallic nanoparticles have been synthesized by co-reduction of precursors of two metals or post-deposition of a second metal on the surface of iron particles.[70] An overview[54] of the use of nanoscale Fe particles and Fe/Pd, Fe/Pt, Fe/Ag, Fe/Ni and Fe/Co in the reduction of a variety of organic pollutants, such as chlorinated alkanes and alkenes, chlorinated benzenes, pesticides, organic dyes, nitro aromatics, PCBs and inorganic anions, *viz.* nitrates, to less toxic byproducts, is given in the

literature. It was evaluated that dehalogenation of trichloroethylene on bimetallic Pd/Fe nanoparticles was higher than on nZVI particles alone.[71]

Pd/Fe and Ni/Fe nanoparticles are used for hydrohalogenation of chlorinated aliphatics,[72] chlorinated aromatics[73] and polychlorinated biphenyls,[74] but
Ni has better corrosion stability and lower cost than Pd, which makes Ni more
efficient for wastewater treatment.[75] Ni/Fe nanoparticles are also found effective for reductive halogenations of brominated methane, when compared with
the commercial microscale iron.[28] Along with this, tetrachloroethylene could
also be removed from water by supported bimetallic Ni/Fe nanoparticles.[76]
Similarly, Pd-coated Au nanoparticles are found both active and selective for
removal of organic contaminants from water. Au particles themselves do not
act as a catalyst and do not react with organic compounds, but when combined
with Pd they increase the catalytic activity of Pd. It is well known in the literature that bimetallic Pd/Au nanoparticles are very active for aqueous phase
tetrachloroethylene hydrochlorination, better than only a Pd catalyst.[30] It was
also found that when Pd was supported on Au the catalytic activity of Pd was
increased by 700 times.[77]

7.3.9   Single Enzyme Nanoparticles

Enzymes have proven to be more effective than synthetic catalysts in many
areas of application. In wastewater treatment, enzymes can be utilized to develop remediation processes that are environmentally less aggressive than
conventional techniques. Their versatility and efficiency even in mild reaction
conditions gives them an advantage over the conventional physicochemical
treatment methods. The biological origin of enzymes reduces their adverse
impact on the environment, thereby making enzymatic wastewater treatment
an ecologically sustainable technique. Specific enzymes are used for different
contaminants. Some examples of enzyme classes which can be used for remediation are the peroxidases, polyphenol oxidases (laccase, tyrosinase), dehalo-
genases and organophosphorus hydrolases.[31] The great choice of applicable
enzymes implies the possibility for successful remediation of a broad range of
organic contaminants in water. Contaminants such as phenols, polyaromatics,
dyes, chlorinated compounds, organophosphorus pesticides and even explosives can be successfully degraded using appropriate enzymes.[78] A new method
for the stabilization of enzymes has been developed which involves the production of chemically stable single enzyme nanoparticles (SENs). The production of SENs can be achieved by using a combination of enzyme technology
and nanotechnology.[31] SENs are resistant to extreme conditions, such as high/
low pH, high contaminant concentration, high salinity and high/low temperatures, and it is much easier to control SENs than microbial organisms.
They need no nutrients to exist and any metabolic byproducts or mass transfer
limitations, due to cellular transport, are avoided.[31] For the synthesis of SENs
the enzyme chymotrypsin has been used in which the enzyme molecule was
"caged" by a silicate shell which was linked to its surface. While the cage

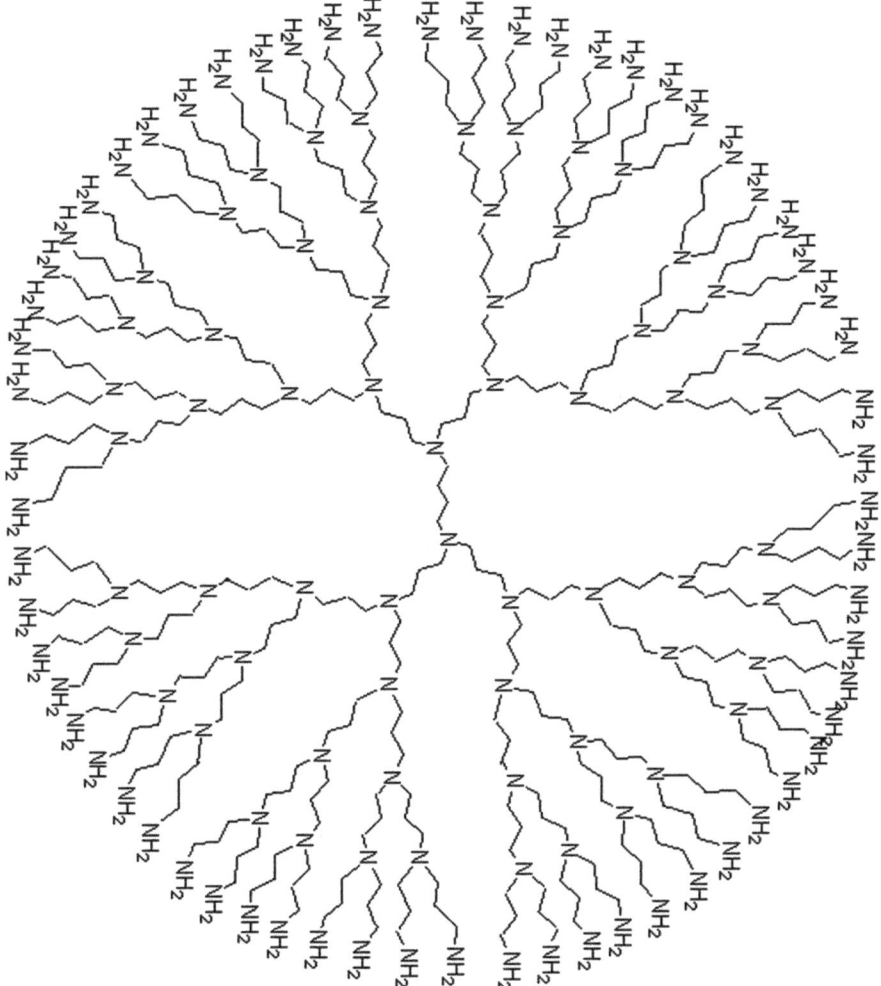

Figure 7.8 Dendrimers.

covered most of the enzyme, the active site was kept chemically accessible to maintain the functionality of the enzyme.[78]

7.3.10 Dendrimers

Dendrimers represent a novel class of three-dimensional, highly branched, globular macromolecules which include hyper-branched polymers, dendrigraft polymers and dendrons (Figure 7.8). Similar to linear polymers, they are composed of a large number of monomer units that are chemically linked together. The size of the dendrimers ranges between 2 and 20 nm, with common shapes including cones, spheres and disc-like structures.[79]

Dendrimers have wide ranges of potential applications, including adhesives and coatings, chemical sensors, medical diagnostics, drug-delivery systems, high-performance polymers, catalysts, building blocks for super molecules, separation agents and water remediation. Recently, poly(amidoamine) (PAMAM) dendrimers have been developed which include functional nitrogen and amide groups repeatedly attached in radially branched layers.[80] They were used in the remediation of wastewater with a variety of transition metal ions such as copper(II). The high concentration of nitrogen ligands within the interior branches makes PAMAM dendrimers useful as chelating agents for copper ions. Similarly, for the remediation of other metals from water, researchers have employed an ethylenediamine core.[81] To expand the initial research, scientists have devised a dendrimer-enhanced ultrafiltration (DEUF) method to recover copper from aqueous solutions. DEUF is a modified form of polymer-enhanced ultrafiltration (PEUF), which is used as a promising technology for metal ion removal from waste streams. PEUF and DEUF work on the same principles, where the binding of metal ions to the polymers or dendrimers allows the removal of contaminants though membrane filtration. The whole filtration process is carried out in two steps:[79] (i) the first step includes the mixing of either linear polymers or dendrimers with contaminated wastewater, where they subsequently bind to metal ions present; the solution is then passed through an ultrafiltration membrane, which prevents passage of the polymer/dendrimer–metal ion complex; (ii) in the second step, the metal ion is detached from the polymer or dendrimer, which can be further reused.

Dendrimers containing bactericide carriers can be used to kill bacteria in a body of water. In this method, after purification of the water the dendrimer can be removed from the body of the water and can be reused by securing an antimicrobial agent onto it again. To improve the effectiveness of water purification materials, two different antimicrobial agents can be secured to the dendrimer. Thus the dendrimers containing a water purification material can work effectively in drinkable fluids such as fruit juices to rid the juices of harmful microbes.

7.3.11 Nanomembranes

Nanoreactive materials have been used to synthesize membranes for use in water treatment. Water filtration membranes fabricated from nanomaterials are gaining interest these days as they are resistant to fouling, durable and cost-effective, and because of unlimited benefits in terms of metals and organic and biological contaminants removed. The selectivity of the membrane is determined by the compound size and permeability. Since sedimentation, flocculation, coagulation and activated carbon each remove a narrow spectrum of water pollutants, membrane filtration has played a significant role in reducing pollutants and producing high-quality pure water.[82] In the last two decades, the development of polymeric and ceramic membranes has positively impacted on the use of membranes. Nanoreactive membranes are able to decompose pollutants such as 4-nitrophenol[83] and bind metal ions in aqueous solution. Nanosilver-impregnated polysulfonate ultrafiltration membranes were found to be effective against bacterial strains and showed a significant improvement in virus removal.[84] Similarly, nanostructure surface-modified microporous ceramics were applied with the aim of efficient virus filtration.[85] In this procedure the internal surface area of highly porous elements was coated with a colloidal nanodispersion of hydrated yttrium oxide that was then heated to obtain an electropositive Y_2O_3 coated surface. Modified nanostructure filters were able to remove bacteriophages from water. The nano-filtration separation process removes turbidity, microorganisms and inorganic ions, reduces water hardness, separates organic impurities from drinking water, and desalinates organic materials. As pressure is added to salt water, it leaves the semi-permeable membrane as soft water, leaving the unwanted particles behind, in this case bivalent and monovalent ions. The nanosilver-impregnated membranes were also resistant to biofouling, mainly because the attachment of bacteria to the membrane surface was prohibited by Ag^+. MWCNTs have also been prepared to form a hollow monolithic cylindrical membrane, which increases the efficiency for the removal of bacteria or hydrocarbons; they can also be easily regenerated. Numerous experiments with CNTs in nano-filtration have been conducted.

7.4 Conclusion

The impact of nanotechnology on the development of tools and techniques for water treatment will be more pronounced in the near future. The advantages of using nanoparticles relies on the direct humanitarian benefit from using nanotechnology and in the promotion of economical viabilities in rural communities. Therefore, the production of nanostructures, nanocomposites and modified nanostructures for water remediation will increase because of the need for producing clean water in fast and low-energy-consumption ways. Nanotechnology should be regarded as the tool to ensure the sustainability of social communities in different places. This is possible through the use of advanced nanomaterials that enable the desalination of seawater, recycling of contaminated water and the reuse of wastewater. Remediation nanotechnologies are different from each other in the way they remove contaminants from the environment. Some of them use chemical conversion mechanisms like oxidation and reduction, while others act as catalysts. Despite the differences, however, the effective operation of most depends on the effective surface area of their active materials. The large surface areas of nanoparticles make them more effective for the removal of contaminants than their bulk alternatives and thus are preferred for environmental remediation. Novel nano-engineered technologies provide more sensitive and reliable water monitoring solutions. In the end, it is likely that it will be the next few years that dictate the direction of nanotechnologies for water remediation. It is suggested that these may be used in the future for large-scale water purification.

References

1. S. Aguayo, M. J. Munoz, A. De la Torre, J. Roset, E. de la Pena and M. Carballo, *Sci. Total Environ.*, 2004, **38**, 69.
2. A. K. M. Kabzhinski, J. Cyran and R. Juszczak, *Pol. J. Environ. Stud.*, 2002, **11**, 695.
3. D. Rickerby and M. Morrison, *J. Sci. Technol. Adv. Mater.*, 2006, **8**, 19.
4. C. Bellona and J. E. Drewes, *Water Res.*, 2007, **41**, 3948.
5. J. Hu, G. Chen and I. M. C. Lo, *Water Res.*, 2005, **39**, 4528.
6. D. Mohan, Jr. and C. U. Pittman, *J. Hazard. Mater.*, 2007, **142**, 1.
7. P. Binks, presented at the Victorian Water Sustainability Seminar, Melbourne, May 2007.
8. D. K. Tiwari, J. Behari and P. Sen, *World Appl. Sci. J.*, 2008, **3**, 417.
9. M. A. Shannon, P. W. Bohn, M. Elimelech, J. G. Georgiadis, B. J. Merinas and A. M. Mayes, *Nature*, 2008, **452**, 301.
10. H. K. Shon, S. Vigneswaran, J. Kandasamy and J. Cho, *Characteristics of Effluent Organic Matter in Wastewater*, Eolss, Oxford, 2007.
11. T. Masciangioli and W. Zhang, *Environ. Sci. Technol.*, 2003, **37**, 102.
12. P. K. Stoimenov, R. L. Klinger, G. L. Marchin and K. J. Klabunde, *Langmuir*, 2002, **18**, 6679.
13. V. L. Colvin, *Nat. Biotechnol.*, 2003, **10**, 1166.
14. C. Lu and H. Chiu, *Chem. Eng. Sci.*, 2006, **61**, 1138.
15. X. Q. Li, D. W. Elliott and W. X. Zhang, *Crit. Rev. Solid State Mater. Sci.*, 2006, **31**, 111.
16. C. Lu, Y. L. Chung and K. F. Chang, *J. Hazard. Mater.*, 2006, **138**, 304.
17. Y. Q. Cai, Y. Cai, S. F. Mou and Y. Q. Lu, *J. Chromatogr. A*, 2005, **1081**, 245.
18. C. L. Mangun, Z. R. Yue, J. Economy, S. Maloney, P. Kamme and D. Cropek, *Chem. Mater.*, 2001, **13**, 2356.
19. T. S. Sreeprasad, S. M. Maliyekkal, K. P. Lisha and T. Pradeep, *J. Hazard. Mater.*, 2011, **186**, 921.
20. A. E. Alvarez, A. G. Sanchez and X. Querol, *Water Res.*, 2003, **37**, 4855.
21. J. Cao, D. Elliott and W. X. Zhang, *J. Nanopart. Res.*, 2005, **7**, 499.
22. X. Q. Li and W. X. Zhang, *Langmuir*, 2006, **22**, 4638.
23. R. Cheng, J. L. Wang and W. X. Zhang, *J. Hazard. Mater.*, 2007, **144**, 334.
24. S. Kar, R. C. Bindal and P. K. Tewari, *Nanotoday*, 2012, **7**, 385.
25. M. E. Pena, G. P. Korfiatis, M. Patel, L. Lippincott and X. Meng, *Water Res.*, 2005, **39**, 2327.
26. S. Nahar, K. Haregawa and S. Kagaya, *Chemosphere*, 2006, **65**, 1976.
27. Y. Chen, J. Crittenden, S. Hackney, L. Sutter and D. Hand, *Environ. Sci. Technol.*, 2005, **39**, 1201.
28. T. T. Lim, J. Feng and B. W. Zhu, *Water Res.*, 2007, **41**, 875.
29. H. L. Lien and W. X. Zhang, *Colloids Surf. A*, 2001, **191**, 97.
30. M. O. Nutt, K. N. Heck, P. Alvarez and M. S. Wong, *Appl. Catal. B*, 2006, **69**, 115.

31. M. Alcalde, M. Ferrer, F. J. Plou and A. Ballesteros, *Trends Biotechnol.*, 2006, **24**, 281.
32. E. R. Birnbaum, K. C. Rau and N. N. Sauer, *Sep. Sci. Technol.*, 2003, **38**, 389.
33. C. W. Tan, K. H. Tan, Y. T. Ong, A. R. Mohamed, S. H. S. Zein and S. H. Tan, *Environ. Chem. Lett.*, 2012, **10**, 265.
34. V. K. K. Upadhyayula, S. Deng, M. C. Mitchell and G. B. Smith, *Sci. Total Environ.*, 2009, **408**, 1.
35. B. Pan and B. Xing, *Environ. Sci. Technol.*, 2008, **42**, 9005.
36. Q. Li, S. Mahendra, D. Y. Lyon, L. Brunet, M. V. Liga, D. Li and P. J. Alvarez, *Water Res.*, 2008, **42**, 4591.
37. R. Q. Long and R. T. Yang, *J. Am. Chem. Soc.*, 2001, **123**, 2058.
38. Y. H. Li, S. Wang, A. Cao, D. Zhao, X. Zhang, C. Xu, Z. Luan, D. Ruan, J. Liang, D. Wu and B. Wei, *Chem. Phys. Lett.*, 2001, **350**, 412.
39. Y. H. Li, S. Wang, J. Wei, X. Zhang, C. Xu, Z. Luan, D. Wu and B. Wei, *Chem. Phys. Lett.*, 2002, **357**, 263.
40. C. Park, E. S. Engel, A. Crowe, T. R. Gilbert and N. M. Rodriguez, *Langmuir*, 2000, **16**, 8050.
41. D. Y. Lyon, D. A. Brown and P. J. J. Alvarez, *Water Sci. Technol.*, 2008, **57**, 1533.
42. P. Kondratyuk and J. T. Yates, *Chem. Phys. Lett.*, 2004, **383**, 314.
43. S. S. Gupta, T. S. Sreeprasad, S. M. Maliyekkal, S. K. Das and T. Pradeep, *ACS Appl. Mater. Interfaces*, 2012, **4**, 4156.
44. H. Li, L. Zou, L. Pan and Z. Sun, *Environ. Sci. Technol.*, 2010, **44**, 8692.
45. K. Zhang, V. Dwivedi, C. Chi and J. Wu, *J. Hazard. Mater.*, 2010, **182**, 162.
46. V. Chandra, J. Park, Y. Chun, J. W. Lee, I. C. Hwang and K. S. Kim, *ACS Nano*, 2010, **4**, 3979.
47. D. Y. Lyon, J. D. Fortner, C. M. Sayes, V. L. Colvin and J. B. Hughes, *Environ. Toxicol. Chem.*, 2005, **24**, 2757.
48. J. Brant, H. Lecoanet and M. R. Wiesner, *J. Nanopart. Res.*, 2005, **7**, 545.
49. D. Y. Lyon, L. K. Adams, J. C. Falkner and P. J. J. Alvarez, *Environ. Sci. Technol.*, 2006, **40**, 4360.
50. E. A. Ayuso, A. G. Sanchez and X. Querol, *Water Res.*, 2003, **37**, 4855.
51. W. Song, L. Gonghu, V. H. Grassian and S. C. Larsen, *Environ. Sci. Technol.*, 2005, **39**, 1214.
52. W. Song, V. H. Grassian and S. C. Larsen, *Chem. Commun.*, 2005, 2951.
53. M. Sprynskyy, B. Boguslaw, A. P. Terzyk and J. Namiesnik, *J. Colloid Interface Sci.*, 2006, **304**, 21.
54. W. Zhang, *J. Nanopart. Res.*, 2003, **5**, 323.
55. K. W. Henn and D. W. Waddill, *Remed. J.*, 2006, **16**, 57.
56. S. Krajangpan, J. J. Bermudez, A. N. Bezbaruah, B. J. Chisholm and E. Khan, *Water Sci. Technol.*, 2008, **58**, 22.
57. T. Li, S. Li, S. Wang, Y. An and Z. Jin, *J. Water Resour. Prot.*, 2009, **1**, 1.
58. W. X. Zhang, *J. Nanopart. Res.*, 2003, **5**, 323.

59. Y. P. Sun, X. Q. Li, J. Cao, W. X. Zhang and H. P. Wang, *Adv. Colloid Interface Sci.*, 2006, **120**, 47.
60. C. Su and R. W. Puls, *Water, Air, Soil Pollut.*, 2008, **193**, 65.
61. J. Kim and J. Grate, *J. Water Resour. Prot.*, 2004, **29**, 220.
62. B. Nowack and T. D. Bucheli, *Environ. Pollut.*, 2007, **150**, 5.
63. S. K. Brar, M. Verma, R. D. Tyagi and R. Y. Surampalli, *Waste Manage.*, 2010, **30**, 504.
64. P. V. Kamat, D. Meisel and C. R. Chimie, *Nanosci. Oppor. Environ. Remed.*, 2003, **6**, 999.
65. Y. Morimoto, N. Kobayashi, N. Shinohara, T. Myojo and I. Tanaka, *J. Occup. Health*, 2010, **52**, 325.
66. C. R. Quetel, E. Vassileva, I. Petrov, K. Chakarova and K. I. Hadjiivanov, *Anal. Bioanal. Chem.*, 2010, **396**, 2349.
67. L. Xiao, L. Erdei, A. McDonagh and M. Cortie, presented at the International Conference on Nanoscience and Nanotechnology, Melbourne, 2008.
68. H. Zhang, X. Quan, S. Chen, H. Zhao and Y. Zhao, *Sep. Purif. Technol.*, 2006, **50**, 147.
69. G. C. C. Yang and C. J. Li, *Desalination*, 2008, **234**, 354.
70. B. Shrick, J. Blough, D. Jones and T. Mallouk, *Chem. Mater.*, 2002, **14**, 5140.
71. W. X. Zhang, C. B. Wang and H. L. Lien, *Catal. Today*, 1998, **40**, 387.
72. X. Xu, M. Zhou, P. He and Z. Hao, *J. Hazard. Mater.*, 2005, **123**, 89.
73. B. W. Zhu, T. T. Lim and J. Feng, *Chemosphere*, 2006, **65**, 1137.
74. J. Wei, X. Xu, Y. Liu and D. Wang, *Water Res.*, 2006, **40**, 348.
75. P. He and D. Y. Zhao, *Environ. Sci. Technol.*, 2005, **39**, 3314.
76. L. Wu and S. M. C. Ritchie, *Chemosphere*, 2006, **63**, 285.
77. M. O. Nutt, J. B. Huges and M. S. Wong, *Environ. Sci. Technol.*, 2005, **39**, 1346.
78. K. Jungbae and J. Grate, *Nano Lett.*, 2003, **3**, 1219.
79. D. Mamadou, S. Christie, P. Swaminathan, J. Johnson and W. Goddard, *Environ. Sci. Technol.*, 2005, **39**, 1366.
80. D. Mamadou, S. Christie, P. Swaminathan, L. Balogh, X. Shi, W. Um, C. Papelis, W. Goddard and J. Johnson, *Langmuir*, 2004, **20**, 2640.
81. X. Yinhui and D. Zhao, *Environ. Sci. Technol.*, 2005, **39**, 2369.
82. D. M. Dotzauer, J. Dai, L. Sun and M. L. Bruening, *Nano Lett.*, 2006, **6**, 2268.
83. K. Zodrow, L. Brunet, S. Mahendra, D. Li, A. Zhang, Q. Li and P. J. J. Alvarez, *Water Res.*, 2009, **43**, 715.
84. H. Strathmann, *J. Am. Inst. Chem. Eng.*, 2001, **47**, 1077.
85. M. Wegmann, B. Michen and T. Graule, *J. Eur. Ceram. Soc.*, 2008, **28**, 1603.

CHAPTER 8

Applications of Ionic Liquids in Metal Extraction

GEETA DURGA,[a] DEEPTI GOYAL[b] AND ANURADHA MISHRA*[b]

[a] Department of Applied Sciences, School of Engineering and Technology, Sharda University, Greater Noida, Gautam Budh Nagar – 201306, India; [b] Department of Applied Chemistry, School of Vocational Studies and Applied Sciences, Gautam Buddha University, Greater Noida, Gautam Budh Nagra – 201310, India
*Email: anuradha_mishra@rediffmail.com

8.1 Introduction

The deleterious effects of many organic solvents combined with serious environmental issues, such as atmospheric emissions and contamination of aqueous effluents, have changed the focus of many researchers to the development of "green engineering", which represents research aimed at finding environmentally benign alternatives to harmful chemicals. Among the neoteric solvents applicable in "green technologies", ionic liquids have garnered increasing attention over others. Room temperature ionic liquids, sometimes called simply ionic liquids, a comparatively new class of solvents, have many fascinating properties, such as low vapor pressure, high viscosity, dual natural polarity, good thermal stability, a wide range of miscibility with water and other organic solvents[1] and no emissions of volatile organic compounds compared with traditional molecular organic solvents. As a result of their unique chemical and physical properties, ionic liquids have aroused increasing

RSC Green Chemistry No. 23
Green Materials for Sustainable Water Remediation and Treatment
Edited by Anuradha Mishra and James H. Clark
© The Royal Society of Chemistry 2013
Published by the Royal Society of Chemistry, www.rsc.org

interest for their promising role as an alternative medium for many industrial applications, such as synthetic, electrochemical, catalytic and extraction processes.[2]

8.2   What Are Ionic Liquids?

Ionic liquids (ILs), also known as molten salts, consist of organic cations and organic/inorganic anions. According to a widely accepted definition, ILs are a class of salts which are liquid below $100\,^\circ\text{C}$, whereas salts that are liquid at or below room temperature (*i.e.* $25\,^\circ\text{C}$) are known as room temperature ionic liquids (RTILs). In general, ionic liquids consist of a salt where one or both of the ions are large, and the cation has a low degree of symmetry. These factors tend to reduce the lattice energy of the crystalline form of the salt, and hence lower the melting point.[3] Ionic liquids come in two main categories, namely simple salts (made of a single anion and cation) and binary ionic liquids (salts where an equilibrium is involved). For example, $[\text{EtNH}_3][\text{NO}_3]$ is a simple salt, whereas mixtures of aluminum(III) chloride and 1,3-dialkylimidazolium chlorides (a binary ionic liquid system) contain several different ionic species, and their melting point and properties depend upon the mole fractions of the two components present. Ionic liquids consisting of a simple salt show simple melting behavior, whereas for the binary systems the melting point depends upon composition.[4]

Structurally, ILs are usually composed of nitrogen-, phosphorus- or sulfur-containing organic cations (*e.g.* imidazolium, pyridinium, ammonium or phosphonium) paired with organic or inorganic counter anions such as halide, tetrafluoroborate (BF_4^-), hexafluorophosphate (PF_6^-) or bis-[(trifluoromethyl)sulfonyl]imide (NTf_6^-). Owing to the fact that a wide variety of cations and anions can be paired to generate a variety of ILs, it is estimated that there could be up to 10^{18} possible combinations of cations and anions.[5] The structures, names and abbreviations of some common cations and anions are shown in Figure 8.1.

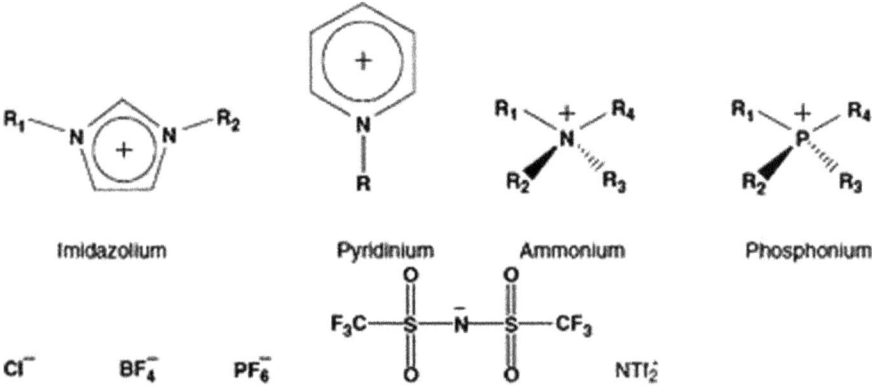

Figure 8.1   Structure of common cations and anions used in ionic liquids.

The first IL was reported by Gabriel and Weiner in 1888.[6] The melting point for this ethanolammonium nitrate IL was reported to be 52–55 °C. The first RTIL, ethylammonium nitrate, was reported by Walden in 1914.[7] The discovery of dialkylimidazolium chloroaluminate RTILs by Wilkes and co-workers in 1982 spearheaded the search for new classes of ILs based on organic cations.[8] However, these RTILs were not very stable in air and water owing to hydrolysis of the chloroaluminate anion. Therefore, Wilkes and Zaworotko were the first to report the synthesis of air- and water-stable ILs with imidazolium or pyridinium cations and acetate or tetrafluoroborate anions.[9] This was a significant discovery and the starting point for current research aimed at synthesizing varying classes of RTILs and ILs as well as exploiting their unique properties.

Since then, researchers in diverse fields of science and engineering have devoted a significant amount of effort into the study and application of ILs. Currently, ILs can be synthesized from relatively cheap starting materials and an increasing number of ILs is now commercially available. However, the cost of ILs is still much greater than the cost of molecular solvents. A trade-off can be obtained if the application requires small amounts of IL and/or the IL can be recycled and reused.[10]

There is a plethora of applications of ILs in various domains of the physical sciences. They are used as "solvents" for organic synthesis and catalysis, as electrolytes, as lubricants, as a stationary phase for chromatography, as matrices for mass spectrometry, as supports for the immobilization of enzymes as liquid crystals, as templates for the synthesis of mesoporous nanomaterials and ordered films, as materials for embalming and tissue preservation, and in separation technologies, *etc.*[11–13]

8.3 Ionic Liquids for Metal Extraction

Typically, a complexing ligand is dissolved in the hydrophobic extracting phase to complex the metal ions. Volatile organic compounds (VOCs) have been widely used as extraction solvents in liquid–liquid extraction (LLE), from synthesis to hydrometallurgy, which has led to health and safety concerns.[14,15] ILs have been regarded as successful alternative extraction solvents for metal ions owing to their extremely low vapor pressure, nonflammability and high thermal stability. Also, extractions at elevated temperature are feasible when ILs are used as the extraction solvent. Moreover, ILs exhibit remarkable solubility for a wide range of compounds, making them excellent solvents for extraction. These unique and exciting properties of ILs make them ideal solvents in extraction techniques and also make them greener and safer. ILs can be custom-designed, thereby varying the physical and chemical properties, and can even be made hydrophobic by simply changing the combination and structural nature of the cations and counter anions, which is why ILs are also termed "designer solvents". The metal ions extractability of the ionic liquid medium, which is sometimes noncoordinating, can be increased by the addition of organic extractants capable of forming organophilic polar

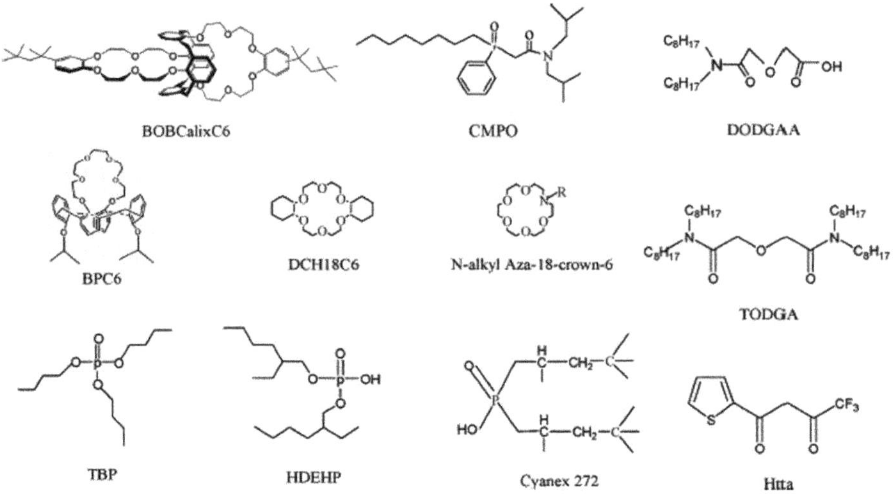

Figure 8.2   Structures of common extractants used in IL-based extraction.

complexes with the metal ions stabilized in the RTIL phase. Thus far, a wide variety of complexing ligands (also called extractants) such as crown ethers, *N*,*N*-dioctyldiglycol amic acid (DODGAA), diglycolamide derivatives, the-noyltrifluoroacetone, *etc.* (Figure 8.2) have been dissolved in ILs for the LLE of targeted metal ions from aqueous solutions. Several investigators have focused on the application of RTILs using extractants for LLE of metal ions. Dai *et al.*,[16] for example, first discovered that highly efficient extraction of strontium ions can be achieved when dicyclohexano-18-crown-6 (DC18C6) is combined with RTILs. Subsequently, Visser *et al.*[17] and Chun *et al.*[18] reported the extraction of various alkali metal ions with crown ethers in RTILs. In the presence of crown ethers in RTIL-based liquid–liquid separations, the resulting metal ion partitioning depends on the hydrophobicity of the crown ether and the composition of the aqueous phase. In comparison to traditional solvent extraction, RTILs show exceptional behavior and the possibility of a significantly complicated partitioning mechanism.

Structural complexing functionality can also be incorporated into the cation and/or anion of ILs to form functionalized ionic liquids (FILs) [also referred to as task-specific ionic liquids (TSILs)], allowing the extraction of metal ions without the presence of additional complexing ligands.[10,19] This approach provides priority advantages, including the increased affinity of a target species for the ionic liquid over a second phase and a great reduction of the chance for loss of the chelating agent to the aqueous phase. FILs are viewed as tunable, multipurpose materials for many applications other than just as diluents. They can behave as the organic phase and extracting agents, suppressing the problems encountered from extractant/diluent miscibility and facilitating species extraction and solvent recovery.[20–24] The incorporation of functionality into an IL is usually accomplished by the grafting of pre-existing groups onto one of the ion structures (Scheme 8.1).[25]

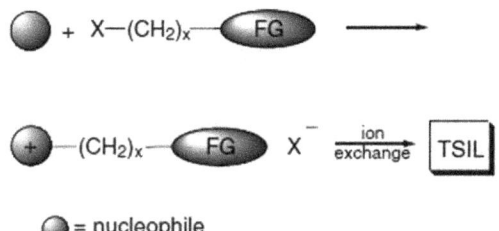

= nucleophile

Scheme 8.1   Grafting of pre-existing groups onto ion structures.

8.4   Types of Ionic Liquids for Metal Extraction

Metal ions, including alkali and alkaline earth metal ions, heavy metal ions and rare earth ions, can be extracted from the aqueous phase by LLE with ionic liquids. The ILs more frequently used in analytical chemistry are organic salts, composed of organic cations (*e.g.*, imidazolium, phosphonium, pyrrolidinium, pyridinium or quaternary ammonium) and appropriate anions (*e.g.*, hexa-fluorophosphate, tetrafluoroborate, alkyl sulfates, alkylsulfonates, chloride or bromide).[26]

8.4.1   Imidazolium Ionic Liquids

ILs derived from imidazole have been frequently utilized in analytical chemistry for elemental determination. Typical IL characteristics and good stability in both oxidative and reductive conditions led imidazolium-class ILs to attract special interest. They offer greater versatility and scope for the design and the application of metal-separation methods.[27] When the alkyl-chain length grows in their structures, the solubility in water diminishes and the viscosity increases, so both parameters have to be considered in the selection of an appropriate extracting phase, since low solubility allows minimal IL consumption, while high viscosity could cause practical drawbacks during microextraction procedures.

1-Alkyl-3-methylimidazolium tetrafluoroborate, $[C_n mim][BF_4]$, and 1-alkyl-3-methylimidazolium bis[(trifluoromethyl)sulfonyl]imide, $[C_n mim][NTf_2]$, have found different applications.[28] Visser *et al.* studied the extraction of Na^+, Cs^+ and Sr^{2+} from aqueous solution into $[C_n mim][PF_6]$, ($n = 4$, 6, 8) by crown ethers.[29] Visser and co-workers[30] also incorporated thiourea, thioethers and urea into derivatized imidazolium cations and used these functionalized ILs as the extractant in LL extraction of Hg^{2+} and Cd^{2+}. In that report, low distri-bution ratios were obtained with less expensive ILs, such as 1-butyl-3-methy-limidazolium hexafluorophosphate, $[C_4 mim][PF_6]$ or $[bmim][PF_6]$, as extractant for Hg^{2+} and Cd^{2+}. The distribution ratios were highly improved when functionalized ILs were mixed with $[C_4 mim][PF_6]$. Therefore, functionalized ILs played the roles of both extractant and the hydrophobic phase. Rogers and co-workers also utilized $[C_4 mim][PF_6]$ with organic and inorganic extractants for the LL extraction of heavy metal ions with a radiotracer technique.[31] Hg^{2+}, Cd^{2+} and Am^{3+} were successfully extracted[32–34] from the aqueous phase into TSILs, or a mixture of the TSILs and a less expensive conventional IL. Ag^+, Hg^{2+}, Cu^{2+}, Pb^{2+}, Cd^{2+} and Zn^{2+} were successfully extracted into [C$_4$mim][PF$_6$] by employing dithizone as a chelator to form neutral metal–dithizone complexes.[35] It was found that the extraction efficiency of IL is higher than that of chloroform at low pH. Furthermore, metal ions can be extracted from the aqueous phase into [C$_4$mim][PF$_6$] and then back-extracted into the aqueous phase with high recovery by manipulating the pH value of the extraction system. Recently, it was reported that Hg^{2+} can be extracted into [C$_n$mim][PF$_6$] ($n = 4$, 6, 8) without the need for any complexing reagent.[36] A new class of imidazolium salts with appended aminodiacetic acid moieties as di-*tert*-butyl esters has been used for the formation of metal chelates with Cu^{2+}, Ni^{2+} and Co^{2+} in aqueous solutions, rendering them useful for metal ion remediation. The increase in the alkyl chain length improves the hydrophobicity of the complexes, making them increasingly organophilic in *n*-butanol–water mixtures, and also easily separable from aqueous solutions as a separate phase.[37]

8.4.2 Quaternary Ammonium Ionic Liquids

Ammonium-containing ILs generally show lower melting points and higher viscosity than their phosphonium analogs, but this strongly depends on the choice of anion. This class of ILs has found some unique applications in various fields, such as phase-transfer catalysts, solvents, lubricants, gas capture agents, coating materials or chemical sensors. Owing to the hydrophobicity of quaternary ammonium ionic liquids (quats), the formation of liquid–liquid biphasic systems with aqueous phases makes them suitable for extraction processes. Tricapryl(methyl)ammonium chloride (methyltrioctylammonium chloride, [MTOA][Cl], Aliquat® 336) has been evaluated as an extracting agent (in most cases dissolved in an appropriate organic solvent) for different metals from acidic aqueous solutions (*e.g.* Cd^{2+}, Fe^{3+}, Pt^{2+} and Hg^{2+} from hydrochloric acid solutions) since 1960.[38] Recently, the complete removal of Zn^{2+}, Cd^{2+} and Fe^{3+} has been achieved using [MTOA][Cl] from hydrochloride aqueous solutions.[39] In general, the extraction mechanism with quats is based on the anion exchange/ion association of metal chloride/sulfate species; it therefore strongly depends on the composition of the aqueous phase (formation of different metal species) and pH. Hence, effective metal stripping is mostly easily achieved by variation of pH, or *via* an efficient complexing agent like thiourea. Owing to ion exchange, the potential leachability of both the IL and organic solvent in the aqueous phase may sometimes pose an environmental risk. Therefore to overcome this problem and attain enhanced extracting ability, it is possible to combine favorable properties of quaternary ammonium ions with different anions containing functional moieties known for their affinity towards selected metals, and forming TSILs as metal scavengers.[40] By anchoring different functional groups onto the anion, it is possible to combine the hydrophobicity of ammonium with the affinity of the functional group towards selected metal species, and hence enhance both the efficiency and selectivity of the extracting agent. Egorov *et al.* successfully applied an Aliquat (2:1 mol/mol mixture of methyltrioctyl- and methyltridecylammonium chlorides)-based TSIL with salicylate anion as extracting agent for iron and copper from a model matrix, suggesting the formation of salicylate complexes with metal species. High efficiency and selectivity were achieved for the extraction of cadmium[41] and chromium[42] from a natural river matrix with methyltrioctylammonium thiosalicylate, [MTOA][TS], a thiol-containing TSIL. On the other hand, [MTOA][SCN] has been proved as an extracting agent for actinides from acid solutions *via* anion exchange.[43] In another study, [MTOA][TS], prepared *via* a halide-free synthesis route, is commercially available and has been evaluated as an extracting agent for heavy metals (*e.g.* Hg^{2+}, Cu^{2+} or Pb^{2+}) from aqueous solutions with high distribution coefficients.[44]

8.4.3 Phosphonium Ionic Liquids

Tetraalkylphosphonium-type ILs have been demonstrated to be thermally and chemically stable, and their viscosities depend on the choice of anion. However, at typical industrial reaction temperatures (*e.g.*, 70–100 °C) their viscosities generally decrease to less than 1 P. Compared to imidazolium ILs, phosphonium ILs have a lower density than water.[43] This property can be beneficial in product work-up steps while decanting aqueous streams containing inorganic salt byproducts.[45] Phosphonium ILs generally have higher electronic polarizabilities than their ammonium counterparts, indicating a difference in their solvation behavior.[46]

Phosphonium ILs have also attracted particular attention due to their capability to extract a variety of metal ions.[47] Trihexyl(tetradecyl)-phosphonium chloride ([$P_{14,6,6,6}$]Cl) and trihexyl(tetradecyl)phosphonium bis[(trifluoromethyl)sulfonyl]imide ([$P_{14,6,6,6}$][NTf_2]) have been employed for Zn extraction,[48] while [$P_{14,6,6,6}$]Cl has also found application for Fe extraction.[49] Stojanovic *et al.* synthesized hydrophobic long-chain quaternary ammonium and phosphonium ILs with functionalized aromatic anions bearing hydroxy, methoxy, thiol and thioether functionalities, as well as tetraphenylborate anions, which resulted in increased chemical stability of the ILs and an alteration of their physicochemical properties. Furthermore, aromatic anions significantly decreased the water solubility and water uptake of both ammonium- and phosphonium-based ILs. Thiol and thioether ILs were used for the extraction of platinum from the aqueous phase and showed a rapid elimination of up to 95% platinum after 30 min.[50]

8.4.4 Pyridinium Ionic Liquids

Among many within this category, 1-alkylpyridinium hexafluorophosphate ([C_npy][PF_6]) has been the most widely used IL. This is due to its particular physicochemical properties, such as high hydrophobicity, water immiscibility and high viscosity, which make it very attractive for performing biphasic separation and preconcentration of metals.[51] Despite the potential benefits of its use, only two microextraction methods have been proposed with 1-hexylpyridinium hexafluorophosphate ([C$_6$py][PF$_6$]) as an extractant solvent for separation and preconcentration of trace Zn and Al.[51,52] Also, [C$_6$py][PF$_6$] has been a reactant for synthesizing novel TSILs, acting as chelating agents for Pd.[52] Nicolas *et al.* performed an extraction of metal ions from water using different ILs containing the cations 4-methyl-1-octylpyridinium, [4MOPYR]$^+$, 1-methyl-1-octylpyrrolidinium, [MOPYRRO]$^+$, or 1-methyl-1-octylpiperidinium, [MOPIP]$^+$, and the anions BF$_4^-$, TfO$^-$ or nonafluorobutanesulfonate, NfO$^-$, and observed that ionic liquids containing pyridinium cations have high extraction efficiencies for mercury removal.[53]

8.4.5 Pyrrolidinium Ionic Liquids

These ILs provide new alternatives to pyridinium and imidazolium ILs. This class of IL has found application in electrochemical development. High thermal stability (decomposition temperatures all above 300 °C) increases safety in applications (*e.g.*, rechargeable lithium-ion batteries and other electrochemical devices). They are sparingly soluble in water but are hygroscopic.[54] Regarding metal determination and speciation, 1-butyl-1-methylpyrrolidinium bis[(trifluoromethyl)sulfonyl]imide has been employed for LLE of Cu in the presence of pyridine-based N^1,N^1,N^4,N^4-tetrakis[2-(pyridin-2-yl)ethyl]butane-1,4-diamine (C$_4$N$_2$Py$_4$), used as an ionophore to set up the extraction system.[53] Lee synthesized ILs containing pyridinium, pyrrolidinium and piperidinium cations with the anion Tf$_2$N$^-$ which were utilized to extract noble metals from water and these functionalized ILs were found to exhibit good extraction properties for silver.[55]

8.5 Extraction of Different Types of Metal Ions

Ionic liquids have emerged as novel media for LLE of metal ions from their aqueous solutions, having the main advantage that they reduce the need to use chlorinated solvents like chloroform, which is highly toxic and is known to be a potential carcinogen. Since ILs are known to be "green solvents", there is high feasibility of recycling the IL used in the extraction. A high extraction efficiency and selectivity can be achieved even after five repeated cycles using ILs.[56] Extraction of metal ions from polluted environmental samples is the main use of ILs. To achieve optimal extraction efficiency, there are several controlling factors:[57]

(1) By altering the structure of the cation/anion to change its hydrophobicity can improve the partition coefficients of the metal ions.[16,58]
(2) By varying the types of extractants or modifying the extractants (such as crown ethers) to achieve optimal selectivity for a specific application.[59]
(3) By controlling the pH of the system the extraction efficiency of metal complexes can be modified.[60]

8.5.1 Alkali Metals and Alkaline Earth Metals

In the pioneering work of using RTILs in the extraction of metal ions, several groups presented the use of crown ethers as extractants for the extraction of Group 1 and Group 2 metal ions from the aqueous phase. Dai *et al.*[16] demonstrated the use of imidazolium-based ILs with PF_6^- and NTf_2^- anions as extraction solvents for Sr^{2+} ions using crown ethers as extractants. Large distribution coefficients for strontium nitrate were obtained. Visser *et al.*[17] further studied the extraction ability of Sr^{2+} by different crown ethers in 1-butyl-3-methylimidazolium hexafluorophosphate ([C_4mim][PF_6]; Figure 8.1) and the result indicated that a crown ether with a higher rigidity structure shows higher extraction efficiency. However, [C_4mim][PF_6] is unstable under conditions of high acidity.

In other work, Yakshin *et al.*[61] studied the extraction of alkali metals and Sr with a crown ether, *cis-syn-cis*-dicyclohexano-18-crown-6 (DC18C6), dissolved in 1-butyl-3-methylimidazolium derivative ILs ([C_4mim][PF_6], [C_4mim][BF_4], [C_4mim][NTf_2]) and confirmed the best physicochemical characteristics (solubility in the aqueous phase, viscosity, hydrophobicity, *etc.*) are exhibited by a solution of DC18C6 in [C_4mim][NTf_2]. ILs as solvents exhibit unusual properties in the extraction of alkali and alkaline earth elements from neutral solutions when compared with DC18C6 in the usual organic solvents.

Chun *et. al*[18] proposed a better method to synthesize 1-alkyl-3-methylimidazolium hexafluorophosphates in which the 1-alkyl group is varied systematically from butyl to nonyl to produce a series of RTILs. They observed the extraction efficiency generally diminished as the length of the 1-alkyl group was increased for competitive solvent extraction of aqueous solutions of alkali metal chlorides with solutions of DC18C6 in these RTILs. However, the extraction with the same extractant was found to be undetectable with conventional organic solvents. The extraction selectivity order for DC18C6 in the RTILs was $K^+>Rb^+>Cs^+>Na^+>Li^+$; with the elongation of alkyl group in the RTILs, the K^+/Rb^+ and K^+/Cs^+ selectivities exhibited general increases, with the larger enhancement for the latter. For DC18C6 in 3-methyl-1-octylimidazolium hexafluorophosphate, the alkali metal cation extraction selectivity and efficiency remained unaffected by varying the aqueous-phase anion from chloride to nitrate to sulfate. A series of imidazolium-based RTILs[62] using DC18C6 as the extractant were analyzed to be used as solvents for Sr^{2+} extraction and a high distribution ratio (10^3) under certain conditions was achieved which was much larger than that in the DC18C6/*n*-octanol system. In the study, E_{Sr} followed the order [C_2mim][NTf_2]>[C_4mim][NTf_2]>-[C_4mim][PF_6]>[C_6mim][NTf_2], and a decrease in E_{Sr} with an increase in the concentration of HNO_3 in the aqueous phase was observed.

In another study, *N*-alkyl-aza-18-crown-6 was synthesized and compared with DC18C6 for the extraction of different metal ions such as Sr^{2+}, Na^+, K^+ and Cs^+.[5,63] It was found that *N*-alkyl-aza-18-crown-6 exhibited higher selectivity for the bivalent ion Sr^{2+} as opposed to DC18C6 which exhibited high selectivity for K^+. Also, the selectivity of ions varied based on the alkyl chain present on the imidazolium-based NTf_2^- ILs. In other work, the extraction of different metal ions (Na^+, K^+, Sr^{2+} and Cs^+) was achieved from the aqueous phase using solutions containing calix[4]arene-bis(*tert*-octylbenzo-crown-6) in *N*-alkyl-*N*-methylimidazolium bromide and *N*-alkyl-*N*-methylimi-dazolium NTf_2^-.[64] Higher distribution coefficients of the metal ions were obtained using imidazolium cations with short alkyl chain substituents but limited the solubility of calix[4]arene-bis(*tert*-octylbenzo-crown-6). In a separate study the Xu group[65] has reported the efficient removal of Cs^+ from aqueous solutions using even low concentrations of the calix crown ether bis(2-propyloxy)calix[4]crown-6 (BPC6) in the ionic liquids $[C_n\text{mim}][NTf_2]$, where $[C_n\text{mim}]$ is 1-alkyl-3-methylimidazolium. Turanov *et al.*[66] studied the effective extraction of micro quantities of Ca^{2+}, Sr^{2+} and Ba^{2+} from aqueous solutions into the organic phase containing the diglycolamide ligand *N,N,N′,N′*-tetraoctyl-3-oxapentanediamide (TODGA) and the ionic liquids 1-butyl-3-methylimidazolium bis[(trifluoromethyl)sulfonyl]imide ($[C_4\text{mim}][NTf_2]$) and 1-butyl-3-methylimidazolium hexafluorophosphate ($[C_4\text{mim}][PF_6]$) and demonstrated the effect of HNO_3 concentration in the aqueous phase, the extractant and IL concentration in the organic phase on the extraction of the metal.

8.5.2   Transition Metals

The heavy metals such as Ni, Cd, Hg, Pb, *etc.*, and their compounds belong to the list of priority substances which have become a serious problem in the aquatic environment due to their toxicity, bioaccumulation and persistence. Their emission causes pollution of the water cycle, leading to poor water quality, insufficient supply of drinking water and complicated drinking water pretreatment. Because of their high affinity for water, it is difficult to remove them using conventional solvents for LLE. Hydrophobic ILs have emerged as potential green solvents for useful extraction processes of these metal ions.

The removal of metal ions from standard solutions, real wastewater samples and activated sewage sludge was investigated by Fuerhacker *et al.*[67] using four ILs, namely trihexyl(tetradecyl)phosphonium thiosalicylate, [PR₄][TS], trihex-yl(tetradecyl)phosphonium 2-(methylthio)benzoate, [PR₄][MTBA], methyl-trioctylammonium thiosalicylate, [A336][TS], and methyltrioctylammonium 2-(methylthio)benzoate, [A336][MTBA], and they were proved to be successful sorbing agents for Zn, Ni, Cu, Cr, Cd and Pb from activated sewage sludge. It was further explained that removal of Cr and Cd decreased with increasing IL dosing rate, but remained nearly constant for the other metals. However, the results of sorption experiments with standard solutions and real wastewater were unsatisfactory except for the efficient removal of Pb and Cu. In another study,[41] the TSIL methyltrioctylammonium thiosalicylate was evaluated as a potential extracting agent for removal of chromium (>90%) from industrial effluents and compared with conventional methods. Reusability of the IL was also mentioned.

Fischer *et al.*[68] evaluated the use of various anion-functionalized ammonium- and phosphonium-based ILs regarding purity and extraction efficiencies for the metal(oid)s Ag, As, Cd, Cr, Cu, Hg, Ni, Pb, Pt, Sn and Zn and the cancerostatic platinum compounds cisplatin and carboplatin from model matrixes and from communal and industrial wastewater samples by applying LLE procedures. Extraction efficiencies >95% could be obtained for Ag, Cu, Hg and Pt with both phosphonium- and ammonium-based ILs bearing sulfur functionality with an extraction time of 120 min in the solutions at environmentally relevant concentrations, whereas other metals were extracted to a lower extent (7–79%). In the case of the cancerostatic platinum compounds, a phosphonium-based IL bearing thiosalicylate functionality showed high extraction efficiency for monoaquacisplatin. The authors recommended the application of the thiol- and thioether-functionalized ILs [A336][TS], [A336] [MTBA], [PR₄][TS] and [PR₄][MTBA] for communal wastewater treatment and the use of [A336][SCN] for industrial wastewater with high levels of Zn contamination. In another study[52] for selective extraction of Cu, Hg, Ag and Pd ions from water, the TSIL-containing cations 4-methyl-1-octylpyridinium, [4MOPYR]⁺, 1-methyl-1-octylpyrrolidinium, [MOPYRRO]⁺, and 1-ethyl-1-octylpiperidinium, [MOPIP]⁺, were used, with the anions BF_4^-, TfO⁻ and NfO⁻. Also, some new TSILs with the cations 1-(cyanopropyl)-4-methylpyridinium, [4MPYRCN]⁺, 1-(cyanopropyl)-1-methylpiperidinium, [MPIPCN]⁺, 1-methyl-1-[4,5-bis-(methylsulfanyl)]pentylpiperidinium, [4MPIPS2]⁺, and 1-methyl-1-[4,5-bis-(methylsulfanyl)]pentylpyrrolidinium, [4MPYRROS₂]⁺, and the anion NTf_2^- were employed. Ionic liquids containing octylpyridinium cations were found to be good for extracting Hg^{2+} ions. Hg^{2+} and Cu^{2+} ions were extracted efficiently and selectively using ILs bearing a disulfide functional group, whereas Ag^+ and Pd^{2+} were removed by those containing a nitrile functional group. However, low distribution coefficients were obtained for all other ions. It was suggested by the authors that, in addition to the functional groups, the heterocyclic cation as well as the anion of an ionic liquid had an influence on the distribution coefficients. Pyridinium or pyrrolidinium cations exhibit distribution co-efficients higher than those for imidazolium or piperidinium cations and the anions TfO⁻ or NfO⁻ instead of NTf_2^- led to enhanced extraction for mercury.

LLE of heavy metal ions (Ag^+, Cu^{2+}, Pb^{2+}, Cd^{2+}, Zn^{2+}) from aqueous solution was achieved using 1-butyl-3-methylimidazolium hexafluorophosphate ([C₄mim][PF₆]) along with dithizone as a metal chelator.[35] The authors explained that high extraction efficiencies were obtained by tailoring the pH value of the aqueous phase and indicated the feasibility of recycling RTILs in the LLE of metal ions. In another study,[69] Cu^{2+} ions were selectively (over Ag^+, Zn^{2+} and Cd^{2+} ions) and efficiently extracted from aqueous solution using the water-immiscible RTIL [C₄mim][PF₆] in the presence of metal chelators, *viz.* dithizone, 8-hydroxyquinoline or 1-(2-pyridylazo)-2-naphthol. The extraction efficiency of copper ion from an aqueous phase was comparable with a traditional organic solvent (CH_2Cl_2) and can be manipulated by tailoring the type of chelator and pH value of the extraction system; also the recycling of the IL for reuse was shown to be possible. González *et al.*[70] also studied the extraction of Cu^{2+} ions from aqueous solutions using a newly synthesized RTIL, 3-butylpyridinium bis(trifluoromethylsulfonyl)imide, [3-BuPyr][NTf$_2$], but without addition of any complexing agent or pH control of the aqueous phase. In the extraction process the assistance of water for the release of the hydrogen atom (attached to the N atom of the pyridine ring) and re-protonation was described to maintain the low pH (2.3) of the solution. The extraction of Cu^{2+} was found to be possible due to a strong interaction between Cu^{2+} and the N atom of the IL cation in the formed complex. The recovery of [3-BuPyr][NTf$_2$] was also feasible from the Cu^{2+}–IL–H_2O system, and could be reused after washing with hydrochloric acid and further purification.

The extraction of Zn^{2+}, Cd^{2+}, Cu^{2+} and Fe^{3+} from hydrochloride aqueous solutions using ILs containing the cations [C$_4$mim]$^+$ and 1-octyl-3-methylimi-dazolium ([C$_8$mim]$^+$) and the anions BF$_4^-$, PF$_6^-$ and NTf$_2^-$ as well as [MTOA]Cl as sole extraction agents were studied.[39] The ionic liquid [MTOA]Cl exhibited extraction percentages higher than 90% for Zn^{2+}, Cd^{2+} and Fe^{3+} at all assayed conditions. Zn^{2+} and Cd^{2+} showed nearly complete extraction with [omim][BF$_4$], whereas the extraction of Cu^{2+} and Fe^{3+} was near zero. There-fore, the use of the latter ionic liquid described the selective separation of Zn^{2+} and Cd^{2+} over Fe^{3+} and Cu^{2+}. Furthermore, the selective recovery of Zn^{2+}/Cd^{2+} and Zn^{2+}/Fe^{3+} was made possible using the ionic liquids [C$_8$mim][PF$_6$], [C$_4$mim][NTf$_2$] and [C$_8$mim][NTf$_2$] due to the low extraction percentage for Zn^{2+}. The increase in the extraction percentages for Zn^{2+}, Cd^{2+} and Fe^{3+} was observed by an increase in HCl concentration, reaching values of 100% at 5 M and suggesting that the predominant mechanism of metal ion removal using imidazolium-based ILs might be the formation of ion pairs with the ionic liquid mediated by HCl.

In another study conducted by Harjani *et al.*,[37] TSILs containing imidazo-lium salts having di-*tert*-butyl esters capable of generating an aminodiacetic acid moiety have been used for the formation of metal chelates with Cu^{2+}, Ni^{2+} and Co^{2+} in aqueous solutions. The increase in the alkyl chain length onto the imidazolium core improved the hydrophobicity of the complexes, making them increasingly organophilic in *n*-butanol–water mixtures, and also easily separ-able from aqueous solutions as a separate phase with an extraction efficiency up to 93%. The coordination ability and nature of the bonding of these imida-zolium salts to the metal ions have been established by X-ray crystallographic analysis. Another TSIL,[40] methyltrioctylammonium salicylate, was used for the highly efficient extraction of transition metal ions (99% for Fe^{3+} and 89% for Cu^{2+}) from their aqueous solutions. The authors suggested that the high effi-ciency was due to the formation of stable metal–salicylate complexes.

The selective and effective extraction[71] of Ag^+ among five different transition metal ions (Ag^+, Cu^{2+}, Zn^{2+}, Co^{2+}, Ni^{2+}) using the extractant pyr-idinocalix[4]arene, ^{t}Bu[4]CH$_2$Py, dissolved in [C$_8$mim][PF$_6$] when compared with extracting systems having ^{t}Bu[1]CH$_2$Py was demonstrated and the results were compared with the extractants in chloroform. The compound ^{t}Bu[4]CH$_2$Py forms a stable 1:1 complex with silver ions and its extraction ability was remarkably higher in the RTILs than in chloroform. Furthermore, the recovery of silver ions from [C$_8$mim][PF$_6$] into a receiving phase can be achieved under mild acidic conditions, compared to those required for the chloroform system. The reusability of the ^{t}Bu[4]CH$_2$Py–RTILs extraction system was confirmed by performing five cycles of forward and back extractions and the extraction ability was maintained at a high level (average degree of extraction = 99.5%). Lertlapwasin *et al.*[72] optimized the conditions for the extraction of Ni^{2+}, Cu^{2+} and Pb^{2+} from water using the synthetic ionic liquid [C$_4$mim][PF$_6$] combined with the 2-aminothiophenol ligand and the results were compared with the ligand in chloroform. High extraction efficiencies of Ni^{2+}, Cu^{2+} and Pb^{2+} were obtained with ionic liquids having the 2-aminothiophenol ligand and were insignificantly affected by competitive ions such as Na$^+$, Ca^{2+}, Mg^{2+}, SO$_4^{2-}$ and Cl$^-$. The optimum pH for the extraction of Ni^{2+} and Pb^{2+} was 4–6 and 5, respectively, whereas extraction of Cu^{2+} was found to be independent of the pH. The stripping of Pb^{2+} and Cu^{2+} could be done by HNO$_3$ while Ni^{2+} could be stripped by H$_2$O$_2$ in HNO$_3$, suggesting the reusability of the ILs.

Visser *et al.*[17] performed LLE of Hg^{2+} and Cd^{2+} from aqueous solutions using derivatized (urea-, thiourea- and thioether-substituted alkyl groups) imidazolium-based ILs. The metal ion distribution ratios increased several fold in these TSILs, regardless of whether the ionic liquids were used as the sole extracting phase or doped into a series of [1-alkyl-3-methylimidazolium][PF$_6$] (alkyl = C$_4$–C$_8$) ionic liquids to form a 1 : 1 solution; however, the metal ion distribution ratios increased with the increase in length of the alkyl chain. Higher distribution ratios (for Hg^{2+} and Cd^{2+}) were observed by increasing the ratio TSIL/[C$_4$mim][PF$_6$] and in the case of the thiourea- and urea-derivatized cations. However, in the latter case the values for Hg^{2+} were higher than those for Cd^{2+}. In another study,[73] an extraction of Hg^{2+} from aqueous solution with [C$_n$mim][PF$_6$] ionic liquids (n = 4, 6, 8) at pH 4.68 was carried out and quantitative transfer of Hg^{2+} ions in all three "classical" imidazolium-based ionic liquids was gained, in the absence of a chelating agent. However, this transfer occurred with some lag time (max. 30 h) which strongly depended on the alkyl chain length on the imidazolium ring and on the temperature.

Both Pt^{4+} and Pd^{2+} were extracted selectively and efficiently from aqueous chloride media using mixtures of the protic ionic liquids trioctylammonium bis(trifluoromethylsulfonyl)amide ([TOAH][NTf$_2$]) and trioctylammonium nitrate ([TOAH][NO$_3$]) as extractants by Katsuta *et al.*;[74] however, under the same conditions, Na$^+$, Mg^{2+}, K$^+$, Ca^{2+}, Mn^{2+}, Fe^{3+}, Co^{2+}, Ni^{2+}, Cu^{2+}, Zn^{2+}, Ru^{3+}, Rh^{3+} and Cd^{2+} were only slightly extracted. It was observed that the extraction of Pd and Pt increased with increasing content of [TOAH][NO$_3$] in the mixture. Ionic liquids (functionalized/unfunctionalized) containing pyridinium, pyrrolidinium and piperidinium cations with the anion bis(trifluoromethylsulfonyl)imide were investigated for their ability to extract Ag$^+$, Pd^{2+} or Au^{3+} metal ions from water at room temperature.[55] Good extraction properties for Pd^{2+} and Au^{3+} with functionalized ionic liquids containing a disulfide group, an alkenyl group or a nitrile group were observed. The extraction process was found to be largely influenced by the structure of the heterocyclic cation ring as well as the mass ratio of water to ionic liquid.

8.5.3 Rare Earth Metals

Extraction and recovery of lanthanide and actinide metals from nuclear and metal-containing industrial waste streams is important both environmentally and economically. Dietz[75] and Huddeleston[76] introduced the use of RTILs as alternatives to *n*-dodecane in solvent extraction, a traditionally used diluent in the PUREX process. Extraction of lanthanides such as Ce, Eu and Y from their aqueous solutions into an ionic liquid, [C_4mim][PF_6], using (*N,N*-diisobutylcarbamoylmethyl)octyl(phenyl)phosphine oxide (CMPO) was studied by Nakashima *et al.*[77] The extraction efficiency and selectivity of CMPO for metal ions was greatly enhanced compared to when *n*-dodecane was used as the extracting solvent. The reusability of ILs in LLE was also described. In another study,[78] extractions of the lanthanide elements Ce^{3+}, Nd^{3+}, Sm^{3+}, Dy^{3+} and Yb^{3+} (in their chloride forms) by bis(2-ethylhexyl)phosphoric acid (DEHPA) with [1-C_n-3-methylimidazolium][PF_6] ($n = 2$, 4) or [1-butyl-4-methylpyridinium][PF_6] were carried out and the results were compared with hexane, in terms of their distribution coefficient values under varying concentrations of DEHPA and HNO_3 and different temperatures (298–333 K). The ionic liquid system exhibited more than three times greater extractability for the lanthanides compared to when hexane was used under different conditions. The distribution coefficients of the lanthanide ions were influenced by the alkyl chain length and it decreased from ethyl to butyl. Moreover, higher distribution coefficients were obtained for the imidazolium cations compared to the pyridinium cation in an ionic liquid. The [C_4mpy][PF_6] system exhibited higher selectivity to Yb compared to the other systems.

The extraction behavior of Na^+, Cs^+, Ca^{2+}, Sr^{2+} and La^{3+} species from aqueous solutions into the hydrophobic ionic liquid 1-C_n-3-methylimidazolium nonafluorobutanesulfonate ([$C_n$mi][NfO]) ($n = 4$, 5, 6) was studied by Kozonoi and Ikeda.[79] The extraction ratios of Li^+, Na^+, Cs^+, Ca^{2+}, Sr^{2+} and La^{3+} species were found to be 39, 24, 5.0, 81, 79 and 98%, respectively. The extraction efficiency of La^{3+} decreased remarkably (from 97% to 38%) with an increase in HNO_3 concentration in aqueous solutions, and with an increase in the hydrophobicities of the ILs, *i.e.* [bmi][NfO] > [pmi][NfO] > [hmi][NfO]. In another study, a series of novel, phosphine oxide functionalized ILs, *i.e.* 1-(alkyldiphenylphosphine oxide)-2,3-dimethylimidazolium cations [$DMImC_nP(O)Ph_2^+$] and the anions NTf_2^- ($n = 2$, 3, 4, 6, 8) and PF_6^- ($n = 2$, 3) were synthesized[80] and demonstrated their tunable metal ligating abilities (with La and Eu). The authors have proposed the potential use of these ILs as extractants for lanthanide, actinide and transition metal ions.[80]

A novel extraction strategy using ionic liquids as both extractants and diluents has been proposed by Sunad *et al.*[81] for the extraction of rare earth elements (REEs). For this purpose, the anion-functionalized ionic extractants tetrabutylammonium di(2-ethylhexyl)phosphate ([TBA][DEHP]), methyl-trioctylammonium di(2-ethylhexyl)phosphate ([TOMA][DEHP]), and trihex-yl(tetradecyl)phosphonium di(2-ethylhexyl)phosphate ([THTP][DEHP]) were synthesized and were used in the ILs 1-hexyl-3-methylimidazolium bis(tri-fluoromethylsulfonyl)imide ([C_6mim][NTf$_2$]) and diisopropylbenzene (DIPB) along with di(2-ethylhexyl)phosphoric acid (HDEHP). The distribution coefficients (D_M) of HDEHP, [TBA][DEHP], [TOMA][DEHP] and [THTP][DEHP] for REEs in [C_6mim][NTf$_2$] were found to be much higher than those in DIPB; also the D_{IL}/D_{DIPB} values of [TBA][DEHP], [TOMA][DEHP] and [THTP][DEHP] were larger than those of HDEHP for light and heavy REEs. The functionalized ILs used as extractants exhibited much higher extractabilities and selectivities for REEs in [C_6mim][NTF$_2$] than in DIPB, which may be attributed to the "like-dissolves-like" principle. Moreover, effective stripping of REEs was also demonstrated using nitric acid.

Other anion-functionalized paraffin-soluble ionic liquids,[82] methyl-trioctylammonium bis(2-ethylhexyl)phosphate ([A336][DEHP]) and methyl-trioctylammonium bis(2-ethylhexyl)diglycolamate ([A336][DGA]), were used for extraction of Eu^{3+} and Am^{3+} and the results were compared with the precursor ILs under different pH conditions. The extraction of Eu^{3+} and Am^{3+} increased with increase of pH and decreased with increase in the chain length of the paraffin diluents and were extracted to the same extent from the aqueous phase. However, separation of Eu^{3+} from Am^{3+} by adding diethylene-triaminepentaacetic acid (DTPA) to the aqueous phase at fixed pH was also found to be possible. The distribution ratio observed in these ionic liquids was higher than that obtained in solutions of the ionic liquid precursors; it was also found to be higher with the use of the [DGA]-based ionic liquid than using the [DEHP] ionic liquid under similar conditions.

Selective extraction[83] of Y^{3+} from heavy lanthanides into 3-methyl-1-octy-limidazolium hexafluorophosphate ([C_8mim][PF$_6$]) containing Cyanex 923 (a mixture of straight-chain dialkylated phosphine oxides) was achieved by adding a water-soluble complexing agent (EDTA) to the aqueous phase. The extraction efficiency of the Y^{3+}/Cyanex923/[C_8mim][PF$_6$] system was found to decrease rapidly as the aqueous acidity increased. In another study of the extraction of trivalent lanthanides (Ln^{3+}) into an ionic liquid,[84] the synergistic effect of 18-crown-6 derivatives, such as 18-crown-6 (18C6), *cis*-dicyclohexano-18-crown-6 (DC18C6) and dibenzo-18-crown-6 (DB18C6), on 1-butyl-3-methylimidazolium bis(trifluoromethylsulfonyl)imide with 2-thenoyltrifluoroacetone (Htta) was investigated. The extractability of lighter Ln^{3+} elements was enhanced by adding 18C6 or DC18C6, whereas DB18C6 did not exhibit any enhancement. Also, the synergistic effect by the crown ether (CE) was increased along with a decrease in the atomic number of the Ln. It was suggested by the authors that the formation of the [Ln(tta)(CE)]$^{2+}$ complex as the extraction species for Ln^{3+} was responsible for the synergistic effect and this effect may have originated in a size-fitting effect of the CE on complexation to Ln^{3+}.

Some of the early work on actinide extraction[85] included extraction of UO_2^{2+} by tributyl phosphate (TBP) in the ILs 1-alkyl-3-methylimidazolium $[C_n mim][PF_6]$, $(n = 4, 6, 8)$ and the extraction mechanism with RTILs was found to be invariant to diluents such as *n*-dodecane. However, in a study conducted by Cocalia *et al.*,[86] identical extraction behavior and coordination of actinides in *n*-dodecane and 1-*n*-decyl-3-methylimidazolium bis-(trifluoromethylsulfonyl)imide ($[C_{10}mim][NTf_2]$) diluents for dialkylphosphoric or dialkylphosphinic acids as extractants has been reported. In another study, liquid–liquid separation of actinide ions[87,88] using the hydrophobic ionic liquid $[C_{10}mim][NTf_2]$ as the extracting phase and Cyanex-923 [a mixture of trialkylphosphine oxides of varying chain lengths (hexyl and octyl)] as ligand was carried out. Distribution ratio measurements for Pu^{4+} and Am^{3+} showed little acid dependency for Pu^{4+}, but strong acid dependency studies seem to support $1:1$ metal-to-ligand binding for Pu^{4+} and significantly the notion that ionic liquids can provide different coordination environments.

A study for the extraction of Am^{3+} from slightly acidic feed solutions and its separation from Eu^{3+} using a variety of complexing agents has been carried out by dissolving the conventional extractant carbamoyl(methyl)phosphine oxide (CMPO) in RTILs.[89] Diglycolamide extractants such as N,N,N',N'-tetra-*n*-octyldiglycolamide (TODGA) have been found to be significantly more effective for minor actinide partitioning compared to CMPO and several test counter-current runs have been carried out using "hot" radioactive waste solutions.[90]

The extraction of several actinide ions,[91] *viz.* Am^{3+}, Pu^{4+} and UO_2^{2+}, using TODGA as extractant in three different RTILs, namely $[C_4mim][NTf_2]$, $[C_6mim][NTf_2]$ and $[C_8mim][NTf_2]$, as the diluents was investigated. The extraction systems containing TODGA in the RTILs were found to have higher distribution ratios for all the actinides (trivalent, hexavalent, tetravalent) when compared to that containing *n*-dodecane. Moreover, the distribution ratio values of the metal ions were found to decrease with increasing feed nitric acid concentration. The extraction mechanism of the actinides was found to be different compared to that in *n*-dodecane and a cation-exchange mechanism was suggested. Also, the extraction was observed to be independent of the anion of the aqueous phase. Stripping of the actinide ions was carried out using complexing agents such as EDTA, DTPA in guanidine carbonate or a buffer mixture, which otherwise was difficult.

Extraction of Am^{3+} from acidic feed solutions[92] was investigated using extraction systems containing a tripodal diglycolamide (T-DGA) in three RTILs, *viz.* $[C_4mim][NTf_2]$, $[C_6mim][NTf_2]$ and $[C_8mim][NTf_2]$. The results obtained were compared with TODGA and significantly higher distribution coefficients were observed in the case of T-DGA in these RTILs. The D_{Am} values decreased with increasing carbon chain length in the RTILs, which was attributed to hydrophobicities of the RTILs. The distribution studies were done on taking the effect of equilibration time, aqueous phase acid concentration variation and T-DGA concentration variation and significantly higher equilibration times were observed for the extraction systems, which were explained partly due to the viscous RTIL phase and partly due to the slow conformational changes of the T-DGA ligand during complexation. Stripping of the metal ion was 99% in three stages using 0.5 M EDTA or DTPA in 1 M guanidine carbonate.

For the extraction of UO_2^{2+}, task-specific ionic liquids with a quaternary ammonium cation and bearing a phosphoryl group have been synthesized by Ouadi *et al*. In another study, Am^{3+} and another TSIL containing 2-hydroxybenzylamine was used.[34,93]

Srncik *et al*.[94] studied the extraction of uranium from a natural water matrix with TSILs, *viz*. methyltrioctylammonium methionate ([A336][Met]), methyltrioctylammonium thiocyanate, ([A336][SCN]) and methyltrioctylammonium thiosalicylate ([A336][TS]). A high uranium extraction efficiency from a matrix-free water sample was obtained using ILs containing an aliphatic thiocyanate, [A336][SCN], or an aromatic thiosalicylate anion, [A336][TS]. However, extraction from a natural water matrix exhibited significantly different results, where the efficiency for [A336][SCN] was poor and a distribution coefficient of $D > 1000$ was obtained with [A336][TS] for the elimination of ^{238}U from natural water with high selectivity compared to Mg^{2+} and Ca^{2+}, which showed no affinity for this IL. Further satisfying results from stripping experiments of uranium from this IL into a 2 M HNO_3 solution were obtained.

In another study,[95] amidoxime-functionalized ILs, [AO₁mim][NTf₂] and [AO₂mim][NTf₂], were synthesized for extraction of UO_2^{2+} from aqueous solutions (Scheme 8.2). The high distribution ratios for UO_2^{2+} provided high selectivity over Th^{4+} and Eu^{3+}. In total, this study provides evidence for the extraction mechanism of the amidoxime polymers that have been successfully employed in the extraction of uranium from seawater. To explore the extraction mechanism, the distribution ratios were measured as a function of pH and ionic strength. The experiments and calculations were conducted in a manner as described above except that the aqueous phase was changed from deionized water to varying concentrations of either HNO_3 or $NaNO_3$. The distribution values decreased consistently as a function of nitric acid concentration for both ILs, suggesting that the extraction involves deprotonation of the amidoxime.

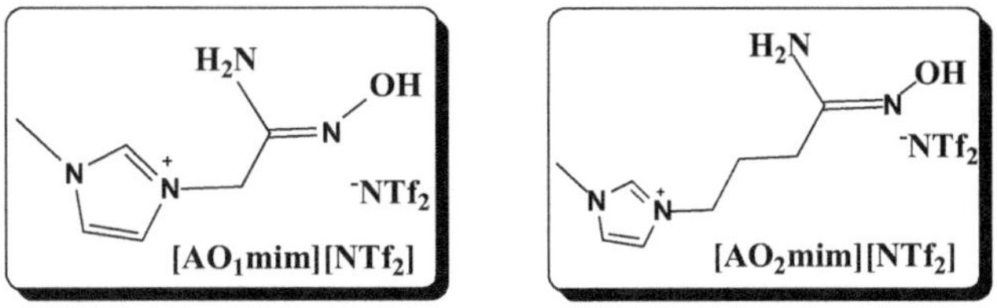

Scheme 8.2 The amidoxime-functionalized ILs [AO₁mim][NTf₂] and [AO₂mim][NTf₂].

8.6 Mechanism for Metal Extraction

Ionic liquids are structurally different from molecular organic solvents and exhibit different behavior towards metal extraction from aqueous solution. However, an IL extraction system will only be considered as "green" if the IL is insoluble in the aqueous phase. In order to utilize them efficiently and selectively, it is crucial to understand their extraction mechanism. Many investigators have revealed these mechanisms as solvent ion pair extraction, ion exchange or a combination of these. However, the ion exchange mode has been widely utilized to understand the separation performance and it was found to be more complex than those with conventional organic solvents as the extracting phases. This complexity in the extraction behavior is attributed to the involvement of either or both cations and anions from the IL to contribute.

8.6.1 Cationic Mechanism

Dietz and Dzielawa[75] studied the extraction behavior of Sr^{2+} by dicyclohexano-18-crown-6 (DC18C6, CE) in 1-alkyl-3-methylimidazolium bis-(trifluoromethylsulfonyl)imide ($[C_n mim][NTf_2]$) and compared the results with the octanol/aqueous system. For the extraction system using *n*-octanol, NO_3^- is co-extracted to the organic phase as a strontium nitrate–crown ether complex (neutral complex mechanism). Hence, an increase in the concentration of NO_3^- in the aqueous phase due to further addition of HNO_3 pushes the equilibrium to the right and increases the extraction of Sr^{2+}; in ILs, Sr^{2+} partitioning does not involve nitrate ion co-extraction and with an increase in HNO_3 a decrease in ion partitioning is observed. An increase in partitioning is observed by transfer of imidazolium cations to the aqueous phase. Thus, the extraction is explained by a cation-exchange mechanism (eqn 8.1):

$$[Sr(CE)]_{aq}^{2+} + 2[C_n mim]_{IL}^{2+} \rightleftharpoons [Sr(CE)]_{IL}^{2+} + 2[C_n mim]_{aq}^{2+} \quad (8.1)$$

However, in a study conducted by Jensen *et al.*[96] for the extraction of Sr^{2+} in the presence of CE, extended X-ray absorption fine structure (EXAFS) measurements were used to examine metal–ligand interactions and found that the two axial-bound nitrates were substituted with two water molecules bound to the metal ion and forming a charged species, $[Sr(CE)(H_2O)_2]^{2+}$ (Figure 8.3), that was transferred to the IL phase (eqn 8.2):

$$Sr_{aq}^{2+} + L_{IL} + 2[C_5 mim]_{IL}^+ + 2H_2O \rightleftharpoons [Sr(L)(H_2O)_2]_{IL}^{2+} + 2[C_5 mim]_{aq}^+ \quad (8.2)$$

The extraction mechanism of Ce^{3+}, Eu^{3+} and Y^{3+} ions has been studied by Nakashima *et al.*[77] and also revealed the occurrence of a cation exchange mechanism, which was confirmed by slope analysis and extraction tests. The extraction of metal ions was not influenced by HNO_3 concentration in the IL system, except for a little decrease in the extraction ability at higher acid

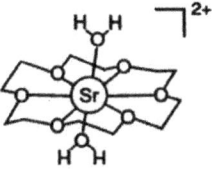

Figure 8.3 Structure of $[Sr(CE)(H_2O)_2]^{2+}$.

concentrations. A metal cation coordinated with CMPO, $[M^{3+}(CMPO)_3]$, was exchanged for a cationic component of the IL, $[C_4mim]^+$ (eqn 8.3):

$$M^{3+}_{aq} + 3CMPO_{org} + 3[C_4mim]^+_{org} \rightleftharpoons [M^{3+}(CMPO)_3]_{org} + 3[C_4mim]^+_{aq} \quad (8.3)$$

In another study,[85] the synergistic extraction system containing CEs and Htta as extractants dissolved in ionic liquid, $[C_4mim][NTf_2]$, were used for extraction of Ln^{3+}. The Ln^{3+} were extracted as cationic ternary complexes, $[Ln(tta)_2(CE)]^+$ and $[Ln(tta)(CE)]^{2+}$, and were exchanged with the IL cation (eqn 8.4):

$$Ln^{3+} + xHtta_{(e)} + yCE_{(e)} + (3-x)[C_4mim]^+_{(e)} \rightleftharpoons [Ln(tta)_x(CE)]^{(3-x)+}_y$$
$$+ xH^+ + (3-x)[bmim]^+ \quad (8.4)$$

Another example of a cation exchange mechanism was studied by Shimoja and Goto[71] for extraction of Ag^+ from aqueous solution by an IL, $[C_8mim][PF_6]$, in the presence of the ligand pyridinocalix[4]arene (eqn 8.5). This compound transfers Ag^+ into RTIL phases *via* complex cation formation (stable 1 : 1 complex):

$$Ag^+ + {}^tBu[4]CH_2Py_{org} + [C_8mim]^+_{org} \rightleftharpoons [Ag^tBu[4]CH_2Py]^+_{org}$$
$$+ [C_8mim]^+ \text{ in } [C_8mim][PF_6] \quad (8.5)$$

Visser *et al.*[97] observed that the high extraction of UO_2^{2+} with the well studied CMPO ligand to an IL phase was due to a cationic mechanism, different from that observed in molecular organic solvents. Ultraviolet and visible (UV/Vis) and EXAFS spectroscopies were used to probe the coordination environment of uranyl nitrate with CMPO and TBP in a series of hydrophobic ILs. In the molecular solvent dodecane, the existence of a hexagonal bipyramidal $[UO_2(NO_3)_2(CMPO)_2]$ complex was revealed by the EXAFS data, while an average equatorial coordination number of 4.5 in $[C_4mim][PF_6]$ and $[C_8mim][NTf_2]$ was suggested by the data, yielding a net stoichiometry of $[UO_2(NO_3)(CMPO)]^+$ (eqn 8.6):

$$UO_2^{2+}_{aq} + [C_nmim]^+_{IL} + NO_3^-_{aq} + L_{IL} \rightleftharpoons [UO_2(NO_3)(L)]^+_{IL} + [C_nmim]^+_{aq} \quad (8.6)$$

Shimojo *et al.*[98] proposed that the transfer of lanthanides with TODGA into ILs proceeded *via* a cation exchange mechanism, in contrast to ion pair extraction in the isooctane system. The effect of the anionic species on the extraction of lanthanides using TODGA was investigated using H_2SO_4 instead of HNO_3, but no substantial effect in the overall extraction behavior of lanthanides with H_2SO_4 in the $[C_2mim][NTf_2]$ system was observed, in contrast to the reduced partitioning of lanthanides into isooctane. However, in the isooctane system it was observed to be reduced considerably using H_2SO_4 compared with the results for HNO_3. It was further suggested that as anionic species are not involved in the transfer of lanthanides into ILs with TODGA, the cation exchange transfer emerges as the most convincing extraction mechanism and this was confirmed by slope analysis. The extraction equilibrium equation of the cation exchange transfer is represented as follows (eqn 8.7):

$$Ln^{3+}_{aq} + 3TODGA_{IL} + 3[C_nmim]^+_{IL} \rightleftharpoons [Ln(TODGA)_3]^{3+}_{IL} + 3[C_nmim]^+_{aq} \text{ in ILs}$$

$$(8.7)$$

Panja *et al.*[91] studied the extraction mechanism of An^{n+} (Am^{3+}, Pu^{4+} and UO_2^{2+}) ions using TODGA as the extractant and it was found to be independent of the anion of the aqueous phase. Hence, it was suggested that a cation exchange mechanism occurred in imidazolium-based RTILs for the extraction of actinides, in contrast to their behavior in *n*-dodecane. Am^{3+} forms a mixture of di- and tri-solvates in all three RTILs, which is in sharp contrast to the tri-solvate species reported by Shimojo *et al.*[97] for the extraction of trivalent lanthanides using TODGA. UO_2^{2+} was found to form a mono-solvate in both *n*-dodecane as well as in the RTILs. On the other hand, Pu^{4+} forms a mixture of mono- and di-solvates in the RTILs. Thus, when *n*-dodecane was replaced by the RTILs as the diluent, Am^{3+} and Pu^{4+} formed different species whereas UO_2^{2+} formed a similar complex (eqns 8.8–8.10):

$$Am^{3+}_{aq} + (2-3)TODGA_{RTIL} + x[C_nmim]^+_{RTIL} \rightleftharpoons [Am(TODGA)_{(2-3)}]^{3+}$$
$$+ x[C_nmim]^+_{aq}$$

$$(8.8)$$

$$UO_2^{2+}{}_{aq} + TODGA_{RTIL} + x[C_nmim]^+_{RTIL} \rightleftharpoons [UO_2(TODGA)]^{2+} + x[C_nmim]^+_{aq}$$

$$(8.9)$$

$$Pu^{4+}_{aq} + (1-2)TODGA_{RTIL} + x[C_nmim]^+_{RTIL} \rightleftharpoons [Pu(TODGA)_{(1-2)}]^{3+}$$
$$+ x[C_nmim]^+_{aq}$$

$$(8.10)$$

8.6.2 Anionic Mechanism

Equilibrium thermodynamics, optical absorption and luminescence spectroscopies, high-energy X-ray scattering, EXAFS and molecular dynamics simulations studies were performed to explore the structure and stoichiometry of the Ln^{3+} complexes with the ligand Htta by Jensen *et al.*,[99] and the formation of Ln^{3+}–tta$^-$ complexes in the biphasic aqueous/[C$_4$mim][NTf$_2$] system was confirmed. The ionic liquid [C$_4$mim][NTf$_2$] behaved as a liquid *anion* exchanger, by promoting formation of an anionic complex, Ln(tta)$_4^-$, in the RTIL phase, in contrast to the cationic complexes previously reported in RTILs under the conditions specified by the authors. The neutrality of the IL was maintained by exchange of Ln(tta)$_4^-$ anions with Tf$_2$N$^-$ anions from the ionic liquid, and the resulting [C$_4$mim][Ln(tta)$_4$] ion pairs become part of the ionic liquid without greatly altering the general structure (eqn 8.11):

$$Ln^{3+} + 4Htta_{org} + [C_4mim][Tf_2N]_{org} \rightleftharpoons [C_4mim][Ln(tta)_4]_{org} + 4H^+ + Tf_2N^-$$

$$(8.11)$$

In another study,[100] the extraction behavior of several metal ions (Cu^{2+}, Mn^{2+}, Co^{2+}, Ni^{2+}, Zn^{2+} and Cd^{2+}) in ionic liquid chelate extraction systems was investigated using several ionic liquids, [C$_n$mim][NTf$_2$] ($n = 4, 6, 8$), as extraction solvent and Htta as extractant and was compared with [C$_n$mim][PF$_6$]. The extracted species for Mn^{2+}, Co^{2+}, Zn^{2+} and Cd^{2+} in the [C$_n$mim][NTf$_2$] systems were anionic, M(tta)$_3^-$, and it was suggested that the alkyl chain length of [C$_n$mim]$^+$ affected the extractability of these metals as anionic M(tta)$_3^-$ (eqn 8.12):

$$M^{2+} + 3Htta_{(e)} + Tf_2N^-_{(e)} \rightleftharpoons M(tta)^-_{3\,(e)} + 3H^+ + Tf_2N^- \qquad (8.12)$$

Ouadi *et al.*[34] synthesized two imidazolium-type FILs containing 2-hydroxybenzylamine entities for Am^{3+} extraction. The authors presented a possible anion exchange mechanism for Am^{3+} partitioning. In the mechanism, the negative charge of the extracting complex in the IL phase is balanced by the transfer of NTf$_2^-$ from the [C$_4$mim][NTf$_2$] phase into the aqueous phase.

8.6.3 Multi-Mode Mechanism

The mixed extraction mechanism involves the occurrence of more than two mechanisms in some of the ionic liquids operating under different extraction conditions in the system. The hydrophobicities of the ILs increase as the length of alkyl group increases in [C$_n$mim]$^+$. It has been observed by Dietz *et al.*[75] that an increase in the hydrophobicity of an IL brings a gradual change in the mode of its partitioning from cationic to a strontium nitrate–crown ether complex during the extraction of Sr^{2+} from nitrate medium to DC18C6/[C$_n$mim][NTf$_2$] ($n = 5, 6, 8, 10$). However, in order to determine the generality of their observations, the authors have also examined the partitioning of a representative alkali metal cation, Na^+, between $[C_nmim][NTf_2]$ ($n = 5$–10) ILs and aqueous nitrate media in the presence of DC18C6. The extraction mechanism was found to follow a "ternary (*i.e.*, three-path) partitioning mechanism": (a) sodium nitrato–crown ether complex partitioning (eqn 8.13), (b) exchange of the 1:1 sodium–crown ether complex for the IL cation (eqn 8.14) and (c) crown ether-mediated Na^+/H_3O^+ exchange (eqn 8.15):[101]

$$[Na \cdot DC18C6]^+ + NO_3^- \rightarrow [Na(NO_3) \cdot DC18C6]_{org} \qquad (8.13)$$

$$[Na \cdot DC18C6]^+ + [C_nmim][Tf_2N_2]_{org} \rightarrow [Na \cdot DC18C6][Tf_2N]_{org} + [C_nmim]^+ \qquad (8.14)$$

$$Na^+ + [H_3O \cdot DC18C6]_{org}^+ \rightarrow [Na \cdot DC18C6]_{org}^+ + H_3O^+ \qquad (8.15)$$

It has further been suggested by the authors that merely by increasing the hydrophobicity of the IL cation (from $[C_5mim]^+$ to $[C_{10}mim]^+$) could not eliminate the possibility of ion exchange as a mode of metal ion partitioning between acidic media and ILs containing neutral extractants and the accompanying solubilization of the ionic liquid. In another study, a dual extraction mechanism of Cs^{2+} by the BPC6/$[C_nmim][NTf_2]$ system ($n = 2, 4, 6$), *i.e. via* exchange of a $[PC6 \cdot Cs]^+$ complex or Cs^+ by $[C_nmim]^+$, has been proposed by Xu *et al.*[65]

8.7   Conclusion and Future Prospects

Ionic liquids have emerged as "green" solvents in separation processes due to many fascinating properties and bearing the potential for replacing conventional VOCs. Moreover, the required properties in the extraction system, *i.e.* the hydrophobicity, polarity, efficiency and selectivity, can be tailor-made using ILs. ILs have been explored in many separation processes as simple biphasic liquid–liquid extractions, liquid-phase micro-extractions, ionic liquid-based solid-phase microextractions (SPME), thin layer chromatographic (TLC) and high-performance liquid chromatographic (HPLC) methods, electro-migration methods, gas–liquid chromatographic (GLC) methods, and supported ionic liquid membrane separations by various investigators. An ionic liquid remains as a "pure" ionic liquid solvent medium in liquid–liquid extractions, whereas in other cases the ionic liquid may be the minor or major phase component, depending on its miscibility with water or organic solvent(s) used to create the two-phase extraction. ILs have been used for the extraction and separation of metal ions, organic and bio-molecules, and gases and have found applications in pharmaceutical, biomedical, environmental and other industries. Many research articles and patents have been published on these applications and many new ILs have been commercialized. However, to increase their applicability, various factors such as cost, stability, purity, acidity, viscosity, solubility, density, electrochemical windows, electrical conductivity, extraction efficiency, selectivity, extraction mechanism, stripping ratio and recycling need to be considered. The potential of improvement of ILs lies in increasing their extraction efficiency, designing "functionalized ILs", reducing the loss of the IL phase in aqueous streams and process intensification. The design of environmentally friendly processes based on ionic liquid technologies will undoubtedly benefit from further toxicity, ecotoxicity and bioaccumulation studies.

References

1. *Ionic Liquids as Green Solvents: Progress and Prospects*, ed. R. D. Rogers and K. R. Seddon, ACS Symposium Series 856, American Chemical Society, Washington, 2003, p. 599.
2. H. Heitzman, B. A. Young, D. J. Rausch, P. Rickert, D. C. Stepinski and M. L. Dietz, *Talanta*, 2006, **69**, 527.
3. K. R. Seddon, in *Proceedings of 5th International Conference on Molten Salt Chemistry and Technology*, ed. H. Wendt, Trans Tech, Zurich, 1998, vol. 5, p. 53.
4. M. J. Earle and K. R. Seddon, *Pure Appl. Chem.*, 2000, **72**, 1391.
5. M. D. Joshi and J. L. Anderson, *RSC Adv.*, 2012, **2**, 5470.
6. S. Gabriel and J. Weiner, *Ber. Dtsch. Chem. Ges.*, 1888, **21**, 2669.
7. P. V. Walden, *Bull. Acad. Imp. Sci. St. Petersburg*, 1914, **8**, 405.
8. J. S. Wilkes, J. A. Levisky, R. A. Wilson and C. L. Hussey, *Inorg. Chem.*, 1982, **21**, 1263.
9. J. S. Wilkes and M. J. Zaworotko, *J. Chem. Soc., Chem. Commun.*, 1992, 965.
10. Q. Zhao and J. L. Anderson, in *Comprehensive Sampling and Sample Preparation*, ed. J. Pawliszyn, Elsevier, Amsterdam, 2012, vol. 2, p. 213.
11. R. Renner, *Environ. Sci. Technol.*, 2001, **35**, 410.
12. J. Dupont, *J. Braz. Chem. Soc.*, 2004, **15**, 341.
13. L. Andreani and J. D. Rocha, *Braz. J. Chem. Eng.*, 2012, **29**, 1.
14. J. Maa and X. Hong, *J. Environ. Manage.*, 2012, **99**, 104.
15. D. M. Roundhill, *Metal Extraction from Soils and Waters*, Kluwer/Plenum, New York, 2001.
16. S. Dai, Y. H. Ju and C. E. Barnes, *J. Chem. Soc., Dalton Trans.*, 1999, 1201.
17. A. E. Visser, R. P. Swatloski, W. M. Reichert, S. T. Griffin and R. D. Rogers, *Ind. Eng. Chem. Res.*, 2000, **39**, 3596.
18. S. Chun, S. V. Dzyuba and R. A. Bartsch, *Anal. Chem.*, 2001, **73**, 3737.
19. C. X. Sun, H. Luo and S. Dai, *Chem. Rev.*, 2012, **112**, 2100.
20. S. G. Lee, *Chem. Commun.*, 2006, 1049.
21. R. Giernoth, *Angew. Chem. Int. Ed.*, 2010, **49**, 2834.
22. J. D. Holbrey, A. E. Visser, S. K. Spear, W. M. Reichert, R. P. Swatloski, G. A. Broker and R. D. Rogers, *Green Chem.*, 2003, **5**, 129.
23. D. Kogelnig, A. Stojanovic, M. Galanski, M. Groessl, F. Iirsa, R. Krachler and B. K. Keppler, *Tetrahedron Lett.*, 2008, **49**, 2782.
24. J. H. Davis, *Chem. Lett.*, 2004, **33**, 1072.

25. *Green Industrial Applications of Ionic Liquids*, ed. R. D. Rogers, K. R. Seddon and S. Volkov, NATO Sci. Ser. 92, Kluwer, Dordrecht, 2002.
26. E. M. Martinis, P. Berton and R. P. Monasterio, *Trends Anal. Chem.*, 2010, **29**, 1185.
27. R. G. Wuilloud, S. Li, S. Cai, W. Hu, H. Chen and H. Liu, *Spectrochim. Acta, Part B*, 2009, **64**, 666.
28. M. R. Rosocka, *Sep. Purif. Technol.*, 2009, **66**, 19.
29. A. E. Visser, R. P. Swatloski, W. M. Reichert, S. T. Griffin and R. D. Rogers, *Ind. Eng. Chem. Res.*, 2001, **39**, 3596.
30. A. N. Visser, R. P. Swatloski, W. M. Reichert, R. Mayton, S. Sheff, A. Wierzbicki, J. H. Davis Jr. and R. D. Rogers, *Chem. Commun.*, 2001, 135.
31. A. E. Visser, R. P. Swatloski, S. T. Griffin, D. H. Hartman and R. D. Rogers, *Sep. Sci. Technol.*, 2001, **36**, 785.
32. N. Li, G. Fang, B. Liu, J. Zhang, L. Zhao and S. Wang, *Anal. Sci.*, 2010, **26**, 455.
33. A. E. Visser, R. P. Swatloski, W. M. Reichert, R. Mayton, S. Sheff, A. Wierzbicki, J. H. Davis and R. D. Rogers, *Environ. Sci. Technol.*, 2002, **36**, 2523.
34. A. Ouadi, B. Gadenne, P. Hesemann, J. E. Moreau, I. Billard, C. Gaillard, S. Mekki and G. Moutiers, *Chem.–Eur. J.*, 2006, **12**, 3074.
35. G. Wei, Z. Yang and C. Chen, *Anal. Chim. Acta*, 2003, **488**, 183.
36. R. Germani, M.V. Mancini, G. Savellia and N. Spreti, *Tetrahedron Lett.*, 2007, **48**, 1767.
37. J. R. Harjani, T. Friscic, L. R. MacGillivray and R. D. Singer, *Dalton Trans.*, 2008, 4595.
38. A. Stojanovic, C. Morgenbesser, D. Kogelnig, R. Krachler and B. K. Keppler, in *Ionic Liquids: Theory, Properties, New Approaches*, ed. A. Kokorin, InTech, Rijeka, Croatia, 2011, p. 657.
39. A. P. de los Ríos, F. J. Hernandez-Fernandez, L. J. Lozano, S. Sanchez, J. I. Moreno and C. Godınez, *J. Chem. Eng. Data*, 2010, **55**, 605.
40. V. M. Egorov, D. I. Djigailo, D. S. Momotenko, D. V. Chernyshov, I. I. Torocheshnikova, S. V. Smirnova and I. V. Pletnev, *Talanta*, 2010, **80**, 1177.
41. A. Rajendran, *Afr. J. Pure Appl. Chem.*, 2010, **4**(6), 100.
42. D. Kogelnig, A. Stojanovic, M. Galanski, M. Groessl, F. Jirsa, R. Krachler and B. K. Keppler, *Tetrahedron Lett.*, 2008, **49**, 2782.
43. F. L. Moore, *Anal. Chem.*, 1964, **36**, 2158.
44. R. S. Kalb, R. Krachler and B. K. Keppler, in *Chemical Industry and Environment V*, ed. W. Höflinger, EMChIE, Vienna, 2006, vol. 1, p. 259.
45. C. J. Bradaric, A. Downard, C. Kennedy, A. J. Robertson and Y. Zhou, *Green Chem.*, 2003, **5**, 143.
46. M. Tariq, P. A. S. Forte, M. F. Costa Gomes, J. N. Canongia Lopes and L. P. N. Rebelo, *J. Chem. Thermodyn.*, 2009, **41**, 790.
47. M. Regel-Rosocka, *Sep. Purif. Technol.*, 2009, **66**, 19.
48. D. Kogelnig, A. Stojanovic, F. Jirsa, W. Korner, R. Krachler and B. K. Keppler, *Sep. Purif. Technol.*, 2010, **72**, 56.

49. A. Stojanovic, D. Kogelnig, L. Fischer, H. Hann, M. Galanski, M. Groessl, R. Krachler and B. Keppler, *Aust. J. Chem.*, 2010, **63**, 511.
50. H. Abdolmohammad-Zadeh and G. H. Sadeghi, *Anal. Chim. Acta*, 2009, **649**, 211.
51. H. Abdolmohammad-Zadeh and G. H. Sadeghi, *Talanta*, 2010, **81**, 778.
52. N. Papaiconomou, J. M. Lee, J. Salminen, M. Von Stosch and J. M. Prausnitz, *Ind. Eng. Chem. Res.*, 2008, **47**, 5080.
53. S. C. N. Hsu, C. J. Su, F. L. Yu, W. J. Chen, D. X. Zhuang, M. J. Deng, I. W. Sun and P.Y. Chen, *Electrochim. Acta*, 2009, **54**, 1744.
54. J. Salminen, N. Papaiconomou, R. A. Kumar, J. M. Lee, J. Kerr, J. Newman and J. M. Prausnitz, *Fluid Phase Equilib.*, 2007, **261**, 421.
55. J. M. Lee, *Fluid Phase Equilib.*, 2012, **319**, 30.
56. D. D. Patel and J. M. Lee, *Chem. Rec.*, 2012, **12**, 329.
57. H. Zhao, S. Xia and P. Ma, *J. Chem. Technol. Biotechnol.*, 2005, **80**, 1089.
58. A. E. Visser, R. P. Swatloski, W. M. Reichert, S. T. Griffin and R. D. Rogers, *Ind. Eng. Chem. Res.*, 2000, **39**, 3596.
59. A. E. Martell and R. D. Hancock, *Metal Complexes in Aqueous Solutions*, Plenum, New York, 1979.
60. G. T. Wei, Z. Yang and C. J. Chen, *Anal. Chim. Acta*, 2003, **488**, 183.
61. V. V. Yakshin, N. A. Tsarenko, A. M. Koshcheev, I. G. Tananaev and B. F. Myasoedov, *Radiochemistry*, 2012, **54**, 54.
62. C. Xu, X. Shen, Q. Chen and H. Gao, *Sci. China, Ser. B*, 2009, **52**, 1858.
63. H. Luo, S. Dai and P. V. Bonnesen, *Anal. Chem.*, 2004, **76**, 2773.
64. H. Luo, S. Dai, P. V. Bonnesen, A. Buchanan III, J. D. Holbrey, N. J. Bridges and R. D. Rogers, *Anal. Chem.*, 2004, **76**, 3078.
65. C. Xu, L. Yuan, X. Shen and M. Zhai, *Anal. Chem.*, 2004, **76**, 3078.
66. A. N. Turanov, V. K. Karandashev and V. E. Baulin, *Solvent Extr. Ion Exch.*, 2010, **28**, 367.
67. M. Fuerhacker, T. M. Haile, D. Kogelnig, A. Stojanovic and B. Keppler, *Water Sci. Technol.*, 2012, **65**, 1765.
68. L. Fischer, T. Falta, G. Koellensperger, A. Stojanovic, D. Kogelnig, M. Galanski, R. Krachler, B. K. Keppler and S. Hann, *Water Res.*, 2011, **45**, 4601.
69. G. T. Wei, J. C. Chen and Z. Yang, *J. Chin. Chem. Soc.*, 2003, **50**, 1123.
70. J. M. Reyna-González, R. Galicia-Pérez, J. C. Reyes-López and M. Aguilar-Martínez, *Sep. Purif. Technol.*, 2012, **89**, 320.
71. K. Shimojo and M. Goto, *Anal. Chem.*, 2004, **76**, 5039.
72. R. Lertlapwasin, N. Bhawawet, A. Imyim and S. Fuangswasdi, *Sep. Purif. Technol.*, 2010, **72**, 70.
73. R. Germani, M. V. Mancini, G. Savellia and N. Spreti, *Tetrahedron Lett.*, 2007, **48**, 1767.
74. S. Katsuta, Y. Yoshimoto, M. Okai, Y. Takeda and K. Bessho, *Ind. Eng. Chem. Res.*, 2011, **50**, 12735.
75. M. L. Dietz and J. A. Dzielawa, *Chem. Commun.*, 2001, 2124.
76. J. G. Huddleston, H. D. Willauer, R. P. Swatloski, A. E. Vissser and R. D. Rogers, *Chem. Commun.*, 1998, 1765.

77. K. Nakashima, F. Kubota, T. Maruyama and M. Goto, *Ind. Eng. Chem. Res.*, 2005, **44**, 4368.
78. S. J. Yoon, J. G. Lee, H. Tajima, A. Yamasaki, F. Kiyono, T. Nakazato and H. Tao, *J. Ind. Eng. Chem.*, 2010, **16**, 350.
79. N. Kozonoi and Y. Ikeda, *Monatsh. Chem.*, 2007, **138**, 1145.
80. J. A. Vicente, A. Mlonka, H. Q. N. Gunaratne, M. Swadzba-Kwasny and P. Nockemann, *Chem. Commun.*, 2012, **48**, 6115.
81. X. Sunad, H. Luob and S. Dai, *Talanta*, 2012, **90**, 132.
82. A. Rout, K. A. Venkatesan, T. G. Srinivasan and P. R. Vasudeva Rao, *Sep. Purif. Technol.*, 2012, **95**, 26.
83. X. Q. Suna, B. Peng, J. Chena, D. Q. Li and F. Luob, *Talanta*, 2008, **74**, 1071.
84. H. Okamura, N. Hirayama, K. Morita, K. Shimojao, H. Naganawa and H. Imura, *Anal. Sci.*, 2010, **26**, 607.
85. P. Giridhar, K. A. Venkatesan, T. G. Srinivasan and P. R. V. Rao, *J. Nucl. Radiochem. Sci.*, 2004, **5**, 21.
86. V. A. Cocalia, M. P. Jensen, J. D. Holbrey, S. K. Spear, D. C. Stepinski and R. D. Rogers, *Dalton Trans.*, 2005, 1966.
87. V. A. Cocalia, J. D. Holbrey, K. E. Gutowski, N. J. Bridges and R. D. Rogers, *Tsinghua Sci. Technol.*, 2006, **11**, 188.
88. K. A. Venkatesan, T. G. Srinivasan and P. R. Vasudeva Rao, *J. Nucl. Radiochem. Sci.*, 2009, **10**, R1.
89. A. Rout, S. Karmakar, K. A. Venkatesan, T. G. Srinivasan and P. R. Vasudeva Rao, *Sep. Purif. Technol.*, 2011, **81**, 109.
90. R. B. Gujar, S. A. Ansari, D. R. Prabhu, P. K. Mohapatra, P. N. Pathak, A. Sengupta, S. K. Thulasidas and V. K. Manchanda, *Solv. Extr. Ion Exch.*, 2012, **30**, 156.
91. S. Panja, P. K. Mohapatra, S. C. Tripathi, P. M. Gandhi and P. Janardan, *Sep. Purif. Technol.*, 2012, **96**, 289.
92. A. Sengupta, P. K. Mohapatra, M. Iqbal, W. Verboom, J. Huskensb and S. V. Godbole, *RSC Adv.*, 2012, **2**, 7492.
93. A. Ouadi, O. Klimchuk, C. Gaillard and I. Billard, *Green Chem.*, 2007, **9**, 1160.
94. M. Srncik, D. Kogelnig, A. Stojanovic, W. Korner, R. Krachler and G. Wallner, *Appl. Radiat. Isot.*, 2009, **67**, 2146.
95. P. S. Barber, S. P. Kelley and R. D. Rogers, *RSC Adv.*, 2012, **2**, 8526.
96. M. P. Jensen, J. Dzielawa, P. Ricket and M. L. Dietz, *J. Am. Chem. Soc.*, 2002, **124**, 10664.
97. A. E. Visser, M. P. Jensen, I. Laszak, K. L. Nash, G. R. Choppin and R. D. Rogers, *Inorg. Chem.*, 2003, **42**, 2197.
98. K. Shimojo, K. Kurahashi and H. Naganawa, *Dalton Trans.*, 2008, 5083.
99. M. P. Jensen, J. Neuefeind, J. V. Beitz, S. Skanthakumar and L. Soderholm, *J. Am. Chem. Soc.*, 2003, **125**, 15466.
100. K. Kidani, N. HIirayama and H. Imura, *Anal. Sci.*, 2008, **24**, 1251.
101. M. L. Dietz and D. C. Stepinski, *Green Chem.*, 2005, **7**, 747.

CHAPTER 9

Periphyton Biofilms for Sustainability of Aquatic Ecosystems

YONGHONG WU

State Key Laboratory of Soil and Sustainable Agriculture, Institute of Soil Sciences, Chinese Academy of Sciences, 71 East Beijing Road, Nanjing 210008, Jiangsu, People's Republic of China
Email: yhwu@issas.ac.cn

9.1 Introduction

9.1.1 Periphyton Biofilm

The term "periphyton" refers to the microfloral community attached to the surfaces of submerged objects in water;[1] it includes algae, bacteria, fungi, protozoa, zooplankton, and other invertebrates.[2] Periphyton accumulates on a dead or living surface/substrate, ranging from clay particles, fine sand, pebbles, and rocks to short-lived filamentous algae, macrophytes, and animal bodies.[1] Like phytoplankton, periphyton can be found in almost every type of water body, from small ponds to large oceans and in trophic conditions that range from the most oligotrophic to the more eutrophic.[3]

9.1.2 Composition and Structure of Periphyton Biofilm

Periphyton communities are very diverse, depending on a range of factors such as habitat, surface types, light intensity, grazing pressure, seasonality, nutrient

RSC Green Chemistry No. 23
Green Materials for Sustainable Water Remediation and Treatment
Edited by Anuradha Mishra and James H. Clark
© The Royal Society of Chemistry 2013
Published by the Royal Society of Chemistry, www.rsc.org

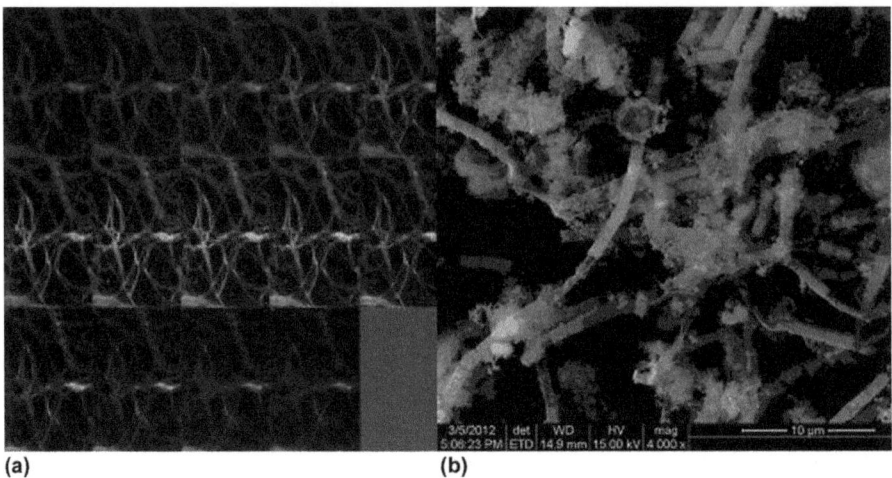

(a)   (b)

Figure 9.1   (a) Periphyton biofilms attached on glass and cultured on BG-11 media, observed by fluorescence confocal microscopy. (b) Periphyton biofilms attached on glass and cultured on BG-11 media, observed by scanning electron microscopy.

availability, and physical disturbances.[1] The most frequently encountered groups are the diatoms (Bacillariophyceae), cyanobacteria (Cyanophyceae and Microcystis), and green algae (Chlorophyceae) (Figure 9.1). Many protozoa and metazoa usually attach at the periphyton biofilms, such as leishmania, nematodes, dinoflagellates, oxytrichida, annelids, amoebae, rotaria, paramecia, vorticella, and rotifers (Figure 9.2).

It is well known that the structure of microbial aggregates such as periphyton biofilms ranges from patchy monolayers to filamentous accretions during different phases of biofilm formation.[4] The basic structure of periphyton biofilms includes at least three conceptual models: (i) heterogeneous mosaic biofilm aggregations, (ii) penetrated water-channel biofilms, and (iii) dense confluent biofilms.[3,4]

The structure of periphyton biofilms is affected by the local environmental conditions, including disturbances, stressors, resources, hydraulic conditions, and biotic interactions.[5] For example, the effects of overlying flow conditions on periphyton structure and denitrification were measured in three laboratory mesocosms (120 cm long and 60 cm wide) under average velocities of 0.05, 0.5 and 5 cm s^{-1}. Periphyton biofilms were cultivated on polyethylene benthic nets overlaying a thin layer of sand. Periphyton structural characteristics were quantified in terms of algal/bacterial biomass, algal species composition, and microbial enumeration. Confocal microscopy was used to investigate the spatial organization of the periphyton. Different benthic microbial communities developed under the three flow conditions, while the total microbial biomass accrual increased monotonically with increasing overlying velocity. The periphytic community that developed under the fastest velocity was characterized

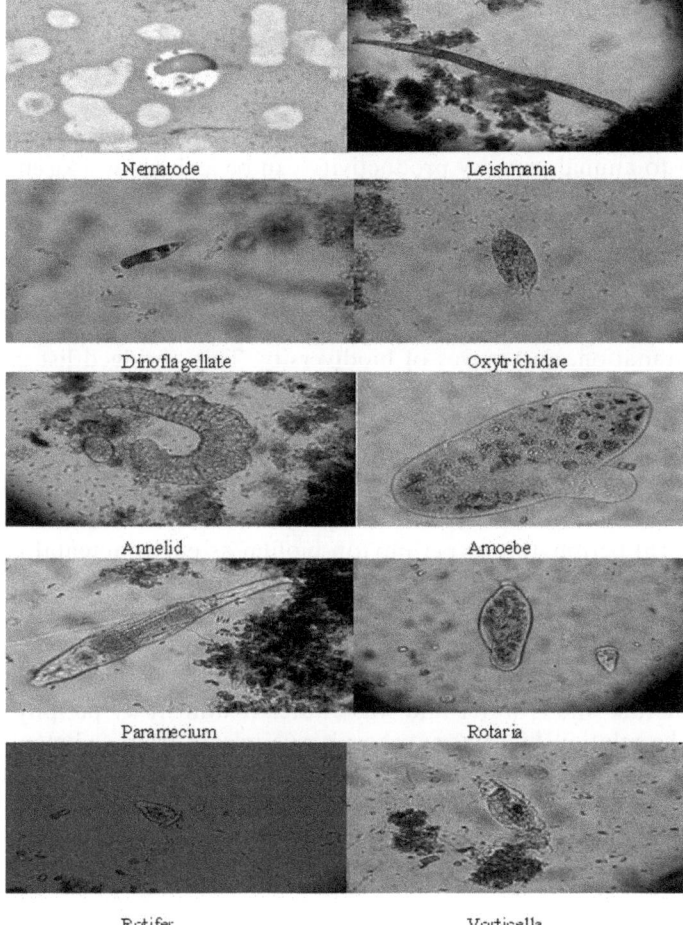

Figure 9.2   Microscopic observation of the protozoa and metazoa in periphyton biofilms.

by the largest fractional biovolume of diatoms (predominantly *Achnanthidium minutissimum*) and promoted establishment of a consortium of bacterial denitrifiers more physiologically active than those that developed under slower overlying velocities.[6]

The size of freshwater periphytic algae ranges from that of a single cell of about one thousandth of a millimeter in diameter to a large alga (*Cladophora*) up to many tens of centimeters in length.[1–3] The shape varies from simple nonmotile single cells to motile, multicellular, and filamentous structures. The colonization of periphyton on a substrate is initiated by a coating of dissolved organic substances and subsequently by bacteria. In the presence of sufficient light, algae attach themselves to the substrates and increase exponentially until reaching a peak, after which they shift to a loss process through sloughing and dislodgment. The time required to reach the peak biomass varies from a few days to several months. The biomass and productivity of periphytic communities vary by several orders of magnitude. A complex interplay of biotic and abiotic factors governs this variation. In shallow lakes, periphyton contributes 42–97% of the total annual productivity and the contribution of periphyton to annual primary productivity can be as high as $1\,kg\,cm^{-2}$.

9.1.3 Periphyton Biofilms Included in this Chapter

Periphyton communities are solar-powered biogeochemical reactors, biogenic habitats, hydraulic roughness elements, early warning systems for environmental degradation, and troves of biodiversity. This abridged list gives some indication of the ecological and cultural importance of periphyton.[5] Therefore, many studies have been published to address the importance of periphyton in aquatic ecosystems. These studies can be classified into 11 broad topics: (1) effects of physical disturbances, (2) effects of exposure to stressors, (3) limiting abiotic factors, (4) competitive interactions, (5) effects of herbivores, (6) periphyton or periphyton biofilm as environmental indicators, (7) the roles of periphyton in nutrient cycling in food webs and between abiotic pools,[5] (8) interspecific interactions,[7] (9) biomonitoring,[8] and (10) nutrient dynamics,[9] as well as (11) influence of hydraulic conditions.[1,10]

In this chapter, two main topics will be presented, based on our previous studies. The first one is water and wastewater treatment by periphytons. The second is the relationship among phosphorus release from sediments, cyanobacterial blooms, and periphyton biofilms. These findings will help the readers to learn more information about the green material—periphyton biofilm—and benefit them to use this kind of green resource of periphyton biofilm appropriately.

9.2 Treatment of Water and Wastewater

9.2.1 Nutrient Removal

It is necessary to "eat" some nutrients during periphyton growth. The common nutrient elements include carbon, nitrogen, and phosphorus. The incorporation of nutrients within microbial biomass is assumed to be very efficient through the use of microorganisms.[11] At the same, the microporous structure of periphyton biofilm facilitates the adsorption of nutrient, improving the removal of nutrients and even dominating the nutrient removal.[12] For example, periphyton and phytoplankton were co-limited by phosphorus and nitrogen, in contrast to previous findings of strong phosphorus limitation in more pristine regions of the marsh, and in the nearby Florida Everglades.[13]

The microorganisms in periphyton biofilm are more apt to "eat" the forms of nitrogen, such as nitrite, nitrate, or ammonium, as the sole nitrogen source for growth.[14] In the first reactor of an intermittently aerated anaerobic–aerobic activated sludge process, nitrification and phosphorus uptake occur during the aeration period, followed by denitrification and phosphorus release during the agitation period.[15] In the second reactor, nitrification and phosphorus uptake occur during aeration and denitrification and weak phosphorus uptake occurs during agitation.[15,16]

The nitrogen removal process by biodegradation of periphyton biofilms has been addressed very well.[16] This process often features two predominant processes: autotrophic nitrification and heterotrophic denitrification.[11] Some minor processes such as heterotrophic nitrification and aerobic denitrification are also involved.[17] Organisms or aggregates that degrade nitrogenous compounds can be divided into three main groups according to the biological nitrogen removal processes they conduct: (i) degradation of nitrogen organic matter and the release of ammonia by various microorganisms, (ii) conversion of ammonia to nitrate by certain autotrophic microorganisms, and (iii) conversion of nitrate to nitrogen gas by a mixed culture that uses nitrate as an electron acceptor (as opposed to free oxygen used by the nitrifiers) in the metabolism of organic carbon.[11,16,18,19]

Another major route for the removal of biological nitrogen from waste liquids involves the removal of nitrogen by a nitrification–denitrification processes.[14] The use of a multi-level bioreactor fosters the coexistence of photoautotrophic and heterotrophic microorganisms, which provide environments for the combination of oxidative and reductive processes.[19] Santos *et al.*[20] investigated a system that uses the oxidative and reductive environments within polymer beads to remove nitrogen *via* a nitrification–denitrification processes. Their results showed that high nitrogen removal rates (up to $5.1 \, mmol \, N \, m^{-3}$ polymer s^{-1}) were achieved under continuous flow and aerobic conditions because the nitrifier *Nitrosomonas europaea* and either of the denitrifiers *Pseudomonas denitrificans* or *Paracoccus denitrificans* were co-immobilized in this system.

When the periphyton biofilms concentrated by soft industrial carriers were applied to remove nitrogen, the results showed that the total nitrogen (TN) concentration in water significantly decreased from 7.9–$7.6 \, mg \, L^{-1}$ to 3.3–$3.5 \, mg \, L^{-1}$ at day 15 after the periphyton biofilms were added, while the TN concentration in water in the control (without the addition of periphyton biofilms) only reached $6.2 \, mg \, L^{-1}$ during the experimental period. Moreover, it was also found that the nitrate concentration in the water decreased and the nitrate removal efficiency was about 50% at day 30 after the addition of periphyton biofilms (Figure 9.3). These results imply that the inclusion of periphyton biofilm can eliminate the nitrogen content in water by converting nitrate nitrogen to N_2.

The excessive input of phosphorus into surface waters often accelerates the eutrophication process.[11] Phosphorus is often regarded as the limiting factor for phytoplankton growth, thereby accelerating the growth of harmful algal blooms.[21] Thus, the removal of phosphorus, especially biological phosphorus, has recently been a subject of great concern. The complex surface structures consist of the microorganisms mentioned above and feature some special properties, such as adhesion and flocculation abilities,[22] which enable

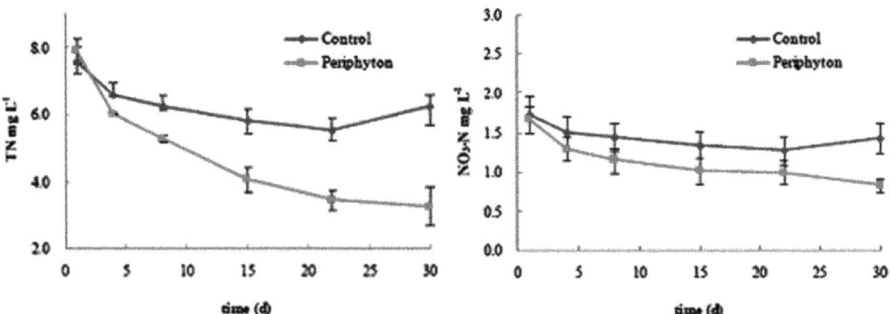

Figure 9.3 Variation of total nitrogen and nitrate in water after the employment of periphyton biofilms.

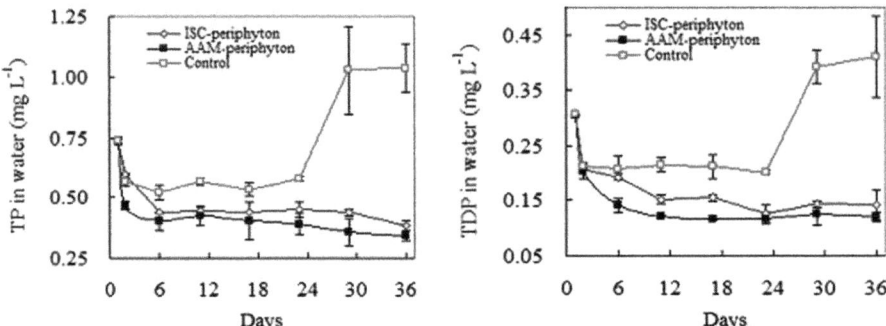

Figure 9.4 Changes in total dissolved phosphorus and total phosphorus concentrations in the overlying water.

microorganism communities to adsorb some phosphorus compounds, heavy metals, dyes, and toxic materials from solutions.[12,22,23]

Phosphorus removal efficiency is associated with periphyton composition. We have investigated periphyton biofilms that consist of diatoms which include the following species: *Melosira varians* Ag., *Gomphonema parvulum* Kütz., *Synedra ulna* Kutz., *Nitzschia amphibia* Grun., and *Fragilaria vaucheriae* Kutz. Most of the bacteria were found to be bacilli and cocci in these periphyton biofilms.[12] It was found that the initial total dissolved phosphorus (TDP) and total phosphorus (TP) concentrations in the overlying water for both treatment and control samples were 0.31 and 0.74 mg L^{-1}, respectively. During the first six days, both TDP and TP decreased gradually in all samples. In the periphyton treatment samples the TDP and TP concentrations in overlying water were relatively steady from the sixth day until the end of the experiment. In contrast, in the control sample, both TDP and TP concentrations in the overlying water were maintained relatively stable between the sixth and 23rd days, after which the TDP and TP concentrations in the overlying water increased drastically (Figure 9.4).

Owing to the ingestion, the decreases in the TP and TDP concentrations in the overlying water in the periphyton samples were partially attributed to phosphorus uptake by the periphytons. As we know, the structure of periphyton biofilm is composed of many microporous surfaces. As a result, the periphytons can be phosphorus sinks[24] and play an important role in the removal, uptake, or transformation of biologically available soluble phosphorus.[25,26]

In periphyton biofilm systems that are used for treating water and wastewater, biological phosphate removal is based on the capacity of some microorganisms to store orthophosphate intracellularly as polyphosphate. These microorganisms store polyhydroxybutyrate (PHB) anaerobically, which is oxidized in a phase with an electron acceptor such as oxygen or nitrate present.[16]

The group of microorganisms in periphyton biofilms that are largely responsible for P removal are known as the polyphosphate accumulating organisms (PAOs).[27] PAOs take up readily biodegradable chemical oxygen demand substrates and store them as polyhydroxyalkanoates (PHAs). The energy required for this anaerobic process is derived from the hydrolysis of intracellular polyphosphate. In the subsequent aerobic stage, PAOs use PHAs to generate energy for growth and phosphorus uptake. In this process the PAOs take up more phosphorus than that released during the anaerobic stage (luxury uptake).[28] Removing phosphate using nitrate instead of oxygen has the advantage of saving energy (oxygen input) and using less organic carbon. It has been shown that it is possible to accumulate denitrifying P-removing bacteria (DPB), which can remove P and N simultaneously in microbial aggregate systems.[29]

During phosphorus removal the periphyton biofilms also respond to the high-loading phosphorus. For example, a field experiment was conducted to determine the effects of increased phosphorus loading on periphyton abundance, productivity, and taxonomic composition in an oligotrophic Everglades slough characterized by abundant metaphyton and epiphyton. The result showed that the added phosphorus was accumulated rapidly by the periphyton at a rate proportional to the load. Phosphorus accumulation caused the loss of the extensive mats of cyanobacteria and diatoms that were abundant in the surrounding slough. This oligotrophic assemblage was replaced by floating mats of eutrophic cyanobacteria and diatoms at the highest loading rates (6.4–$12.8 \, g \, P \, m^{-2}$ per year), and by diffuse masses of filamentous chlorophytes at intermediate loads (1.6–$3.2 \, g \, P \, m^{-2}$ per year). Metaphyton and epiphyton biomass-specific productivity increased in proportion to the loading rate and remained elevated at higher loads until the end of the wet season. Despite higher productivity rates, both epiphyton biomass and floating mat coverage declined at higher loads compared to controls. Periphyton changes induced by phosphorus enrichment may affect wetland function by reducing (1) periphyton dominance, (2) the food quality of the periphyton for herbivores, and (3) the nutrient storage capacity of the wetland. Many of these changes also have been documented in other wetlands, thereby implicating phosphorus as a principal factor affecting wetland periphyton structure and function.[30]

9.2.2   Organic Matter Removal

9.2.2.1   COD Removal

In environmental chemistry, the chemical oxygen demand (COD) value is often used to reflect the level of organic compounds in water or wastewater. Usually, ignoring the organic compound composition, COD removal is associated with the content of "organisms" in water. The oxygen amount depleted by these "organisms" is named as biological oxygen demand (BOD). When the ratio of BOD to COD is relatively high, the organic compounds are easily degraded by the microorganisms in the periphyton biofilm.

In recent years, some researchers have found that low-biodegradable wastewater with a low BOD/COD ratio also can be purified by some specific biofilms. For instance, a biofilm configured system with a sequencing/periodic discontinuous batch mode operation was evaluated for the treatment of low-biodegradable composite chemical wastewater (low BOD/COD ratio 0.3, high sulfate content $1.75\,\mathrm{g\,L^{-1}}$) in aerobic metabolic function. The results showed a COD removal of 88%, accounting for a substrate degradation rate of 0.81 kg COD/cum-day at steady state conditions. The reactor attained stable conditions within 15 days and remained more or less constant thereafter (Figure 9.5b). Subsequently, after achieving stable performance, the reactor was operated at higher organic loading rates to understand the performance (1.5, 3.07 and 4.76 kg COD/cum-day), keeping all other operating conditions the same (Figure 9.5).[31] It is evident from the results that the low-biodegradable wastewater with a low BOD/COD ratio can be purified consistently at a higher organic loading rate and stabilized within relatively less time.

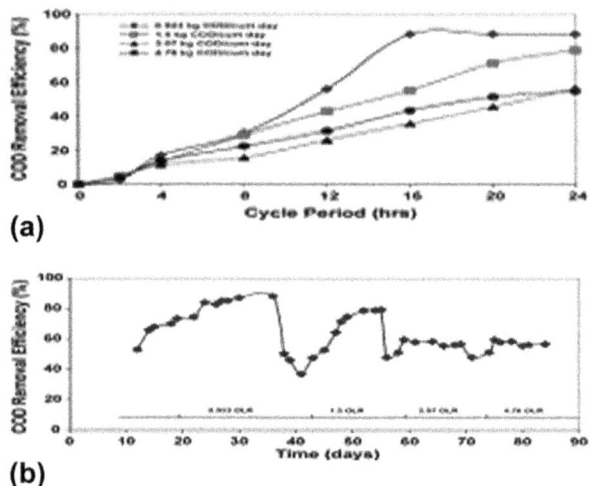

Figure 9.5   COD removal efficiency during sequence phase operation of a batch mode reactor.[29]

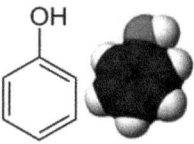

Figure 9.6 Chemical structure of phenol.

9.2.2.2 *Removal of Phenolic Compounds*

Phenol, also known as carbolic acid and phenic acid, is an organic compound with the chemical formula C_6H_5OH. It is a white crystalline solid at room temperature. The molecule consists of a phenyl group (C_6H_5) bonded to a hydroxyl group (OH) (Figure 9.6). It is mildly acidic, but requires careful handling due to its propensity to cause burns.

Because of phenolic compound toxicity to humans and fauna and flora, increasingly stringent restrictions have been imposed on the concentrations of these compounds in wastewater for safe discharge and are regarded as priority pollutants in the USEPA list.[32] In the natural environment, the biodegradation rate of phenol is slow. Consequently, phenols accumulate in the environment and persist for a long time, threatening the safety of flora and fauna as well as human beings.[33] From a practical standpoint, it is therefore important to study phenol removal.[11]

According to the literature, several processes are used to remove phenolic compounds, like granular or biologically activated carbon, H_2O_2/UV processes, O_3/UV processes, Fenton processes (Fe^{2+}/H_2O_2), solvent extraction, membrane processes, *etc.* Conventional processes have been mostly physicochemical processes, but since they cause secondary problems in the effluents (for example, phenol becomes chlorophenols if chlorination is used), biological treatments are preferred for large-scale removal of this type of pollutant.[34]

The periphyton biofilms are often fixed in a reactor (called a biofilm reactor), which has been widely applied to treat water and/or wastewater containing phenolic compounds. For example, a new biofilm-electrode method was used for phenol degradation, because of its low current requirement. The biofilm-electrode reactor consisted of immobilized degrading bacteria on a Ti electrode as cathode and Ti/PbO_2 electrode as anode. With the biofilm-electrode reactor in a divided electrolytic cell, the phenol degradation rate could achieve 100% at 18 h, which was higher than using traditional methods, such as biological or electrochemical methods. The COD removal rate of the biofilm-electrode reactor was also greater than that using biological and electrochemical methods, and could reach 80% at 16 h. The results suggested that the biofilm-electrode reactor system can be used to treat wastewater containing phenol.[35]

Many phenol-degrading microorganisms, including bacteria, fungi, yeast, and periphyton, have been identified in aqueous solutions.[36–39] For example, a new phenol-degrading bacterium with high biodegradation activity and high tolerance of phenol, strain PD12, was isolated from the activated sludge of the Tianjin Jizhuangzi Wastewater Treatment Facility in China. This strain was capable of removing 500 mg phenol L^{-1} in liquid minimal medium by 99.6% within 9 h and metabolizing phenol at concentrations up to 1100 mg L^{-1}. DNA sequencing and homologous analysis of the 16S rRNA gene identified PD12 as an *Acinetobacter* spp.[38] Fungi strains (*Graphium* spp. and *Fusarium* spp.) have high percentages of phenol degradation, with 75% degradation of 10 mM phenol in 168 h.[40] Yeast *C. tropicalis* could degrade 2000 mg L^{-1} phenol alone and 280 mg L^{-1} *m*-cresol alone within 66 and 52 h, respectively. The capacity of the strain to degrade phenol was obviously higher than that to degrade *m*-cresol. The presence of *m*-cresol intensely inhibited phenol biodegradation. Only 1000 mg L^{-1} phenol can be completely degraded in the presence of 280 mg L^{-1} *m*-cresol. On the other hand, phenol of low concentration (from 100 to 500 mg L^{-1}) supplied a sole carbon and energy source for *C. tropicalis* in the initial phase of biodegradation and accelerated the assimilation of *m*-cresol, resulting in the fact that the *m*-cresol biodegradation velocity was higher than that without phenol. Besides, the capacity of *C. tropicalis* for *m*-cresol biodegradation was increased up to 320 mg L^{-1} in the presence of 60–100 mg L^{-1} phenol.[39]

In some cases, phenolic compounds can be effectively removed by plank-tonics such as *Pseudomonas pseudoalcaligenes* in constructed wetland systems. This is because the biofilms form on the planktonic surface and the biofilms may have some specific microorganisms that are capable of degrading phe-nols.[37] It was also found that some bacteria, such as *Pleurotus* spp. strains, have a significant ability to degrade high concentrations of phenols. For instance, bacteria such as *Pleurotus* spp. strains have the ability to remove phenolic compounds from olive mill wastewater that contains high concentrations of phenols and high antibacterial activity.[11,41,42]

The biodegradation of phenols by microorganisms is mainly carried out by lignin-degrading enzymes. For instance, biphenyl (BP), biphenylene (BN), dibenzofuran (DF), dibenzo-*p*-dioxin (DD), and diphenyl ether (DE) were utilized as diphenyl substrates for the lignin-degrading basidiomycete *Phanerochaete chrysosporium*, and its extracellular enzyme, lignin peroxidase. Among these compounds, only BN and DD were oxidized by lignin peroxidase. Cyclic voltammetry measurements revealed that BN and DD possess lower redox potentials than the other diphenyl substrates utilized, in accordance with the reactivity of lignin peroxidase. Although the degradations of BP, DF and DE were not extracellularly initiated by lignin peroxidase, they were metabolized *via* intermediate formation of hydroxylated products by intracellular enzymes.[43]

In addition, manganese peroxidase and laccase, which are produced by white rot basidiomycetes microorganisms, also contribute to the biodegradation of phenols (*e.g.* bisphenol A or BPA).[36] Manganese peroxidase is a heme peroxidase that oxidizes phenolic compounds in the presence of Mn(II) and H_2O_2, while laccase is a multi-copper oxidase that catalyzes the one-electron oxidation of phenolic compounds by reducing oxygen to water.[36] Manganese peroxidase and laccase can degrade BPA and disrupt its estrogenic activity.[36] In the case of laccase, BPA metabolism is faster in the presence of mediators, such as 1-hydroxybenzotriazole (HBT) and 2,2′-azobis(3-ethylbenzothiazoline-6-sulfonate), than with laccase alone.[36]

Environmental conditions, *i.e.* whether aerobic or anaerobic, can significantly affect the efficiency of phenol biodegradation.[11] BPA in river waters is biodegraded under aerobic conditions but not under anaerobic conditions.[36] BPA in spiked samples was rapidly biodegraded under aerobic conditions (>90%), with little decrease in BPA observed under anaerobic conditions (<10%) over 10 days.[44] Under anaerobic conditions, such as those in anaerobic marine sediments, BPA was not biodegraded even after 3 months of incubation. These results suggest that anaerobic bacteria have little or no ability to degrade BPA.[11,36,45]

Phenol biodegradation by microorganisms is also influenced by temperature and microbe counts. For example, the effect of temperature on the efficiency of degradation of the aromatic compounds benzoic acid, 2,3,5-trichlorobenzoic acid, phthalic acid, methyl and ethyl phthalate, resorcinol, phenol, pentachlorophenol, 4-nitrophenol, 4-ethylphenol, and *o*-, *m*-, and *p*-cresol by a mesophilic or a thermophilic anaerobic community was examined at 37 °C and 55 °C. Benzoic acid, phthalic acid, methyl phthalate, phenol, and *m*- and *p*-cresol were mineralized by the mesophilic cultures incubated at 37 °C. Apart from benzoic acid, no aromatic compounds were mineralized by the thermophilic community incubated at 55 °C, suggesting that channeling reactions to the central intermediate benzoyl-CoA were inoperative in this microbial community.[46]

The half-lives for phenol biodegradation in 15 river water samples averaged 4 and 7 days at 30 and 20 °C, respectively, but only about 20% (0.04 mg L^{-1}) of the spiked phenol was biodegraded at 4 °C over a period of 20 days.[47] With respect to bacterial counts, owing to the greater bacterial biomass in the subsurface flow of constructed wetlands, the phenol removal rate is higher than that at the surface.[37] It has also been reported that phenol biodegradation does not correlate with bacterial counts.[48] These differences may be due to the size of bacterial populations that can execute fast and complete phenol biodegradation or mineralization.[11,48]

Moreover, phenol removal is associated with the biomass growth rate and the substrate initial concentration. Some studies have used the Haldane model to fit the relationship between biofilm masses and phenol initial concentrations, which is summarized in Table 9.1.[49–62]

9.2.2.3   Removal of Microcystins

Toxic blooms of cyanobacteria occur worldwide in eutrophic lakes, ponds, and reservoirs. Microcystins (MCs) are the most commonly detected cyanobacterial toxins. These toxins have the general structure cyclo(D-Ala-X-D-MeAsp-Z-Adda-D-Glu-Mdha-), in which X and Z represent variable L-amino acids, and Adda refers to the β-amino acid residue of 3-amino-9-methoxy-2,6,8-trimethyl-10-phenyldeca-4,6-dienoic acid. Over 70 structural analogues of MCs have been identified. These structural variants differ primarily in the X–Z amino acids and in methylation or demethylation on MeAsp and MDha.[63–65] When the L-amino acids at positions 2 and 4 are both arginine, the MC is indicated as MC-RR (Figure 9.7).[65]

Table 9.1 Haldane parameters for different bacteria grown on phenol.[54]

Bacterial strain	Conc. range (mg L^{-1})	Phenol conc. at max. (mg L^{-1})	Haldane's model			T($°C$)/pH	Ref.
			μ_{max}	K_s	K_i		
Pseudomonas putida MTCC 1194	0–1000	64.6	0.305	36.3	129.8	29.9 ± 0.3/7.1	49
P. putida ATCC 49451	25–800	50	0.9	6.93	284.3	30/6.5	50
P. putida DSM 548	1–100	25	0.436	6.19	54.1	26/6.8–6.3	51
P. putida CCRC 14365	0–600	65.4	0.33	13.9	669	30/6.8	52
P. putida CCRC 14365	0–400	45.5	0.245	12.1	1185.8	30/7.0	53
P. putida LY1	0–800	50	0.217	24.4	121.7	25/7.1–7.3	54
Acinetobacter	60–350	60	0.80–0.85	1.2–1.5	188–315	30/7.4	55
Yeast *Trichosporon cutaneum* R57	500	500	0.42	110	380	28–30/7.6	56
Yeast *Candida tropicalis*	0–2000	50	0.48	11.7	207.9	30/6.0	57
Ewingella americana	0–1000	116.9	0.29	5.16	1033.7	37/7.5	58
Alcaligenes faecalis	10–1400	38.9	0.15	2.22	245.4	30/7.2	59
Candida albicans PDY-07	0–1800	68.9	0.315	19.5	208.6	35/7.0	60
Mixed culture	23.5–659	188.2	0.309	74.6	648.1	25 ± 2/7.2	61
Bacillus brevis	750–750	≤50	0.026–	2.2–	868.0–	34 ± 0.1/8.0	62
			0.078	29.3	2434.7		

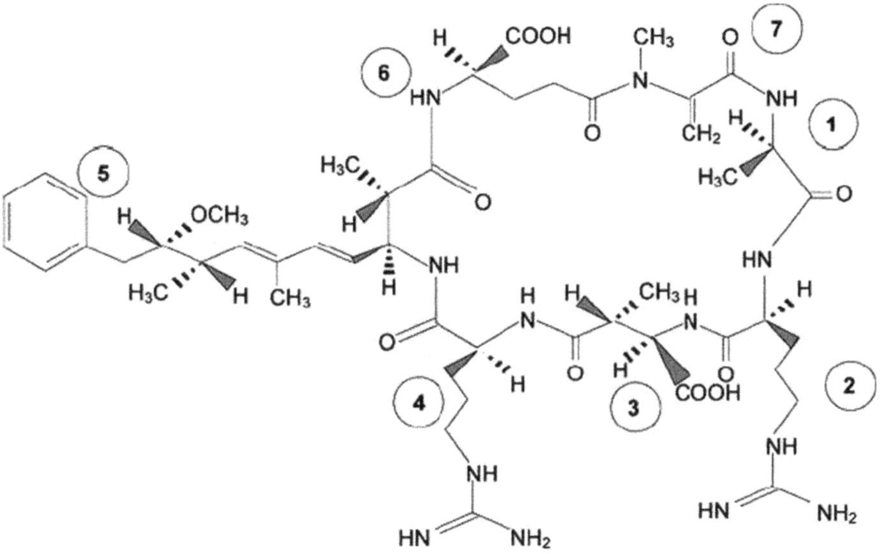

Figure 9.7 Chemical structure of microcystin-RR showing the two variable amino acids (arginine) at positions 2 and 4.[65]

MCs are toxic to mammals, fish, plants, and invertebrates.[4,65] At the molecular level, MCs bind irreversibly to and inhibit the serine/threonine protein phosphatases 1 and 2A (PP1, PP2A). Microcystin-LR has been shown to be a liver-tumor promoter.[66,67] MCs are known to be chemically stable compounds,[65,68] so it is not surprising that conventional drinking water treatments have only limited efficacy in removing dissolved MCs.[65,69] Many methods have been put forward to remove microcystins.[4,69] Among these methods, the technique of using micro-organisms (or/and biofilm) that are capable of degrading microcystins is widely accepted. Methods employing microcystin-degrading microorganisms can be classified into two groups. One is entirely ascribed to the service of biofilms that grow on the surface of substrates within bioreactors such as biological sand,[70,71] granular activated carbon filters,[72] biofilm reactors based on immobilized microorganisms,[73] or practical biological treatment facilities combined with conventional treatment processes.[74,75] The other depends on specific micro-organisms that possess effective microcystin-degrading activities. For instance, bacteria such as *Sphingpoyxis* spp. LH21 from reservoir water treated with microcystins[76] and *Sphingomonas* spp. isolated from lakes[77] and reservoir water[65] were found to be capable of degrading microcystin-LR and -RR.

However, these methods may not be sufficient to treat surface waters due to a number of limitations.[4] Firstly, drawing surface water with cyanocystins through bioreactors (*e.g.* sand filters) to remove the microcystins is very expensive with high running costs, especially when used for large-scale treatment of surface waters, so *ex situ* microcystin removal is not practical. In addition, bioreactors based on biofilms generally require a maturation phase, the time taken for a microorganism population to establish. This maturation phase may take weeks or even months, thus making the method inconvenient for the application.[78,79] Similarly, methods employing specific microcystin-degrading microorganisms may require an acclimation period prior to degradation of the microcystins, by incubating these microorganisms under appropriate conditions after isolating them from surface waters. The longer the acclimation phase, the greater the risk of exposure of animals and humans to the dangers of these toxins.[70] Moreover, some microcystin-degrading microorganisms are unable to entirely adapt to the actual conditions of the surface waters, and thus eventually die.[4]

Periphyton communities refer to the attached floral and faunal microorganisms that grow on submerged surfaces. Periphyton is essentially a kind of biofilm growing in surface waters, mainly consisting of algae, epiphytes, bacteria, protozoa, metazoans, rotifera, crustaceans, eelworms, mollusks, and insect larvae.[80] In freshwater, periphyton biofilms are also more commonly found than other biofilms.[80,81] They are native to surface waters and freshwaters, so will require shorter maturation and acclimation times in such a water matrix. Furthermore, the production site of microcystins is generally in surface waters during harmful cyanobacterial blooms. Thus, using periphyton originating from surface waters to remove microcystins has more practical significance. A similar material, phototrophic biofilm from river ecosystems, has been used to remove microcystins and shows positive effects on microcystin removal.[82] The composition of this biofilm was relatively simple, mostly dominated by phototrophic microorganisms such as the coccal green alga *Oocystis lacustirs* or composed of the colonial green alga *Protoderma* spp. and the diatoms *Navicula* spp. and *Achnanthes* spp.[82] The results from Babica's study[82] showed that the presence of cyanobacterial cells and biofilms is crucial for removal of dissolved MCs (Figure 9.8).

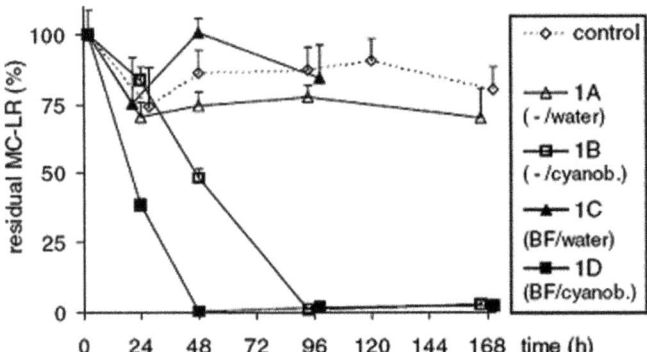

Figure 9.8   The effects of cyanobacterial biomass and biofilms (BF) on the kinetics of MC-LR removal in a microcosm study: (1A) microcosm containing water without cyanobacteria; (1B) microcosm containing water with cyanobacteria; (1C) microcosm with biofilms cultivated in the water without cyanobacteria; (1D) microcosm with biofilms cultivated in the presence of cyanobacteria; (control) aquaria with tap water. Data represent mean of parallel experiments; error bars represent standard deviation.[82]

Such simple biofilm compositions investigated by Babica *et al.*[82] may be susceptible to conflicts with variable conditions of surface waters and hard to form a stable and self-recycling micro-ecosystem in surface waters.[4] Therefore, in our experiments we examined the MC removal effectiveness of a more complex biofilm, periphyton, composed of heterotrophic and phototrophic microorganisms. The objectives of our study[4] were to determine whether MC-RR can be removed by periphyton biofilms effectively, and to provide a clearer insight into the individual adsorption and biodegradation removal mechanisms of MC-RR by discriminating the removal achieved by each process. The study showed that the periphyton community comprised mostly the bacteria and diatoms *Melosira varians* Ag., *Gomphonema parvulum* Kütz., *Synedra ulna* Kütz., *Nitzschia amphibia* Grun., and *Fragilaria vaucheriae* Kütz. The removal rates of MC-RR by different biomasses of periphyton through adsorption and biodegradation are shown in Figure 9.9. The removal rates over 19 days were 64.9, 82.8, 94.8, and 99.7% for the treatment groups with 1.32, 3.96, 6.60, and 9.24 g of periphyton biomass, respectively. The slope of the dynamic curves of MC-RR removal for each treatment group in each experimental period (every day) was calculated. The slope of the change curve of MC-RR removal for each treatment group in the period from 0 to 1 day was greater than that of any other experimental period (see Figure 9.9).[4] This showed that the average removal rate of MC-RR in the first day was the highest. The overall average removal rates of MC-RR in the first day were 18.1, 27.9, 36.3, and 43.5% of the initial MC-RR content for the treatment groups with 1.32, 3.96, 6.60, and 9.24 g of periphyton biomass, respectively, individually accounting for 27.9, 33.7, 38.3, and 43.6% of the total MC-RR removal amount.

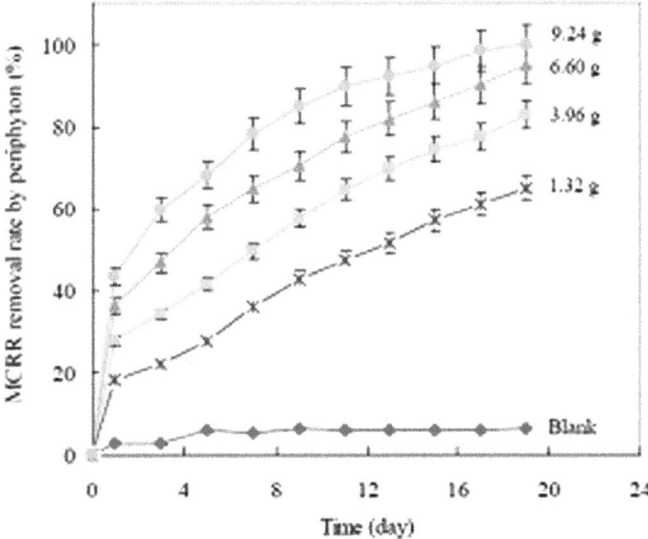

Figure 9.9 Removal rates of MC-RR by different biomasses of periphyton without added NaN₃ (including adsorption and biodegradation processes).[4]

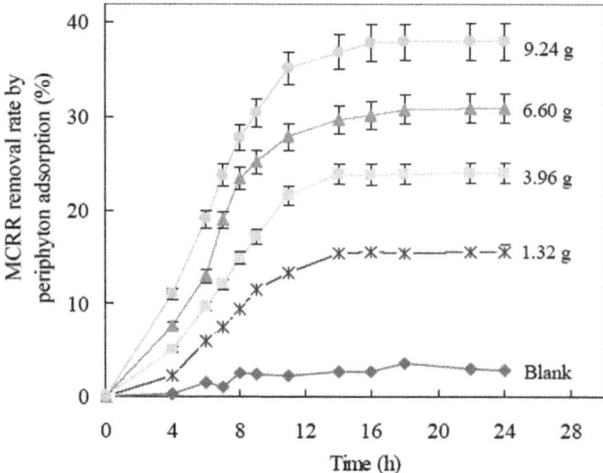

Figure 9.10   Removal rates of MC-RR adsorbed by different biomasses of periphyton with added NaN$_3$ (periphyton respiration was kept at 0 during the experimental period).[4]

The removal rates of MC-RR by different biomasses of periphyton through adsorption[4] are shown in Figure 9.10. The average removal rates of MC-RR through adsorption within 24 h were 15.5, 23.4, 30.3, and 37.6% for the treatment groups with 1.32, 3.96, 6.60, and 9.24 g of periphyton biomass, respectively, which accounted for 85.2, 73.3, 83.5, and 86.5% of the total MC-RR removal amount (through adsorption and degradation) in the first day. This indicated that MC-RR removal was dominated by periphyton adsorption in the first day. Thereafter, the removal of MC-RR by periphyton was largely dependent on biodegradation.

9.2.2.4   Removal of Heavy Metals

Many studies have shown that periphyton can be monitors for heavy metal pollution in water.[83–85] However, heavy metals affect the periphyton biofilm, so the biofilm adapts to the heavy metal pollution and accumulates these heavy metals. For example, periphyton has high Cd, Cr, and Pb accumulation capacities. Both Cr and Pb reduced the levels of Cd sequestrated by periphyton communities. The closer the frequency and duration of the pulse to continuous exposure, the greater the effects of the contaminant on periphyton growth and metal bioaccumulation (Figure 9.11).[83] In addition, the interaction of heavy metals has significant potential to affect the heavy metal bioaccumulation by periphyton biofilm. For instance, Cr and Pb decrease the toxic effects of Cd on periphyton communities, leading the bioaccumulation amount to decrease.[83]

Metal removal by microorganisms is a complex process that depends on the chemistry of the metal ions, the cell wall composition of the microorganisms, the cell physiology, and physicochemical factors such as pH, temperature,

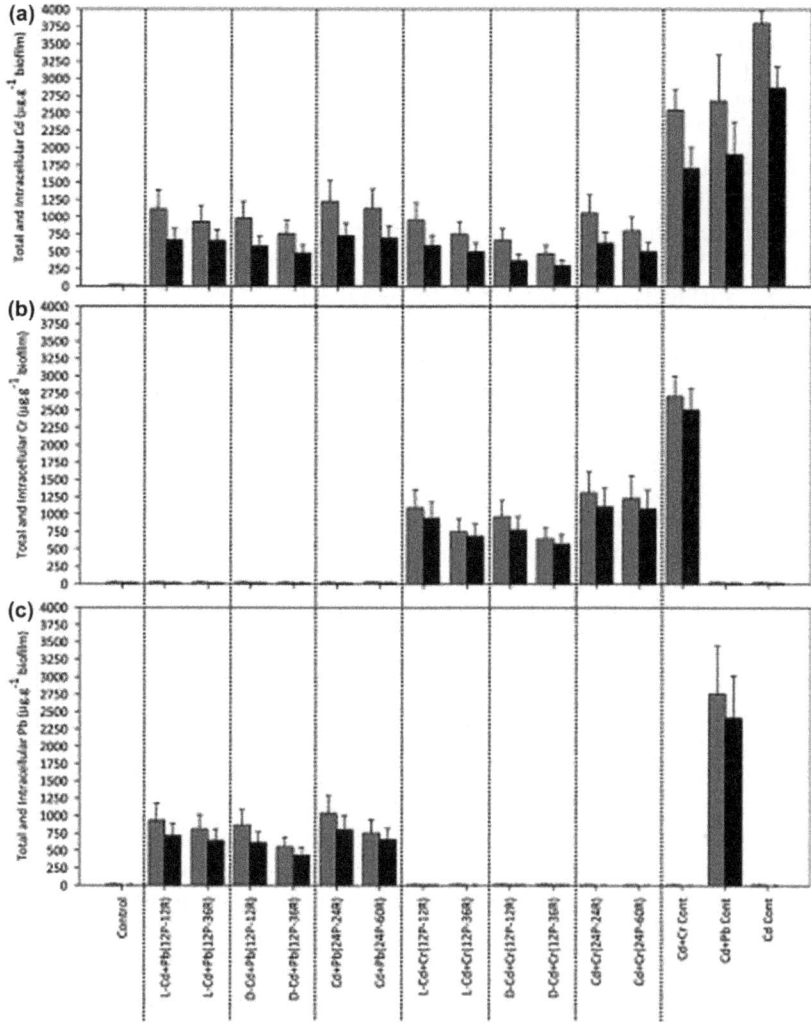

Figure 9.11 Periphyton accumulation of total (*gray bars*) and non-exchangeable (*black bars*) for (a) Cd, (b) Cr(III), and (c) Pb recorded for 16 experimental treatments. Error bars are standard deviation of three replicates; L = light, D = dark, cont = continuous, P = pulse, R = recovery, and numbers before P and R represent time (h).[83]

contact time, ionic strength, and metal concentration.[86–88] Heavy metal removal is probably related to the structure of microbial aggregates such as periphyton biofilms.[11] Owing to the special porous structure of periphyton biofilms, the dynamics of pollutants adsorbed onto or desorbed from the active sites of an aggregate surface can occur concomitantly;[4] this is the case with Cd, Cu, and Pb ions that are freely shuttled into and out of *Ralstionia* spp. and *Bacillus* spp. aggregates.[89]

Among the mechanisms for removing heavy metals by periphyton biofilms, complexation plays an important role. Many functional groups in the extracellular polymeric substances (EPS) of periphyton biofilms, such as carboxyl, phosphoric, thiol, phenolic, and hydroxyl groups, can complex with heavy metals.[23,90] For example, when the biofilms mainly consist of filamentous cyanobacteria (*Phormidium morphotype*) and heterotrophic bacteria (including sulfate-reducing bacteria) and there is calcite supersaturation at the biofilm surface, there is a potential amount of Ca^{2+} bound by EPS *via* complexation.[91] When *Pseudomonas aeruginosa* PAO1 biofilms (attached to *Sepharose* surfaces) were subjected to dissolved Fe^{3+}, most of the iron was removed from solution within 25 h by surface complexation with negatively charged functional groups on the bacterial cell wall *via* a nucleation and mineralization process. Chemical formation of hydrate iron oxides was partially responsible for the dissolved Fe removal.[92] The molecular speciation of Zn within the biofilm was examined with Zn K-edge extended X-ray absorption fine structure (EXAFS) spectroscopy. Zinc sorption to the biofilm was attributed to predominantly Zn–phosphoryl (85 ± 10 mol%) complexes, with a smaller contribution to sorption from carboxyl-type complexes (23 ± 10 mol%). The results of this study spectroscopically confirmed the importance of phosphoryl functional groups in Zn sorption by a bacterial biofilm at neutral pH.[93]

During the complexation period, the microbial composition of the periphyton biofilm employed also plays an important role to remove heavy metals.[11] Some Gram-negative bacterial strains, such as *Acinetobacter calcoaceticus*, *Erwinia herbicola*, *Pseudomonas aeruginosa*, and *P. maltophilia*, have a high affinity for gold biosorption, as do *P. maltophilia* cells immobilized with polyacrylamide gel.[94] *P. aeruginosa* ASU 6a (Gram negative) aggregates and *Bacillus cereus* AUMC B52 (Gram positive) aggregates are inexpensive and efficient biosorbents for Zn(II) removal from aqueous solutions.[75] *Escherichia coli* is an effective bacterial biosorbent used for the removal of multiple heavy metals, such as Pb, Cu, Cd, and Zn.[95]

The ion exchange mechanism is the main mode of interaction between some divalent cations and the EPS.[11,23] It has been reported that the binding between the EPS and divalent cations, such as Ca^{2+} and Mg^{2+}, is one of the main intermolecular interactions supporting microbial aggregate structures. During the removal of metals by microbial aggregate, Ca^{2+} and Mg^{2+} are simultaneously released into solution, indicating that ion exchange is involved.[11,96] Solid–liquid separation mechanisms, such as flocculation and/or precipitation, are important processes employed by periphyton biofilms in removing heavy metal ions from wastewater.[11,89] For example, a brewer's yeast strain (*Saccharomyces cerevisiae*) was used to remove heavy metals (Cu^{2+}, Ni^{2+}, Zn^{2+}, Cd^{2+}, and Cr^{3+}) from a synthetic effluent. The solid–liquid separation process was carried out using the flocculation ability of the strain. The results demonstrated that flocculation by yeast strains can be used as an inexpensive and natural separation process to remove heavy metals from a wide range of industrial effluents.[11,97]

During heavy metal adsorption, ions can be isolated by adsorption onto the EPS from microorganisms and periphyton biofilms.[11,89] Natural and extreme acidic eukaryotic biofilms have a strong binding capacity for heavy metals, such as Hg(II), Zn, Cu, Co, Ni, As, Cd, Cr, and Pb, by releasing colloid materials such as proteins, or by affecting the ion value (*e.g.* the transformation of Hg^{2+} to Hg^0).[11,89,98,99] This indicates that the EPS play an important role during the removal of heavy metals by microbial aggregate adsorption. The EPS characteristics may significantly affect the chemical forms, mobility, bioavailability, and ecotoxicity of heavy metals in aqueous solutions.[11]

The efficiency of heavy metal adsorption by periphyton biofilms is affected by many factors, including biological composition, chemical composition, functional groups, and pH.[11] For example, *Escherichia coli* hosts able to over-express metal-binding proteins (MerP) originating from Gram-positive (*Bacillus cereus* RC607) and Gram-negative (*Pseudomonas* spp. K-62) bacterial strains were used to adsorb Ni^{2+}, Zn^{2+}, and Cr^{3+} from aqueous solutions.[100] Mammalian and fish metallothioneins expression in *E. coli* aggregates leads to a significant increase (5–210%) in the overall efficiency of biosorption of Pb, Cu, Cd, and Zn.[82] Recombinant *E. coli* expressing human metallothionein protein was immobilized with poly(vinyl alcohol) (PVA) for the removal of Cd from solution. The adsorption ability was strongly affected by pH, with optimal performance at pH 5.0, while it was less sensitive to temperature over the range of 20–42 °C (Figures 9.12, 9.13).[101]

The metabolism-independent process of metal binding to the cell walls of bacterial biomass and external surfaces is the main mechanism present in the

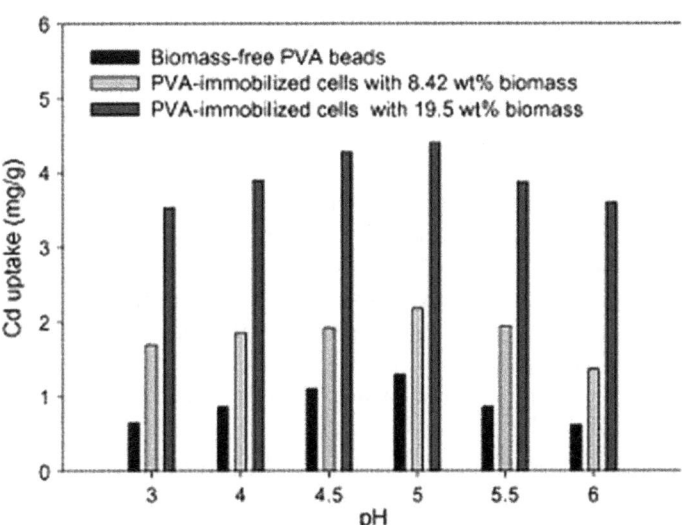

Figure 9.12 Effect of pH on biosorption of Cd^{2+} using cell-free and cell-entrapped PVA gel beads (biomass loading was 8.42 and 19.5 wt%, respectively). Initial Cd^{2+} concentration was 100 mg L^{-1}. The incubation temperature and agitation rate was 30 ± 1 °C and 75 rpm, respectively.[101]

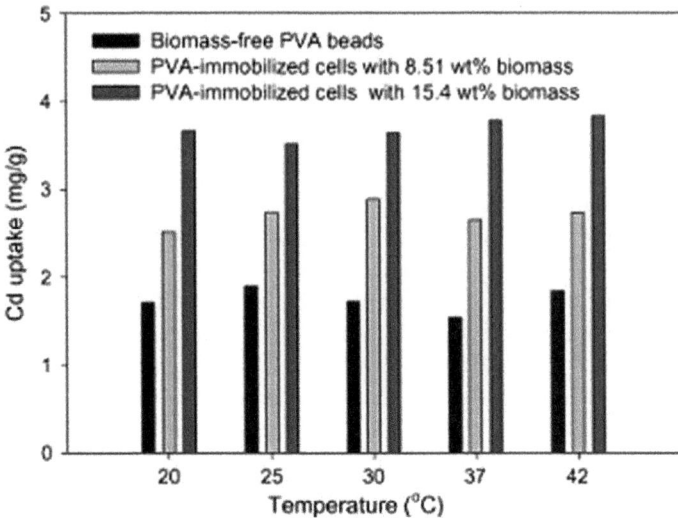

Figure 9.13 Effect of temperature on biosorption of Cd^{2+} using cell-free and cell-entrapped PVA gel beads (biomass loading was 8.51 and 15.36 wt%, respectively). Initial Cd^{2+} concentration was 100 mg L^{-1}. The incubation pH and agitation rate was 5.0 and 75 rpm, respectively.[101]

case of nonliving biomass; it involves an adsorption process, such as ionic and physicochemical adsorption.[88] A variety of functional groups located on the bacterial cell wall are known to be included in metal biosorption. These include carboxyl, amine, hydroxyl, phosphate, and thiol groups.[88] The mechanism of metal biosorption by bacterial biomass occurs through complexation, co-ordination, physical adsorption, chelation, ion exchange, inorganic precipitation, and/or a combination of these processes.[102–104] Such processes include the active participation of several anionic ligands present on the biomass, like phosphoryl, carboxyl, carbonyl, thiol, and hydroxyl groups, to immobilize the metal ions.[88,105] For instance, the presence of amino, carboxyl, hydroxyl, and carbonyl groups led to greater zinc biosorption by a Gram-negative bacterium (*P. aeruginosa*) with respect to that of a Gram-positive bacterium (*B. cereus*).[88]

The major advantages of biosorption by periphyton biofilms are their high effectiveness in reducing heavy metal ions and the use of inexpensive biosorbents. Microbial aggregate biosorption processes are particularly suitable for treating dilute heavy metal wastewater[106] and may be used under various conditions owing to the complex composition of heterotrophic and photo-autotrophic microorganisms. Microbial biomass heavy metal biosorbents characteristically exhibit environmental safety, broad sources, low cost, short maturation and acclimation periods, and rapid adsorption. These periphyton biofilms have vast potential in removing heavy metals under various conditions because of the hierarchical and self-maintained micro-ecosystem that is established in periphyton biofilms. Moreover, these microbial biomasses can be cultivated and fostered in wastewater treatment systems such as bioreactors, which not only improves the heavy metal removal efficiency but also maintains the microbial aggregate micro-ecosystems in a steady state.[19]

9.3 Relationship of Phosphorus Release, Cyanobacterial Bloom, and Periphyton Biofilms

9.3.1 Inhibition of Phosphorus Release from Sediments

Eutrophication of most freshwater systems is limited by nutrients such as the phosphorus concentration. High P concentrations originate from external and internal sources.[107] Therefore, information on the nutrient status, mainly phosphorus, is essential to predict the development of a lake under present or future P loading and to forecast the effect of additional restoration measures. This nutrient status consists of the nutrients available in the water and in the top layer of the sediment. The amount of P in the overlying water is generally small compared to the amount of P in the top layer of sediment.[108] Thus, it is practically significant to control the phosphorus release from sediments.

To test the effects of periphyton biofilms on phosphorus release from sediments, an outdoor experiment was conducted. The phosphorus release process at the interface of the sediment and overlying water was simulated in field microcosms. The microcosm was a $120 \times 100 \times 100$ cm rectangular glass tank. The sediment thickness was *ca.* 15 cm while the overlying water depth was *ca.* 85 cm. The microcosms were installed in the open air to obtain natural light. The sediments and water were collected from a eutrophic lake. The sediments were laid at the bottom of the microcosms and then the water sample was added to the microcosms. The periphyton was cultured onto the two types of substrates (with a diameter of 12 cm and length of 55 cm), namely artificial aquatic mats (AAM) and industrial soft carriers (ISC). After the microcosms were left static for one day, the periphyton together with its substrate was fixed to the water surface under 30 cm. The periphytons on the AAM and ISC surfaces were designated as AAM-periphyton and ISC-periphyton, respectively. Each microcosm had 1600 g of fresh periphyton (weight at 25–30 °C, wetness $85 \pm 5\%$) for both the AAM-periphyton and ISC-periphyton samples. The control sample did not contain any periphyton. The dissolved oxygen was maintained in the range from 8.5 to 9.5 mg L^{-1} by continuous aeration. The temperature was coincident with the ambient temperature, ranging from 22 to 38 °C.

Figure 9.14 shows the change of the average TP and Exch-P concentrations of the control and periphyton samples in the sediments. The TP and Exch-P concentrations of the control samples were lower than those of the periphyton samples. The TP and Exch-P concentrations dropped continuously from the beginning of the experiment. They decreased more dramatically from the 17th day. The average TP concentrations in the sediments were reduced from 11.8 mg g^{-1} of original concentration to 10.8 and 9.3 mg g^{-1} for the 17th and 36th days, respectively (Figure 9.14). The TP and Exch-P concentrations in the

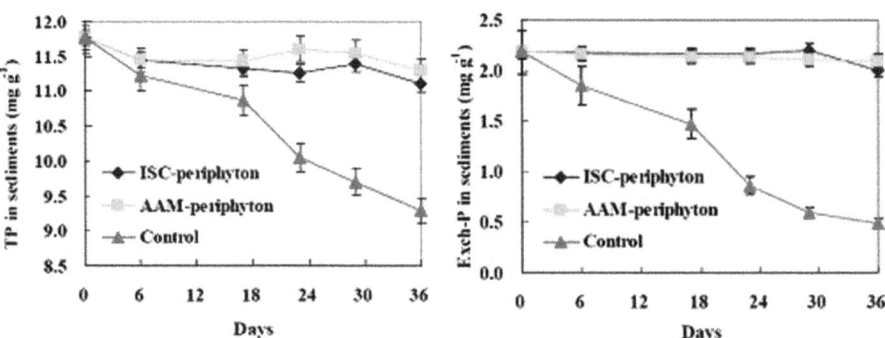

Figure 9.14   Changes of TP and Exch-P with time in the sediments.[12]

AAM- and ISC-periphyton treatments also decreased with experimental time, but in a very insignificant fashion.

The whole average reduced concentrations of Exch-P in the sediments for each treatment (AAM- and ISC-periphyton) were significantly different from that of the control ($p < 0.05$), which indicated that the introduction of the periphytons could positively delay the Exch-P release from sediments into the overlying water. The Fe/Al-P was the main composition of Exch-P, accounting for 86.4% at the beginning of the experiment. These results showed that the periphytons decreased the phosphorus content in the overlying water through delaying phosphorus release.

9.3.2   Control of Cyanobacterial Bloom

Many studies have been conducted to evaluate the allelopathic effects of single phototrophic species or single-species biofilm formed by phytoplankton such as cyanobacteria (*Phormidium, Lyngbya, Oscillatoria, Microcoleus, Aphanothece,* and *Scytonema*),[109,110] diatoms (*Diatoma*),[111] benthic/epiphytic alga (*Uronema confervicolum*),[112] Alexandrium (*Alexandrium tamarense*), Gymnodiniales (*Karenia mikimotoi*), and Prymnesiophyceae (*Chrysochromulina polylepis*).[113] In addition, many studies have been done on the allelopathic effects of heterotrophic microorganisms (or a biofilm consisting of heterotrophic microorganisms) in aquatic ecosystems. For instance, algal-lysing fungi were isolated from the different aquatic habitat types that were sampled. The remaining isolates comprised four bacteria and four streptomycetes. All these isolates lysed *Anabaena flos-aquae* and, in most cases, several other filamentous and unicellular cyanobacteriae.[114] Bacteria, namely *Alphaproteobacteria, Betaproteobacteria, Gammaproteobacteria, Bacteroidetes, Firmicutes,* and *Actinobacteria,* isolated from the Baltic Sea, Germany, showed positive anti-microbial activity.[110,115]

However, the integrated power of all microbial aggregates comprising the periphyton biofilm has often been underestimated or ignored.[110] The organisms in biofilms create their own microhabitats with pronounced gradients of biological and chemical parameters, which enable the efficient and effective use of substrates and energy.[116] Therefore, considering the integrated ecological impact of periphyton biofilms such as allelopathic effects on harmful bacteria, phytoplankton should have more practical significance.

Unfortunately, little information is available on the allelopathic effects of multi-community biofilms on harmful algal growth. Our research group conducted a study[110] to determine whether allelo chemicals released from periphyton biofilm inhibit the growth of cyanobacterial blooms (*Microcystis aeruginosa*) and the detailed inhibitory mechanism on cyanobacterial growth. In this study, the chlorophyll-*a* concentration in the control simulated pond was $250 \, \mu g \, L^{-1}$ on day 17 of the experiment. Microscopic analysis of water samples from this pond revealed that the algae were cyanobacteriae. After 36 days, the water in the control pond was completely covered with a thick layer of cyanobacteriae (Figure 9.15). The statistics showed that the chlorophyll-*a* concentrations in the control pond were significantly different from the ponds with the periphyton biofilm, indicating that the absence of cyanobacterial bloom occurred during the growth of the periphyton biofilm. Despite this decrease of the average concentrations of nitrate, ammonia, and total dissolved phosphorus, the average concentrations of ammonia, nitrate, and total dissolved phosphorus in the overlying water were 0.54, 0.67, and $0.12 \, mg \, L^{-1}$ at the end of the experiment, respectively, which is sufficient to support the growth of cyanobacteriae.[117,118]

Figure 9.16 shows that the growth of cyanobacteria was inhibited when cyanobacteria were cultured in BG11 medium prepared using the extract of periphyton biofilm instead of distilled water. However, the cyanobacteria in the cyanobacterial monoculture grew well. This implied that there are some compounds, named allelochemicals, in the extract of the periphyton biofilms responsible for inhibiting cyanobacterial growth.

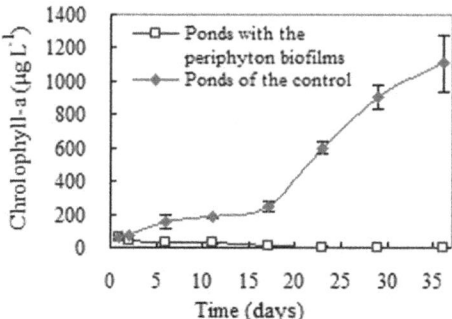

Figure 9.15   Changes in chlorophyll-*a* concentrations in the control and treatment overlying water. An extract of the periphyton biofilm was used to culture cyanobacteria to determine whether some compounds from the periphyton biofilm extract had inhibited cyanobacterial growth.[110]

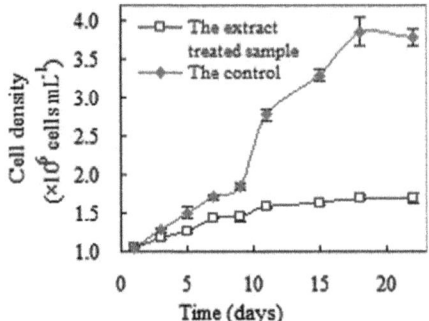

Figure 9.16   Changes in the cell counting of cyanobacteria in control (cyanobacterial monoculture in BG11 medium prepared using distilled water) and biofilm extract treatment (cyanobacteria cultured in BG11 medium prepared using the extract from periphyton biofilm).[110]

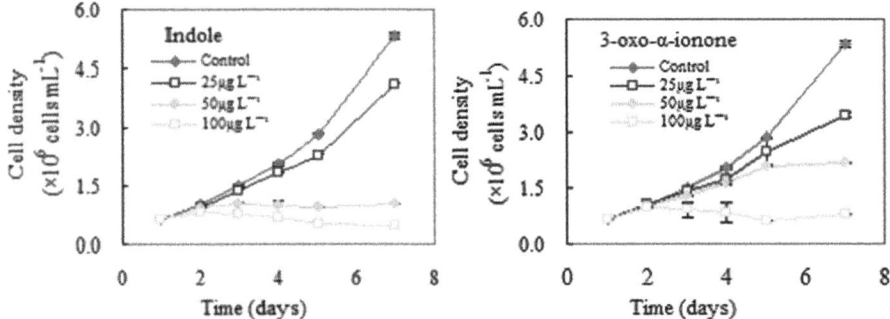

Figure 9.17   Inhibitory effects of indole and 3-oxo-α-ionone on the growth of cyanobacteria.[110]

A total of 30 identified and two unidentified compounds were isolated from the water extract of the periphyton biofilm. Indole and 3-oxo-α-ionone are two micro-water-soluble organic compounds.[119] Both organic compounds are widespread materials in nature that are often found in terrestrial plants, phytoplankton, and macrophytes.[120–122] In addition, types of indole and ionone had exhibited toxicity against certain plankton,[123] such as cyanobacteria.[124] Therefore, a bioassay of 3-oxo-α-ionone and indole on cyanobacterial growth was conducted, which showed that both compounds positively affected the growth of cyanobacteria. The growth of cyanobacteria began to be markedly inhibited when the allelochemical concentration was more than 50 and 100 μg L^{-1} for indole and 3-oxo-α-ionone, respectively (Figure 9.17).

To investigate the parts of cyanobacterial cells stressed by indole and 3-oxo-α-ionone, the cyanobacterial cells were characterized using TEM. The thylakoid membrane separated from the cyanobacterial cytoplasm and other inner materials were lost with a higher dose and with an increase in contact time (Figure 9.18).

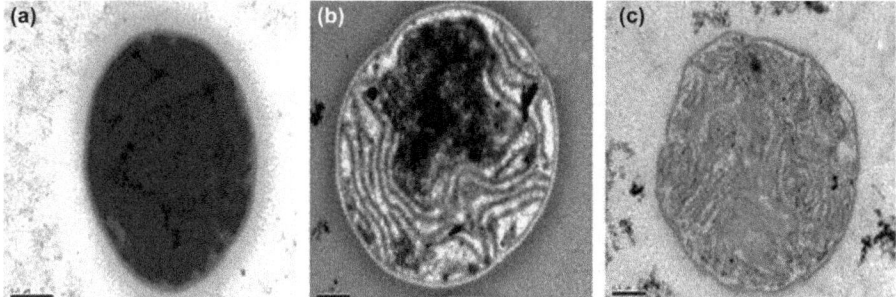

Figure 9.18 (a) Control for cyanobacterial monoculture; the cell structure of the cyanobacteria was intact. (b) Bioassay with 3-oxo-α-ionone; the thylakoid membrane separated from the cyanobacterial cytoplasm. (c) Bioassay with indole; many inner materials in the cyanobacterial cell and thylakoid membrane were lost.[110]

Additionally, the PS II of cyanobacterial cells was investigated during the addition of 3-oxo-α-ionone and indole at a concentration of 50 μL L^{-1} (or 50 μg L^{-1}). The results showed that both effective quantum yields and electron transport rates were significantly decreased with an increase in 3-oxo-α-ionone and indole concentrations. These findings indicate that electron transport in PS II was interrupted by 3-oxo-α-ionone and indole, which leads to the disfunction of PS II in cyanobacterial cells and to the death of the cyanobacteria. This study confirmed that the allelopathy between natural biofilms and cyanobacteria does actually exist, which can control cyanobacterial growth by certain allelochemicals produced by periphyton biofilms, and paves the way for research on allelopathy in aquatic ecosystems from single-single species (community) to multi-community/single-species fields. The use of environmentally friendly and cost-effective ubiquitous periphyton biofilms has an even higher probability to prevent the adverse effects associated with harmful algal blooms in eutrophic surface waters.

9.4 Potential of Periphyton Biofilm Applications

The severity of the worldwide water crisis has prompted the United Nations to conclude that it is water scarcity, not a lack of arable land, that will be the major constraint to increased food production over the next few decades.[125] For instance, Australia is one of the major food producing and land abundant countries, but recent drought has reduced its agricultural and food production substantially.[126] At the same time, water, especially surface water, is facing pollution problems all over the world. For example, water pollution from small rural industries is a serious problem throughout China. Over half of all river sections monitored for water quality are rated as being unsafe for human contact, and this pollution is estimated to cost several per cent of GDP.[127] This problem creates a huge market for newer and safer devices to make water clean.

Periphyton biofilm, a green material, serves as an environmentally friendly means to clean water. It is well known that bioremediation using micro-organisms in aquatic ecosystems is much safer compared to physical and chemical methods. The microorganisms for periphyton biofilms can be collected from local surface waters. Under certain artificial conditions, these collected microorganisms are inoculated and cultured to be periphyton biofilms and then immobilized on substrates in polluted water. Periphyton biofilms show high efficiency in removing COD, UV$_{254}$ nm matter, Cu, Zn, Fe, P, and N.[128,129] The complex composition of periphyton biofilms enables the eco-system of periphyton communities in a relatively steady and balanced status because the self-recycling food web in periphyton communities maintains its material recycle and energy transformation.[130,131] This pyramidal composition structure improves the living abilities of microorganisms in periphyton biofilms and makes them able to adapt to different conditions. As a result, the application of periphyton biofilms needs less maturation and acclimation time,[4] saving the operation time and cost.

Periphyton biofilms can easily be immobilized on their substrates in polluted aquatic ecosystems or reactors.[12,128,129] Therefore it is possible that bioremediation based on periphyton biofilms can be widely applied in some underdeveloped countries or areas having lack of enough professionals. Overall, the application of periphyton biofilms to improve aquatic ecosystem health is to meet the "4-E" concepts: efficient, economical, environment friendly, and easily produced. It is very much possible that periphyton biofilms, an environmentally benign green material, may be employed widely and may dominate the market of water and wastewater bio-treatment in the near future.

References

1. M. E. Azim, in *Encyclopedia of Inland Waters*, ed. G. E. Likens, Academic, Oxford, 2009, p. 184.
2. M. E. Azim, M. C. J. Verdegem, A. A. V. Dam and M. C. M. Beberidge, in *Periphyton: Ecology, Exploitation and Management*, ed. M. E. Azim, M. C. J. Verdegem, A. A. V. Dam and M. C. M. Beberidge, CABI, Wallingford, UK, 2005, p. 1.
3. M. E. Azim and T. Asaeda, in *Periphyton: Ecology, Exploitation and Management*, ed. M. E. Azim, M. C. J. Verdegem, A. A. V. Dam and M. C. M. Beberidge, CABI, Wallingford, UK, 2005, p. 14.
4. Y. Wu, J. He and L. Yang, *Environ. Sci. Technol.*, 2010, **44**, 6319.
5. S. T. Larned, *J. N. Am. Benthol. Soc.*, 2010, **29**, 182.
6. S. Arnon, A. I. Pakman, C. G. Peterson and K. A. Gray, *J. Geophys. Res.*, 2007, **112**, G01002; doi: 01010.01029/02006JG000235.
7. J. R. Holomuzki, J. W. Feminella and M. E. Power, *J. N. Am. Benthol. Soc.*, 2010, **29**, 220.
8. S. Dolédec and B. Statzner, *J. N. Am. Benthol. Soc.*, 2010, **29**, 286.

9. P. J. Mulholland and J. R. Webster, *J. N. Am. Benthol. Soc.*, 2010, **29**, 110.

10. B. Statzner, J. A. Gore and V. H. Resh, *J. N. Am. Benthol. Soc.*, 1988, **7**, 307.

11. Y. Wu, T. Li and L. Yang, *Bioresour. Technol.*, 2012, **107**, 10.

12. Y. Wu, S. Zhang, H. Zhao and L. Yang, *Bioresour. Technol.*, 2010, **101**, 9681.

13. K. E. Havens, T. L. East, J. A. Rodusky and B. Sharfstein, *Aquat. Bot.*, 1999, **63**, 267.

14. C. Yariv, *Bioresour. Technol.*, 2001, **77**, 257.

15. K. Sasaki, Y. Yamamoto, K. Tsumura, S. Ouchit and Y. Mori, *Water Sci. Technol.*, 1996, **34**, 111.

16. S. Villaverde, *Rev. Environ. Sci. Biotechnol.*, 2004, **3**, 171.

17. M. C. M. Van Loosdrecht and M. S. M. Jetten, *Water Sci. Technol.*, 1998, **38**, 1.

18. Y. Wu, J. He, Z. Hu, L. Yang and N. Zhang, *J. Hazard. Mater.*, 2011, **194**, 1.

19. Y. Wu, Z. Hu, P. G. Kerr and L. Yang, *Bioresour. Technol.*, 2011, **102**, 736.

20. V. A. P. M. dos Santos, M. Bruijnse, J. Tramper and R. H. Wijffels, *Appl. Microbiol. Biot.*, 1996, **45**, 447.

21. V. H. Smith, G. D. Tilman and J. C. Nekola, *Environ. Pollut.*, 1999, **100**, 179.

22. Z. Aksu, *Process Biochem.*, 2005, **40**, 997.

23. G. P. Sheng, H. Q. Yu and X. Y. Li, *Biotechnol. Adv.*, 2010, **28**, 882.

24. P. V. McCormick, R. B. E. Shuford III and M. J. Chimney, *Ecol. Eng.*, 2006, **27**, 279.

25. L. J. Sainto and K. R. Reddy, *Aquat. Bot.*, 2003, **77**, 203.

26. R. J. Chiou and Y. R. Yang, *Bioresour. Technol.*, 2008, **99**, 4408.

27. A. Oehmen, P. C. Lemos, G. Carvalho, Z. Yuan, J. Keller, L. L. Blackall and M. A. M. Reis, *Water Res.*,**41**, 2271.

28. S. Zhou, X. Zhang and L. Feng, *Bioresour. Technol.*, 2010, **101**, 1603.

29. S. Tsuneda, T. Ohno, K. Soejima and A. Hirata, *Biochem. Eng.*, 2006, **27**, 191.

30. P. V. McCormick, M. B. O'Dell, R. B. E. Shuford, J. G. Backus and W. C. Kennedy, *Aquat. Bot.*, 2001, **71**, 119.

31. S. V. Mohan, N. C. Rao and P. N. Sarma, *J. Hazard. Mater.*, 2007, **144**, 108.

32. G. S. Veeresh, P. Kumar and I. Mehrotra, *Water Res.*, 2005, **39**, 154.

33. J. Dosta, J. M. Nieto, J. Vila, M. Grifoll and J. Mata-Álvarezlvarez, *Bioresour. Technol.*, 2011, **102**, 4013.

34. B. Marrot, A. Barrios-Martinez, P. Moulin and N. Roche, *Biochem. Eng. J.*, 2006, **30**, 174.

35. X. Zhang, W. Huang, X. Wang, Y. Gao and H. Lin, *J. Environ. Sci.*, 2009, **21**, 1181.

36. J. H. Kang, Y. Katayama and F. Kondo, *Toxicology*, 2006, **217**, 81.

37. E. Kurzbaum, F. Kirzhner, S. Sela, Y. Zimmels and R. Armon, *Water Res.*, 2010, **44**, 5021.
38. Y. Wang, Y. Tian, B. Han, H. Zhao, J. Bi and B. Cai, *J. Environ. Sci.*, 2007, **19**, 222.
39. Y. Jiang, J. Wen, J. Bai, D. Wang and Z. Hu, *Biochem. Eng. J.*, 2006, **29**, 227.
40. V. L. Santos and V. R. Linardi, *Process Biochem.*, 2004, **39**, 1001.
41. G. Bleve, C. Lezzi, M. A. Chiriatti, I. D'Ostuni, M. Tristezza, D. Di Venere, L. Sergio, G. Mita and F. Grieco, *Bioresour. Technol.*, 2011, **102**, 982.
42. A. Tsioulpas, D. Dimou, D. Iconomou and G. Aggelis, *Bioresour. Technol.*, 2002, **84**, 251.
43. N. Hiratsuka, M. Oyadomari, H. Shinohara, H. Tanaka and H. Wariishi, *Biochem. Eng. J.*, 2005, **23**, 241.
44. J. H. Kang and F. Kondo, *Arch. Environ. Contam. Toxicol.*, 2002, **43**, 265.
45. J. W. Voordeckers, D. E. Fennell, K. Jones and M. M. Haggblom, *Environ. Sci. Technol.*, 2002, **36**, 696.
46. L. Leven and A. Schnürer, *Int. Biodeterior. Biodegrad.*, 2005, **55**, 153.
47. J. H. Kang and F. Kondo, *Chemosphere*, 2002, **49**, 493.
48. G. M. Klecka, S. J. Gonsior, R. J. West, P. A. Goodwin and D. A. Markham, *Environ. Toxicol. Chem.*, 2001, **20**, 2725.
49. A. Kumar, S. Kumar and S. Kumar, *Biochem. Eng. J.*, 2005, **22**, 151.
50. S. J. Wang and K. C. Loh, *Enzyme Microb. Technol.*, 1999, **25**, 177.
51. A. A. M. G. Monteiro, R. A. R. Boaventura and A. E. Rodrigues, *Biochem. Eng. J.*, 2000, **6**, 45.
52. T. P. Chung, H. Y. Tseng and R. S. Juang, *Process Biochem.*, 2003, **38**, 1497.
53. R. S. Juang and S. Y. Tsai, *Water Res.*, 2006, **40**, 3517.
54. Y. Li, J. Li, C. Wang and P. Wang, *Bioresour. Technol.*, 2010, **101**, 6740.
55. O. J. Hao, M. H. Kim, E. A. Seagren and H. Kim, *Chemosphere*, 2002, **46**, 797.
56. Z. Alexievaa, M. Gerginova and P. Zlateva, *Enzyme Microb. Technol.*, 2004, **34**, 242.
57. Y. Jiang, J. P. Wen, H. M. Li, S. L. Yang and Z. D. Hu, *Biochem. Eng. J.*, 2005, **24**, 243.
58. K. M. Khleifat, *Process Biochem.*, 2006, **41**, 2010.
59. J. Bai, J. P. Wen, H. M. Li and Y. Jiang, *Process Biochem.*, 2007, **42**, 510.
60. G. Y. Wang, J. P. Wen, G. H. Yu and H. M. Li, *World J. Microb. Biotechnol.*, 2008, **24**, 2685.
61. M. Bajaj, C. Gallert and J. Winter, *Biochem. Eng. J.*, 2009, **46**, 205.
62. V. Arutchelvan, V. Kanakasabai, R. Elangovan, S. Nagarajan and V. J. Muralikrishnan, *J. Hazard. Mater.*, 2006, **129**, 216.
63. G. A. Codd, S. G. Bell, K. Kaya, C. J. Ward, K. A. Beattie and J. S. Metcalf, *Eur. J. Phycol.*, 1999, **34**, 405.

64. J. Fastner, G. Codd, J. Metcalf, P. Woitke, C. Wiedner and H. Utkilen, *Anal. Bioanal. Chem.*, 2002, **374**, 437.
65. A. M. Valeria, E. J. Ricardo, P. Stephan and W. D. Alberto, *Biodegradation*, 2006, **17**, 447.
66. A. R. Humpage and I. R. Falconer, *Environ. Toxicol.*, 1999, **14**, 61.
67. E. Ito, A. Takai, F. Kondo, H. Masui, S. Imanishi and K. I. Harada, *Toxicon*, 2002, **40**, 1017.
68. K. Lahti, M. Niemi, J. Rapala and K. Sivonen, in *Harmful Algae*, ed. B. Reguera, J. Blanco, M. L. Fernandez and T. Wyatt, Xunta de Galicia and Intergovernmental Oceanographic Commission of UNESCO, Paris, 1998, p. 363.
69. C. Svrcek and D. W. Smith, *J. Environ. Eng. Sci.*, 2004, **3**, 155.
70. D. G. Bourne, R. L. Blakeley, P. Riddles and G. J. Jones, *Water Res.*, 2006, **40**, 1294.
71. L. Ho, T. Meyn, A. Keegan, D. Hoefel, J. Brookes, C. P. Saint and G. Newcombe, *Water Res.*, 2006, **40**, 768.
72. H. Wang, L. Ho, D. M. Lewis, J. D. Brookes and G. Newcombe, *Water Res.*, 2007, **41**, 4262.
73. K. Tsuji, M. Asakawa, Y. Anzai, T. Sumino and K. I. Harada, *Chemosphere*, 2006, **65**, 117.
74. J. Li, K. Shimizu, H. Maseda, Z. Lu, M. Utsumi, Z. Zhang and N. Sugiura, *Bioresour. Technol.*, 2012, **106**, 27.
75. J. Li, K. Shimizu, M. Utsumi, T. Nakamoto, M. K. Sakharkar, Z. Zhang and N. Sugiura, *J. Biosci. Bioeng.*, 2011, **111**, 695.
76. L. Ho, D. Hoefel, C. P. Saint and G. Newcombe, *Water Res.*, 2007, **41**, 4685.
77. T. Saito, K. Okano, H. D. Park, T. Itayama, Y. Inamori, B. A. Neilan, B. P. Burns and N. Sugiura, *FEMS Microbiol. Lett.*, 2003, **229**, 271.
78. A. M. Bruce and H. A. Hawkes, *Biological Filters*, Academic, London, 1983.
79. S. Zhang and P. M. Huck, *Water Res.*, 1996, **30**, 456.
80. A. Sladeckova, *Bot. Rev.*, 1962, **28**, 286.
81. F. Juttner, in *Biofilms in Aquatic Environment*, ed. C. W. Keevil, A. Godfree, D. Holt and C. Dow, Royal Society of Chemistry, Cambridge, 1999, p. 43.
82. P. Babica, L. Blaha and B. Marsalek, *Environ. Sci. Pollut.*, 2005, **12**, 369.
83. T. Bere, M. A. Chia and J. G. Tundisi, *Environ. Pollut.*, 2012, **163**, 184.
84. G. J. Ramelow, R. S. Maples, R. L. Thompson, C. S. Mueller, C. Webre and J. N. Beck, *Environ. Pollut.*, 1987, **43**, 247.
85. A. Serra, N. Corcoll and H. Guasch, *Chemosphere*, 2009, **74**, 633.
86. R. M. Gabr, S. H. A. Hassan and A. A. M. Shoreit, *Int. Biodeterior. Biodegrad.*, 2008, **62**, 195.
87. S. H. A. Hassan, S. J. Kim, A. Y. Jung, J. H. Joo, S. E. Oh and J. E. Yang, *J. Gen. Appl . Microbiol.*, 2009, **55**, 27.
88. J. H. Joo, S. H. A. Hassan and S. E. Oh, *Int. Biodeterior. Biodegrad.*, 2010, **64**, 734.

89. A. Choi, S. Wang and M. Lee, *Geosci. J.*, 2009, **13**, 331.
90. D. Zhang, X. Pan, K. M. G. Mostofa, X. Chen, G. Mu, F. Wu, J. Liu, W. Song, J. Yang, Y. Liu and Q. Fu, *J. Hazard. Mater.*, 2010, **175**, 359.
91. F. Shiraishi, A. Bissett, D. de Beer, A. Reimer and G. Arp, *Geomicrobiol. J.*, 2008, **25**, 83.
92. J. U. Lee and T. J. Beveridge, *Chem. Geol.*, 2001, **180**, 67.
93. B. Toner, A. Manceau, M. A. Marcus, D. B. Millet and G. Sposito, *Environ. Sci. Technol.*, 2005, **39**, 8288.
94. T. J. Tsuruta, *Gen. Appl. Microbiol.*, 2004, **50**, 221.
95. W. C. Kao, Y. P. Chiu, C. C. Chang and J. S. Chang, *Biotechnol. Prog.*, 2006, **22**, 1256.
96. B. Yuncu, F. Sanin and U. Yetis, *J. Hazard. Mater.*, 2006, **137**, 990.
97. M. D. Machado, M. S. F. Santos, C. Gouveia, H. M. V. M. Soares and E. V. Soares, *Bioresour. Technol.*, 2008, **99**, 2107.
98. S. George, A. Kishen and K. P. Song, *J. Endod.*, 2005, **31**, 867.
99. T. R. Neu and J. R. Lawrence, in *Microbial Glycobiology*, ed. O. Holst, P. J. Brennan and M. Itzstein, Academic, San Diego, 2010, p. 733.
100. W. C. Kao, C. C. Huang and J. S. Chang, *J. Hazard. Mater.*, 2008, **158**, 100.
101. W. C. Kao, J. Y. Wu, C. C. Chang and J. S. Chang, *J. Hazard. Mater.*, 2009, **169**, 651.
102. D. Kratochvil and B. Volesky, *Trends Biotechnol.*, 2008, **16**, 291.
103. X. C. Chen, Y. P. Wang, Q. Lin, J. Y. Shi, W. X. Wu and Y. X. Chen, *Colloids Surf. B.*, 2005, **46**, 101.
104. N. Akhtar, M. Iqbal and S. I. J. ZafarIqbal, *J. Environ. Sci.*, 2008, **20**, 231.
105. B. Volesky, *Hydrometallurgy*, 2003, **71**, 179.
106. F. Fu and Q. Wang, *J. Environ. Manage.*, 2011, **92**, 407.
107. M. E. Bechmann, D. Berge, H. O. Eggestad and S. M. Vandsemb, *J. Hydrol.*, 2005, **304**, 238.
108. D. T. Van der Molen, R. Portielje, P. C. M. Boers and L. Lijklema, *Water Res.*, 1998, **32**, 3281.
109. F. Juttner and J. T. Wu, *Arch. Hydrobiol.*, 2000, **147**, 13.
110. Y. Wu, J. Liu, L. Yang, H. Chen, S. Zhang, H. Zhao and N. Zhang, *Environ. Microbiol.*, 2011, **13**, 604.
111. F. Juttner, *J. Phycol.*, 2001, **37**, 12.
112. J. Leflaive, G. Lacroix, Y. Nicaise and L. Ten-Hage, *Environ. Microbiol.*, 2008, **10**, 11.
113. G. O. Flstarol, C. Legrand, K. Rengefors and E. Granell, *Environ. Microbiol.*, 2004, **6**, 8.
114. K. Redhead and S. J. Wright, *Appl. Environ. Microbiol.*, 1978, **35**, 8.
115. W. Jutta, T. Vera, N. Kerstin, S. Tim and I. F. Johanner, *Marine Biotechnol.*, 2009, **11**, 14.
116. L. A. Meyer-Reil and M. Köster, *Marine Pollut. Bull.*, 2000, **41**, 14.
117. D. W. Schindler, *Science*, 1977, **196**, 3.
118. F. M. Yusoff and C. D. McNabb, *Aquaculture Res.*, 2008, **28**, 7.

119. P. Z. Ni, *Organic Chemistry*, People's Republic Sanitation Press, Beijing, 2007, p. 170.
120. A. N. Seal, T. Haig and J. E. Pratley, *J. Chem. Ecol.*, 2004, **30**, 6.
121. L. Blazevic and J. Mastelić, *Flavour Fragr. J.*, 2008, **23**, 7.
122. Y. Hong, H. Y. Hu, X. Xie, A. Sakoda, M. Li and F. M. Sagehashi, *Aquat. Toxicol.*, 2009, **9**, 8.
123. P. G. Becher and F. Juttner, *Environ. Toxicol.*, 2005, **20**, 10.
124. Q. M. Xian, H. D. Chen, H. L. Liu, H. X. Zou and D. Q. Yin, *Environ. Sci. Pollut. Res. Int.*, 2006, **13**, 5.
125. UNDP, *Beyond Scarcity: Power, Poverty and the Global Water Crisis*, United Nations Development Programme, New York, 2007.
126. T. Goesch, A. Hafi, M. Oliver, S. Page, D. Ashton, S. Hone and B. Dyack, *Aust. Commodities*, 2007, **14**, 10.
127. M. Wang, M. Webber, B. Finlayson and J. Barnett, *J. Environ. Manage.*, 2008, **86**, 648.
128. Y. Wu, Z. Hu, P. G. Kerr and L. Yang, *Bioresour. Technol.*, 2011, **102**, 736.
129. Y. Wu, Z. Hu, L. Yang, B. Graham and P. G. Kerr, *Bioresour. Technol.*, 2011, **102**, 2419.
130. T. A. Frankovich and J. C. Zieman, *Aquat. Bot.*, 2005, **83**, 14.
131. K. C. Pietro, M. J. Chimney and A. D. Steinman, *Ecol. Eng.*, 2006, **27**, 290.

CHAPTER 10

Remediation of Dye Containing Wastewater Using Viable Algal Biomass

SEEMA DWIVEDI*[a] AND TANVI VATS[b]

[a] School of Biotechnology, Gautam Buddha University, Greater Noida, Gautam Budh Nagar – 201310, India; [b] Department of Applied Chemistry, School of Vocational Studies and Applied Sciences, Gautam Buddha University, Greater Noida, Gautam Budh Nagar – 201310, India
*Email: seema@gbu.ac.in

10.1 Introduction

With an ever-increasing population, rapid urbanization, high industrial growth and climatic mutations, the supply of fresh water has become a limited natural resource in several parts of the world (Figure 10.1).[1] The water resources are stressed not only due to an increased demand but also because of increasing water pollution. Water is typically referred to as polluted when it is impaired by anthropogenic contaminants and either does not support a human use and/or undergoes a marked shift in its ability to support its constituent biotic communities.[2]

Water pollution occurs when pollutants are discharged directly or indirectly into water bodies without adequate treatment to remove harmful compounds. The increasing application of water in every sphere of life has not only raised water consumption but has also lead to greater contamination and generation of large amounts of wastewater.[3] We can thus justify positioning water

RSC Green Chemistry No. 23
Green Materials for Sustainable Water Remediation and Treatment
Edited by Anuradha Mishra and James H. Clark

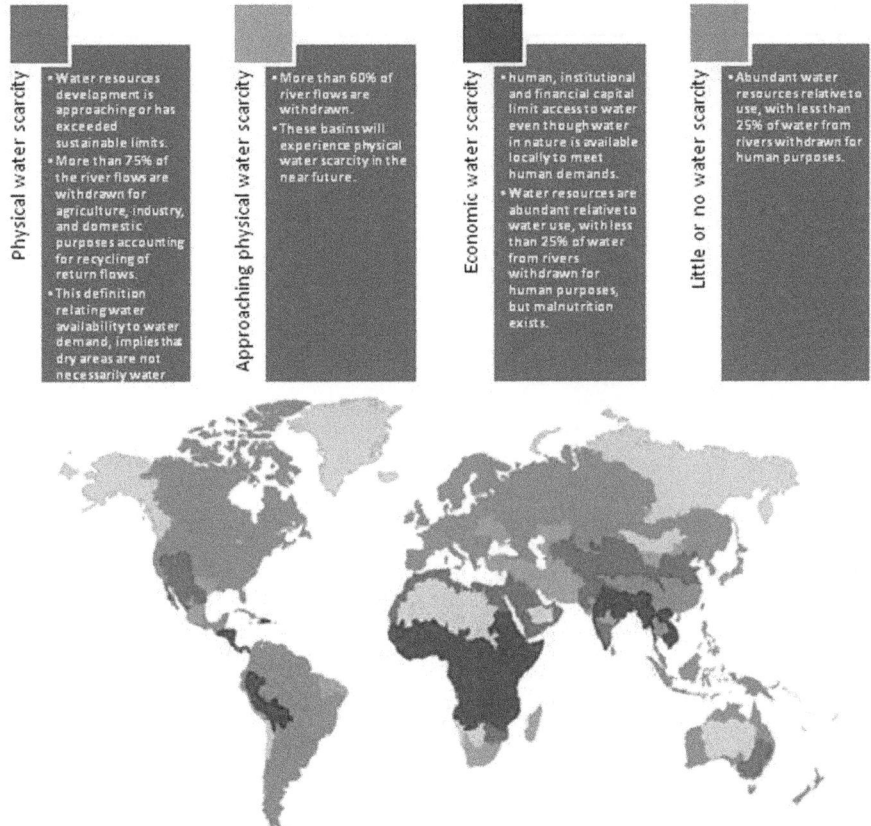

Figure 10.1 Areas of physical and economic water scarcity.

remediation as a key concern for environmentalists all over the globe. There are several sources of water pollution and, among them, synthetic dyes are one of the major contributors.

10.1.1 Dyes as Water Polluters

Mankind's fascination with color dates back to 2600 BC, when the earliest use of dyestuffs in China was reported. All of the initially used dyes were of natural origin and could be labeled as environmentally benign. It was in 1856 that William Henry Perkin serendipitously discovered the first synthetic dyestuff, "mauve" (a basic aniline dye), while searching for a cure for malaria, and this heralded the synthesis of synthetic dyes.[4] Chemically, a dye molecule is built up of two key components: the chromophore, or the color imparting part, and the supplementary auxochrome, which apart from aiding the chromophore imparts water solubility to the dye and in turn enhances fiber affinity. Table 10.1 gives a list of various types of synthetic dyes used in the textile industry based on their properties and applications.[5]

Table 10.1   Types of dyes.

Type of dye	Properties	Principle Chemical Classes	Applications
•Acid	•water soluble	•azo (including premetallized) •polyazo compounds, along with some •stilbenes, phthalocyanines and oxazines •anthraquinone	•used for nylon, wool, silk, modified acrylics • paper •leather • ink-jet printing • food •cosmetics
•Cationic(Basic)	•water soluble dyes • yield coloured cations in solution	•triphenylmethane azine •xanthene •nitro and nitroso	•modified nylons, modified polyesters,silk, wool, and tannin-mordanted cotton •paper •polyacrylonitrile • cation dyeable polyethylene terephthalate •medicine
•Disperse	•water-insoluble nonionic dyes • used for hydrophobic fibers from aqueous dispersion	•diazahemicyanine •triarylmethane • cyanine • hemicyanine •thiazine •oxazine • acridine	•polyester ,nylon • cellulose •cellulose acetate •acrylic fibers
•Direct	•water-soluble •anionic •high affinity for cellulosic fibers in presence of electrolytes	•azo •anthraquinone • styryl • nitro benzodifuranone	•cotton and rayon • paper •leather
•Reactive	•form a covalent bond with the fiber •simple chemical structures •narrow absorption bands •bright dyeing	•azo • anthraquinone •triarylmethane •phthalocyanine •formazan • oxazine	•cotton and other cellulosics
•Solvent	•solvent soluble (water insoluble) •nonpolar or little polar •low cost • good wash fastness	• azo anthraquinone, •phthalocyanine •triarylmethane	•plastics • gasoline • lubricants • oils •waxes
•Sulfur		•intermediate structures	• cotton and rayon •paer •leather
•Vat	•water-insoluble dyes •used as soluble leuco salts	• anthraquinone (including polycyclic quinones) •indigoids	• cotton,rayon and wool

Dye houses, paper printers, textile dyers, pharmaceuticals, food, photography, cosmetics and leather are the major industries which use synthetic dyes.[6–8] As given in the Table 10.1, the chromophores in anionic and nonionic dyes are mostly azo groups or anthraquinone types. The reductive cleavage of azo linkages is responsible for the formation of toxic amines in the effluent. Anthraquinone-based dyes are more resistant to degradation because of their fused aromatic structures. Reactive dyes are typically azo-based chromophores combined with different types of reactive groups. They differ from all other classes because they bind through covalent bonds. They are used extensively in textile industries because they impart bright colors, are water-fast, and have simple application techniques with low energy consumption. Removal of water-soluble reactive and acid dyes is very difficult, as they tend to pass through conventional treatment systems unaffected. Basic dyes impart brilliant and intense colors and are highly visible even in a very low concentration. Disperse dyes do not ionize in an aqueous medium and some disperse dyes have also been shown to have a tendency to bioaccumulate.[9] The binding capacities of dyes depend on their structure and hence their loss in wastewater varies from 2% for basic dyes to as high as 50% for reactive dyes. This loss leads to severe contamination of surface and ground waters in the vicinity of dyeing industries.[10] The textile industry is one of the most water-intensive industries, consuming 80–200 m^3 of water per ton of product and producing 1650 m^3 of wastewater per day.[11] It is estimated that 280 000 tons of textile dyes are discharged in textile industrial effluent every year worldwide.[12] Effluents containing dyes from the textile industries are highly colored and are therefore visually identifiable.[13] The disposal of the highly colored effluents from these industries into the environment is extremely hazardous. The wastewater contaminated by these effluents is difficult to treat and reuse. The presence of dyes deteriorates the water quality in terms of color, salinity, total organic carbon (TOC), biological oxygen demand (BOD), chemical oxygen demand (COD) and suspended solids, has an inhibitory effect on aquatic photosynthesis, is toxic to flora, fauna and humans, and, above all, several commonly used dyes (specifically azo dyes) are reported to be carcinogenic and mutagenic.[14–18] Therefore, remediation of dye-containing wastewater has become an important aspect of environmental research.

10.1.2 Methods of Dye Removal

Gupta and Suhas reviewed the conventional methods used for dye removal and categorized them into the following main components:[5] (a) physical; (b) chemical; (c) acoustic, radiation and electrical processes; and (d) biological.

Physical methods include sedimentation and various types of filtration process. Sedimentation is the part of the primary treatment of municipal and industrial waste based on gravity settlement of waste. Filtration techniques include microfiltration, ultrafiltration, nanofiltration and reverse osmosis. The main drawbacks of these membrane-assisted processes are high working pressures, significant energy consumption, the high cost of membranes and a relatively short membrane life, which makes their use limited for treating dye wastewater. The reverse osmosis process forces water, under pressure, through a membrane that is impermeable to most contaminants.[19] The membrane is somewhat better at rejecting salts than it is at rejecting non-ionized weak acids and bases and smaller organic molecules generally with a molecular weight below 200. Reverse osmosis is an effective decoloring and desalting process for the most diverse range of dye wastes, and has been successfully employed for recycling.

Chemical methods of treatment of dye wastewater mainly deal with flocculation and oxidation. The flocculation method is very effective way of water treatment. The process involves addition of flocculating agents, such as aluminum (Al^{3+}), calcium (Ca^{2+}) or ferric (Fe^{3+}) (taken alone or in combined form) ions, to the dye effluent to induce flocculation.[20,21] The major drawbacks of the process include the presence of large quantities of concentrated sludge in the final product, pH dependence and poor results with highly soluble azo, reactive, acid and basic dyes.[22,23]

The oxidation method is a remediation method by which wastewater is treated using oxidizing agents. The common oxidizing agents used for dye removal are chlorine,[24,25] hydrogen peroxide,[26] Fenton's reagent,[27] ozone or potassium permanganate.[28] Oxidation is the most commonly used method for dye removal because it requires low quantities and short reaction times. The method partially or completely degrades the dyes to lower molecular weight species such as aldehydes, carboxylates, sulfates and nitrogen, though complete oxidation of the dye can theoretically reduce the complex molecules to carbon dioxide and water. Catalysts and pH play a crucial role in the oxidation process.

In addition to the above-mentioned methods, the adsorption process has been widely used for color removal. Adsorption is one of the processes which, besides being widely used for dye removal, also has wide applicability in wastewater treatment.[29] The term *adsorption* refers to a process wherein a material is concentrated at a solid surface from its liquid or gaseous surroundings. Where the attraction between the solid surface and the adsorbed molecules is physical in nature, the adsorption is referred to as physical adsorption (physisorption). The attractive forces between adsorbed molecules and the solid surface in physisorption are weak van der Waals forces and result in reversible adsorption. On the other hand, if the attraction forces are due to chemical bonding, the adsorption process is called chemisorption. Because of the strong bonding in chemisorption, it is difficult to remove chemisorbed species from the solid surface.

Ion exchange is basically a reversible chemical process wherein an ion from solution is exchanged for a similarly charged ion attached to an immobile solid particle. Ion exchange shares various common features along with adsorption, and they can be grouped together as "sorption processes" for a unified treatment to have high water quality. Ion exchange has been successfully employed for dye removal.[30,31]

The most important characteristic of an adsorbent is undoubtedly the quantity of adsorbate it can accumulate. Dyes that are difficult to break down biologically can often be removed using adsorbents. A good adsorbent should possess a porous structure that leads to a high surface area and the time taken for adsorption equilibrium to be established should be as small as possible so that it takes lesser time to remove the dye. The common adsorbates used for dye removal are alumina, silica gel, zeolites and activated carbon.[5] Although these adsorbates effectively remove dyes, their high cost bars them from being very popular materials.

Electrochemical methods can be applied as a tertiary treatment to remove color.[32] This can be achieved either by electrooxidation with insoluble anodes or by electrocoagulation using consumable materials. Commonly used anode materials for electrodegradation of dyes are iron, conducting polymers and boron-doped diamond.[33–35] This technique effectively removes soluble and insoluble dyes with reduction of COD. The anode material and the working potential greatly affect the rate and efficacy of dye removal. The major limitations of the method are the high electricity cost, sludge production and also pollution from chlorinated organics and heavy metals due to indirect oxidation.[5]

Advanced oxidation processes (AOPs) are those processes which involve simultaneous use of more than one oxidation process, as at times a single oxidation system is insufficient for the total decomposition of dyes. These processes commonly comprise (a) photo-Fenton processes, (b) UV photolysis and (c) sonolysis.

The combination of the Fenton reaction in UV light is called the photo-Fenton reaction. Bandala *et al.* have reported that UV radiation enhances the efficiency of the Fenton process in treating dye wastewaters.[36] Muruganandham and Swaminathan presented a case study of photochemical decolorization of a chlorotriazine reactive azo dye by Fenton and photo-Fenton processes. They inferred that on carrying out the reaction under optimized conditions the latter is more efficient than the former.[37]

Photocatalysis is an AOP in which light energy from a light source excites an electron from the valence band of the catalyst to the conduction band with a series of reactions which result in the formation of hydroxyl radicals.[38,39] Having high oxidizing potential, these hydroxyl radicals can attack almost all organic structures, causing oxidation. Common photocatalysts used include oxides such as TiO_2, ZnO, ZrO_2 and CeO_2 and sulfides such as CdS and ZnS. The process is found suitable for a wide range of dyes, including direct, reactive, vat and disperse dyes.[40]

Sonolysis is the use of ultrasonic waves for the decolorization and degradation of dyes. The proposed mechanism for the sonochemical process is based on the formation of short-lived radical species generated in violent cavitation events.[5] Hong *et al.*[41] were perhaps the first to report the accelerated photocatalytic degradation of the dyes alizarin and procion blue in the presence of ultrasonic irradiation. The authors reported the process to be dependent on ultrasound power, total solution volume and the gas phase used.

The major drawbacks of AOPs include the generation of undesirable byproducts, incomplete mineralization and, depending on the type, the process may be influenced by pH, light intensity, radical concentration and the chemical structure of the dye.[5]

With the growing awareness for environmental remediation, biological methods have gained a lot of impetus. They are considered environmentally friendly as they can lead to complete mineralization of organic pollutants at low cost.[42] The use of microorganisms for the removal of synthetic dyes from industrial effluents offers considerable advantages: the process is relatively inexpensive, the running costs are low, and the end products of complete mineralization are not toxic.[43] It is now known that several microorganisms, including fungi, bacteria, yeasts and algae, can decolorize and even completely mineralize many dyes under certain environmental conditions.[42] Biosorption of dyes depends upon basically three parameters, namely the nature of the dye, the specific surface properties of the biomass and the operating conditions (*e.g.* pH, temperature, ionic strength).

The availability of algae in both fresh and saline waters makes them potential candidates as biosorbants. The biosorption capacity of algae is attributed to their relatively high surface area and the high binding affinity by algae has been demonstrated to have good removal rates for dyes.[44] The binding mechanisms of dye to algae vary from physical to chemical.[45] The various functional chemical groups such as carboxyl, hydroxyl, amino, thiol, sulfate and phosphate, present on the algal cell wall, are responsible for the sequestration of dyes, metals and salts from wastewaters. We can thus infer that adsorption of dyes on algae depends on the cell wall properties, which change from one species to another and, depending on the composition of the cell wall, their activities may be different for each dye.[46] Biosorption is hypothesized to occur in a two-step process: an initial rapid process taking minutes to hours, followed by a slow process ranging from several hours to a day. The proposed mechanism is usually an ion exchange process involving exchange of an ion from the solution for a similarly charged ion attached to an immobile particle.[5] The dye molecules diffuse from the aqueous phase onto the solid phase of the algal biopolymers present on the cell wall and accumulate there.[47,48] Mohan *et al.* proposed that the bio-removal of dye using viable algae proceeds *via* a three-step mechanism, as represented in Figure 10.2.

Biosorption does not involve metabolic energy or transport and can occur in either living or dead biomass, whereas bioaccumulation involves metabolism- and temperature-independent and metabolism-dependent mechanism steps. Biodegradation is an energy-dependent process and involves the breakdown of dye into various byproducts through the action of various enzymes. When biodegradation is complete, the process is called mineralization.[49]

The prerogative of this platform is used to give an insight into the removal of dyes from water by viable algal biomass. This chapter deals in detail with the methodology used, the factors effecting and the mechanism involved in the removal of dyes form wastewater. In order to discuss and describe the

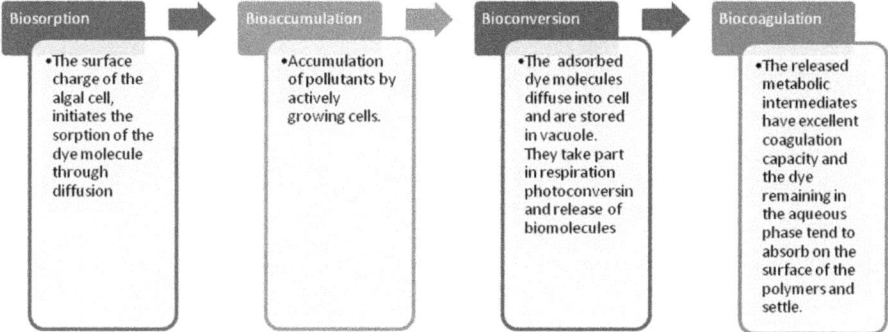

Figure 10.2   Steps involved in bioremediation of textile water using algae.

aforementioned points, we selected the case study of dye removal by live green algae and immobilized algal biomass.[49]

10.2   Biosorption of Dyes by Live Algal Biomass

Use of growing cultures in bioremediation has the advantage over nonliving cells as simultaneous removal of dye is obtained during growth of the organism and separate biomass production can be avoided.[49] Viable microbial species are abundantly available in nature and can be adopted in conventional treatment to enhance their application. In this realm, this section investigates the ability/ potential of commonly available algal species in viable form for removal of the dye color.

10.2.1   Removal of Dyes by *Spirogyra* Species

Spirogyra is a genus of filamentous green algae of the order Zygnematales, named for the helical or spiral arrangement of the chloroplasts (Figure 10.3) . It is commonly found in freshwater areas. *Spirogyra* is unbranched with cells connected end-to-end in long male reproductive system filaments. The cell wall has two layers: the outer wall is composed of pectin that dissolves in water to make the filament slimy to touch, while the inner wall is of cellulose. The cytoplasm forms a thin lining between the cell wall and the large vacuole it surrounds.[50]

In this section we review dye removal using live *Spirogyra* species taking the diazo dye Direct Brown 2 and the monoazo Reactive Yellow 22 as examples.[51,52]

10.2.1.1   Experimental Methodology

The methodology design using live *Spirogyra* species includes collection and identification of the algal biomass as the first and foremost step. The test algae species is then dried by blotting and weighed for further studies. The algal

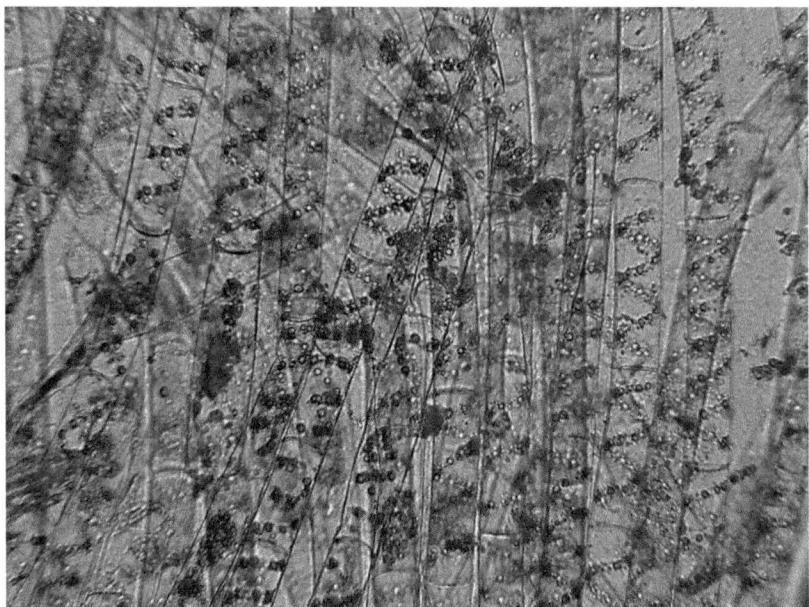

Figure 10.3   Patch of green *Spirogyra* freshwater algae under the microscope.

biomass is then added to the dye solution of required concentration. This
mixture is further kept in diffused sunlight and is exposed to the atmosphere.
The reaction mixture is supplemented by glucose. The extent of dye removal is
usually measured in terms of color removal (which in turn is measured by
calculating the optical density using a spectrophotometer). Apart from this, the
pH, redox potential, conductivity and turbidity are also measured according to
the procedures specified by the Standard Methods (APHA, 1995).[51,52]

10.2.1.2   *Effect of Contact Time, Initial Dye Concentration and Dose of Algae*

Mohan *et al.* reported the removal of Direct Brown 2 by taking different
concentrations of test dye solution by algae biomass and by differing doses of
algae biomass at various time intervals.[51] The rate of dye removal/uptake was
initially less but accelerated with contact time, and finally reached saturation
after a period of four days. The rate of dye removal remained more or less
constant thereafter. The authors attributed the initial low uptake to the accli-
matization of the algae biomass to the dye-bearing aqueous environment.
A similar reaction profile was reported for bio-removal of the Reactive Yellow
dye using *Spirogyra* cultures.[52]

   It was also established that the rate and percentage of dye adsorption was
inversely proportional to the dye concentration. The constant uptake of dye at
lower dye concentration could be attributed to the availability of vacant or
unsaturated sites which become saturated with higher dye concentrations. The

rate of dye uptake was higher in the case of high doses of algae biomass, which provided a greater surface area for adsorption.

10.2.1.3 Redox Potential and pH as Indicators of Dye Removal

During dye sorption, the pH initially decreased and then increased and became constant (almost neutral) on attainment of the saturation level. This variation of the pH values may be attributed to a surface interaction between the cell surface and the dye molecule. The decrease in pH indicates the low net (zero) surface electric charge.[53] On the other hand, the increase in pH may be attributed to the release of CO_2, subsequent formation of H_2CO_3 and thereby establishment of equilibrium with the atmosphere.[54] The probability of the contact and interaction between cells increases due to mutual electric repulsion forces and the interaction becomes saturated slowly when the pH increases. The increase in pH can also be accounted for by the accumulation of dyes (especially acidic ones) by the *Spirogyra* cells where tannin compounds are present.

The redox potential acts as an indicator of biochemical reactions involved in the dye removal by live *Spirogyra* algae. The values were negative at the initiation of the experiment, gradually reached zero and then approached a maximum positive value on saturation and subsequently decreased to near zero values. The increase of redox potential from negative values to zero was correlated to slow and gradual assimilation of dye and stored in vacuoles with tannins (pentagalloyl-*O*-glucose) and subsequent secretion of extracellular polymers and related organic compounds due to biomineralization by algal biomass. These polymers comprise surface functional groups, which enhance the sorption of the dye molecules onto the surface of the polymer (floc) and then settling during the dye removal processes. This process results in a decrease of redox potential.[52]

10.2.1.4 Turbidity as a Parameter of Dye Removal

Turbidity refers to the degree to which light is scattered by suspended particles. The turbidity values showed a minimum value in the initial stages of the experimentation and increased gradually before a slight decline and a final constant reading. The increase of the turbidity gave an idea of the release of extracellular polymers and the subsequent decrease could be an indicator of coagulation of biopolymers. The mechanism of color removal by biological flocculation in the case of algae biomass can be described by metabolic transformation of coloured molecules to non-colored molecules.[55]

10.2.2 Removal of Dyes by Live Microalgal Species

In this section we investigate the potential of widely available, easily cultured microalgae species, *Cosmarium* and *Vaucheria*, for dye removal. The section deals in detail with the method used and the factors influencing dye removal by the aforementioned species in live form.

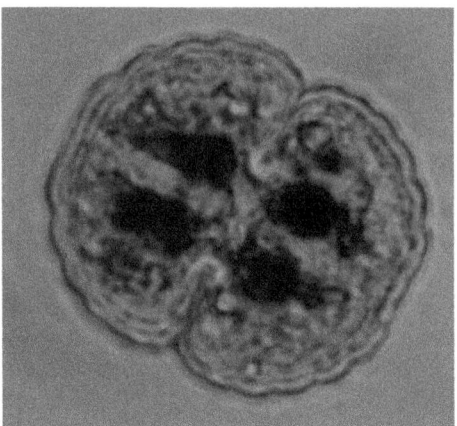

Figure 10.4   Live *Cosmarium* algae.

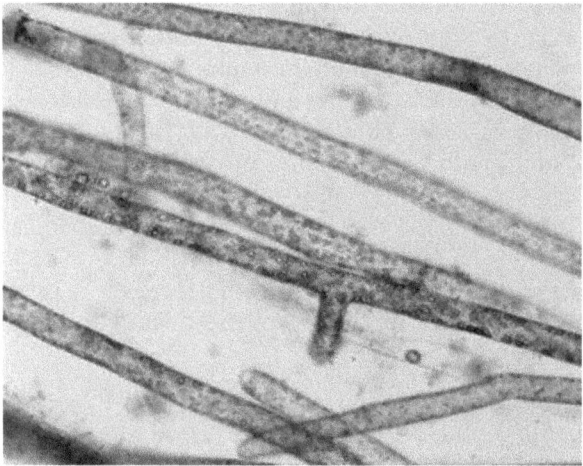

Figure 10.5   Yellow-green algae *Vaucheria*.

Cosmarium is a freshwater member of the Chlorophyta green algae (Figure 10.4). The genus *Cosmarium* is one of the largest genera of desmids with over 2000 species that have been described. They are most abundant in water that is poor in calcium salts, usually acid water with a pH ranging from 7.0 to 5.0. The cells are deeply incised with a clear distinction into two halves. The cell wall is ornamented in different patterns, according to the species.[56]

Vaucheria is a genus of Xanthophyceae or yellow-green algae. *Vaucheria* exhibits apical growth from the tip of filaments, forming mats in either terrestrial or freshwater environments (Figure 10.5). They lack chlorophyll *b* and instead have chlorophyll *c*; this gives them a characteristic yellowish-green color. Many xanthophytes produce a cell wall, though it is not composed of cellulose (as in plants) or of chitin (as in fungi). In fact, the cell wall composition of these protists is still completely unknown, though it is known that cysts in this group often contain silica in their walls. The walls are often, but not always, composed of two overlapping cylindrical halves which fit together, one slightly inside the other. Its filaments form coenocytes with a large central vacuole pushing against the surrounding cytoplasm; the vacuole extends along the entire filament except for the growing tip. The chloroplasts are located on the periphery of the cytoplasm, with the nuclei aggregating toward the center near the vacuole.[57,58]

10.2.2.1 Experimental Methodology

Daneshvar *et al.* reported the removal of malachite green by live *Cosmarium* species.[59] After identification of the algae, the cells were placed in a culture medium for static incubation at 25 °C. The pH of the medium was adjusted as well as maintained using dilute H_2SO_4 and NaOH. The algae culture was exposed to direct sunlight and the growth was measured by counting the number of cells using optical microscopy. A similar batch preparation method was used by Khataee *et al.* in biotreatment of malachite green using *Vaucheria* species.[60]

10.2.2.2 Effect of Dye Concentration and Dose of Biomass

Dye concentration has a tremendous effect on the efficiency of color removal. Daneshvar and co-workers predicted the enhancement of dye removal by *Cosmarium* species on the increment of the initial concentration, as higher concentrations provide an important driving force to overcome all mass transfer resistance of the dye between the aqueous and solid phases.[59] However, the contradictory case was presented for the macroalga *Vaucheria*.[60] Based on toxicity of the dye at higher concentrations, and the ability of the enzyme to recognize the substrate efficiently at very low concentrations, they predicted a decrease in the biological removal of dye.[61] Similar results were reported by Wang *et al.* for the biological decolorization of Reactive Black 5 by *Enterobacter* spp. EC3.[62] Dye removal significantly increased along with an increase in the amount of biomass until it reached a saturation point. The reason for this observation is thought to be the fact that an increase of alga biomass gives more surface area for sorption of the dye molecule.

10.2.2.3 Effect of pH and Temperature

The decolorization efficiency of the algal biomass for a cationic dye, malachite green, increased rapidly on increasing the initial pH of the dye solution from 1.5 to 7.5.[60] The phenomenon was explained on the basis of a zero point of discharge for the biomass. In general, all algal species have an isoelectric point of around pH 3–4.[61] As per the zero point of charge of algae species, their surfaces are presumably positively charged in acidic solution and negatively charged in alkaline solution. As malachite green is a cationic dye, the alkaline solution favors adsorption of it onto the surface of the algae species, which in turn increases the efficiency of dye removal.

On observing the effect of temperature, the decolorization rate increased as the temperature rose for both the algae discussed in this section. The results showed no thermal deactivation of activity under the operational temperatures. Hence, the biomass could be easily used over a broad range of temperatures.

10.2.2.4   Dye Removal and Absorption and IR Spectra

The dye molecules give very good absorption spectra in the visible range and hence a decrease in peak intensity could be an important parameter in predicting the magnitude of dye removal. Apart from this, a close observation of the absorption spectra could help in differentiating between bioadsorption or biodegradation. In adsorption examination, the absorption spectrum will reveal that all peaks decrease approximately in proportion to each other. If the dye removal is attributed to biodegradation, either the major visible light absorbance peak should disappear and/or a new peak should appear.[63] Based on these results the color removal by *Cosmarium* spp. may be largely attributed to biodegradation.[59]

Besides absorption spectra, FTIR could be a useful indicator of dye removal. Changes in the fingerprint region clearly indicate the biochemical process involved in water remediation. These techniques can also be used to predict the reusability of the algal biomass.[59,60]

10.2.3   Removal of Dyes by *Chlorella vulgaris*

Chlorella is a genus of single-cell green algae, belonging to the phylum Chlorophyta (Figure 10.6). It is spherical in shape, about 2–10 μm in diameter, and is without flagella. *Chlorella* contains the green photosynthetic pigments chlorophyll-*a* and -*b* in its chloroplast.[64,65]

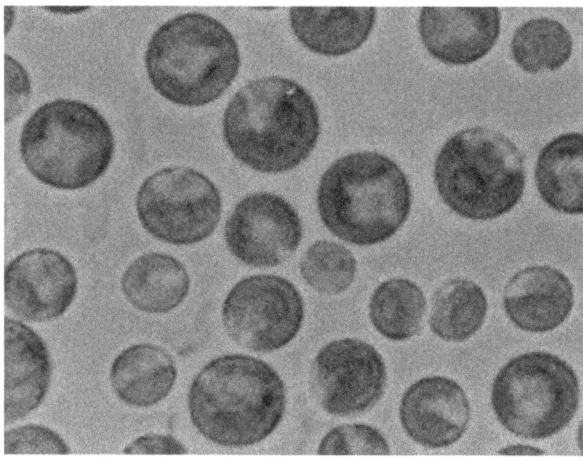

Figure 10.6   Live *Chlorella* spp. under the microscope.

This section deals with the use of live *Chlorella vulgaris* for dye removal. As in the previous sections, the methodology of live culture, as well as other details, is discussed.

10.2.3.1 Experimental Methodology

Lim *et al.* studied the potential application of viable *Chlorella vulgaris* for bioremediation of textile wastewater using four batches of cultures in high-rate algae ponds (HRAP) containing a textile dye (Supranol Red 3BW).[66] The authors screened 10 microalgae and placed them in a culture medium. They further monitored their growth and color removing capacity in textile wastewater. *C. vulgaris* UMACC 001 was selected for further studies, as the preliminary screening showed that it grew best in textile wastewater.

Apart from culture studies in the laboratory, cultures of *C. vulgaris* were grown in HRAP conditions. The nutrients were weighed accordingly and mixed in the HRAP. Growth of *Chlorella* was monitored daily by cell count and chlorophyll-*a* determination. Pollution parameters such as COD, ammonium nitrogen and phosphate phosphorus of textile water were determined. The dry weight was measured at the end of the studies, while physical parameters such as pH, temperature and solar irradiance were monitored throughout.

10.2.3.2 Dye Removal by Live C. vulgaris

C. vulgaris UMACC 001 exhibited remarkable growth in the presence of dyes and efficiently treated the wastewater. Color removal by *Chlorella* decreased with increasing concentration of dyes in the medium. This could be attributed to the factors mentioned in Section 10.2.1.3.

The HRAP system using *Chlorella* demonstrated a practical application for bioremediation of dyes as it could remove up to 50% of the color, besides reducing pollutants such as COD, ammonium nitrogen and phosphate phosphorus. Adsorption equilibrium studies showed that color removal by *Chlorella* followed the Langmuir and Freundlich models.

10.3 Conclusion and Future Perspectives

Treatment of dye-containing wastewater using algae as a biosorbent could be adopted as a cost effective and efficient approach for decolorization of textile industry effluents. Some advantages of this process over conventional wastewater treatments could be summarized as follows.

(a) Cost effective. It has been shown that this process requires low maintenance and thus is more cost effective in removing BOD, pathogens, phosphorus and nitrogen than activated sludge processes.
(b) Low energy requirements. Traditional wastewater treatment processes require high energy costs of mechanical aeration to provide oxygen; on the other hand, algae act as a source of oxygen for aerobic bacteria, thus

eliminating the energy requirement. Aeration, being an energy intensive process, accounts for 45–75% of a wastewater treatment plant's total energy costs, whereas algae provide the needed oxygen through photosynthesis.

(c) Reduction in sludge formation. Industrial effluents are conventionally treated using a variety of hazardous chemicals for pH correction, sludge removal, color removal and odor removal. Extensive use of these chemicals for water treatment results in huge amounts of sludge. This sludge forms solid industrial waste and is disposed of by depositing in landfills. Using algae for wastewater treatment produces sludge with algal biomass which is energy rich and can processed to make valuable products like biofuels, fertilizers, *etc.* Not only is there considerable reduction in sludge formation, but algal technology avoids the use of chemicals and the whole process of remediation is greener and simplified.

(d) Greenhouse gas emission reduction. The US Environmental Protection Agency has specifically identified conventional wastewater treatment plants as major contributors to greenhouse gases. Like other processes, algal biosorbent-based treatment also releases CO_2 but it consumes more CO_2 during its growth than is released by the plant. This makes the entire system carbon negative.

With the advent of interdisciplinary research, the development of models based on kinetic studies, absorption and other techniques can help researchers to predict the reaction path and rate of dye removal more accurately. Besides, these studies can help us gain a better insight into the actual reaction mechanisms involved in bioremediation by live microorganisms. With the aid of neural networks we can easily know the effect of different variables involved in the process and it can lead to path-breaking research in this area which could eventually lead to large-scale industrial applications of live algae in dye removal from wastewater. Development of systems like HRAP in the laboratory is the clear indication of the readiness of this technology for large-scale, safe and cost-effective application.

References

1. *Water for Food, Water for Life: A Comprehensive Assessment of Water Management in Agriculture*, Earthscan, London, and International Water Management Institute, Colombo, 2007.
2. en.wikipedia.org/wiki/Water_pollution.
3. R. Helmer and I. Hespanhol, *Water Pollution Control – A Guide to the Use of Water Quality Management Principles*, Spon, London, 1997.
4. R. M. Christie, *Environmental Aspects of Textile Dyeing*, Woodhead, Cambridge, 2007.
5. V. K. Gupta and Suhas, *J. Environ. Manage.*, 2009, **90**, 2313.
6. H. Ali, *Water, Air, Soil Pollut.*, 2010, **213**, 251.

7. D. C. Kalyani, A. A. Telke, R. S. Dhanve and J. P. Jadhav, *J. Hazard. Mater.*, 2009, **163**, 735.

8. J. Gruwez, J. Mortelmans and S. Deboosere, *Forum for Applied Technology*, Trevi, Gentbrugge, Belgium,1999.

9. Z. Aksu, *Process Biochem.*, 2005, **40**, 997.

10. C. O'Neill, F. R. Hawkes, D. L. Hawkes, N. D. Lourenco, H. M. Pinheiro and W. Delee, *J. Chem. Technol. Biotechnol.*, 1999, **74**, 1009.

11. S. R. Couto, *Biotechnol. Adv.*, 2009, **27**, 227.

12. X. Jin, G. Liu, Z. Xu and W. Yao, *Appl. Microbiol. Biotechnol.*, 2007, **74**, 239.

13. N. K. Kilic, J. P. Nielson, M. Yuce and G. Donmez, *Chemosphere*, 2007, **67**, 826.

14. I. I. Savin and R. Butnaru, *Environ. Eng. Manage. J.*, 2008, **7**, 859.

15. J. C. Akan, F. I. Abdulrahman, J. T. Ayodele and V. O. Ogugbuaja., *Aust. J. Basic Appl. Sci.*, 2009, **3**, 1933.

16. T. Kuberan, J. Anburaj, C. Sundaravadivelan and P. Kumar, *Int. J. Environ. Sci.*, 2011, **1**, 1760.

17. R. Faryal and A. Hameed, *Pak. J. Bot.*, 2005, **37**, 1003.

18. P. Cooper, *Colour in Dyehouse Effluent*, Alden, Oxford, 1995.

19. S. Sostar-Turk, M. Simonic and I. Petrinic, *Dyes Pigments*, 2005, **64**, 147.

20. Q. Y. Yue, B. Y. Gao, Y. Wang, H. Zhang, X. Sun, S. G. Wang and R. R. Gu, *J. Hazard. Mater.*, 2008, **152**, 221.

21. J. P. Wang, Y. Z. Chen, X. W. Ge and H. Q. Yu, *Colloids Surf. A*, 2007, **302**, 204.

22. J. W. Lee, S. P. Choi, R. Thiruvenkatachari, W. G. Shim and H. Moon, *Dyes Pigments*, 2006, **69**, 196.

23. F. I. Hai, K. Yamamoto and K. Fukushi, *Crit. Rev. Environ. Sci. Technol.*, 2007, **37**, 315.

24. C. G. Namboodri, W. S. Perkins and W. K. Walsh, *Am. Dyestuff Rep.*, 1994, **83**, 17.

25. D. Rajkumar and J. G. Kim, *J. Hazard. Mat.*, 2006, **136**, 203.

26. R. Hage and A. Lienke, *Angew. Chem. Int. Ed.*, 2006, **45**, 206.

27. S. Wang, *Dyes Pigments*, 2008, **76**, 714.

28. J. Wu, H. Doan and S. Upreti, *Chem. Eng. J.*, 2008, **142**, 156.

29. R. C. Bansal and M. Goyal, *Activated Carbon Adsorption*, Taylor & Francis, Boca Raton, FL, 2005.

30. C.-H Liu, J.-S Wu, H.-C. Chiu, S.-Y Suen and K. H Chu, *Water Res.*, 2007, **41**, 1491.

31. S. Raghu and C. A. Basha, *J. Hazard. Mater.*, 2007, **149**, 324.

32. V. K. Gupta, R. Jain and S. Varshney, *J. Colloid Interface Sci.*, 2007, **312**, 292.

33. D. Dogan and H. Turkdemir, *J. Chem. Technol. Biotechnol.*, 2005, **80**, 916.

34. A. M. Faouzi, B. Nasr and G. Abdellatif, *Dyes Pigments*, 2007, **73**, 86.

35. F. H. Oliveira, M. E. Osugi, F. M. M. Paschoal, D. Profeti, P. Olivi and M. V. B. Zanoni, *J. Appl. Electrochem.*, 2007, **37**, 583.

36. E. R. Bandala, M. A. Pelaez, A. J. Garcia-Lopez, M. D. J. Salgado and G. Moeller, *Chem. Eng. Process.*, 2008, **47**, 169.
37. M. Muruganandham and M. Swaminathan, *Dyes Pigments*, 2004, **63**, 315.
38. I. Arslan-Alaton, *Color. Technol.*, 2003, **119**, 345.
39. T. K. Ghorai, D. Dhak, S. K. Biswas, S. Dalai and P. Pramanik, *J. Mol. Catal. A*, 2007, **273**, 224.
40. S. K. Kansal, M. Singh and D. Sud, *J. Hazard. Mater.*, 2007, **141**, 581.
41. Q. Hong, J. L. Hardcastle, R. A. J. McKeown, F. Marken and R. G. Compton, *New J. Chem.*, 1999, **23**, 845.
42. A. Pandey, P. Singh and L. Iyengar, *Int. Biodeterior. Biodegrad.*, 2007, **59**, 73.
43. E. Forgacs, T. Cserhati and G. Oros, *Environ. Int.*, 2004, **30**, 953.
44. A. Srinivasan and T. Viraraghavan, *J. Environ. Manage.*, 2010, **91**, 1915.
45. A. Çelekli, M. Yavuzatmaca and H. Bozkurt, *Text. Light Ind. Sci. Technol.*, 2012, **1**, 29.
46. G. Crini, *Bioresour. Technol.*, 2006, **97**, 1061.
47. A. Ozer, G. Akkaya and M. Turabik, *J. Hazard. Mater.*, 2005, **126**, 119.
48. H. Ali, *Water, Air, Soil Pollut.*, 2010, **213**, 251.
49. D. Charumathi and N. Das, *Int. J. Eng. Sci. Technol.*, 2010, **2**, 4325.
50. http://en.wikipedia.org/wiki/Genus.
51. S. V. Mohan and J. Karthikeyan, *Toxicol. Environ. Chem.*, 2000, **74**, 147.
52. S. V. Mohan, N. C. Rao, K. K. Prasad and J. Karthikeyan, *Waste Manage.*, 2002, **22**, 575.
53. W. Stumm and J. J. Morgan, *Aquatic Chemistry*, Wiley, New York, 1981.
54. R. E. Mackenney, *Microbiology for Sanitary Engineers.* McGraw-Hill, New York, 1962.
55. E. G. H. Lee, J. C. Mueller and C. C. Walden, *Tappi*, 1978, **61**, 59.
56. http://www.hib.no/avd_al/naturfag/plankton/english/plankton/plankton-algae/conjugate_algae/cosmarium.html.
57. www.ucmp.berkeley.edu/chromista/xanthophyta.html.
58. en.wikipedia.org/wiki/Vaucheria.
59. N. Daneshvar, M. Ayazloo, A. R. Khataee and M. Pourhassan, *Bioresour. Technol.*, 2007, **98**, 1176.
60. A. R. Khataee, M. Zarei, G. Dehghan, E. Ebadi and M. Pourhassan, *J. Taiwan Inst. Chem. Eng.*, 2011, **42**, 380.
61. C. I. Pearce, J. R. Lloyd and J. T. Guthrie, *Dyes Pigments*, 2003, **58**, 179.
62. R. Henderson, S. A. Parsons and B. Jefferson, *Water Res.*, 2008, **42**, 1827.
63. K. C. Chen, J. Y. Wu, D. J. Liou and S. C. J. Hwang, *J. Biotechnol.*, 2003, **101**, 57.
64. en.wikipedia.org/wiki/Chlorella.
65. http://www.naturalways.com/graphics/back.gif.
66. S. L. Lim, W. L. Chu and S. M. Phang, *Bioresour. Technol.*, 2010, **101**, 7314.

CHAPTER 11

Factors Affecting Surfactant Modification of Solid Media for Removal of Oxo Ions

KARIKA K. BRIDGERS AND KIRIL D. HRISTOVSKI*

College of Technology and Innovation, Arizona State University –
Polytechnic Campus, 7171 Arroyo Mall, Mesa, AZ 85212, USA
*Email: kiril.hristovski@asu.edu

11.1 Introduction

In search of new, inexpensive, and green materials that exhibit ion-exchange properties suitable for removal of oxo anions from water, researchers have examined approaches that involve surface modification of existing commercially available sorbent media with positively charged surfactants.[1–35] Considering the low cost of surfactants when compared to ion-exchange media, these approaches have demonstrated satisfactory promise in removal of oxo anions for a fraction of the cost of commercially available ion-exchange media. The relatively straightforward concept involves reversing the negative surface charge to a positive one using surfactants with positively charged groups similar to those found in ion-exchange resins. Once the positively charged surfactant groups are properly arranged on the surface of the media, the ion-exchange process can occur and oxo anions can be removed *via* the same mechanism as in polymer-based ion-exchange resins. However, the efficiency of the ion-exchange process is closely related to the properties of the sorbent media and the surfactants used in the surface modification process.

RSC Green Chemistry No. 23
Green Materials for Sustainable Water Remediation and Treatment
Edited by Anuradha Mishra and James H. Clark
© The Royal Society of Chemistry 2013
Published by the Royal Society of Chemistry, www.rsc.org

During research, the importance of some of these factors is often neglected, which may lead to unrealistic performance results of the media. Therefore, it is important to identify critical factors that could potentially lead to unrealistic experimental results and other misconceptions related to the performance of oxo anion removal from water by surfactant-modified media. Although the same principles apply for all surfactants and solid sorbent media, this chapter focuses only on peer-reviewed journal articles pertaining to the two most commonly used sorbent media, granular activated carbon and zeolites, and the most commonly used surfactant, hexadecyltrimethylammonium bromide (HDTMA-Br). Based on their predominant presence in the research literature, these materials could be considered good model representatives to describe the existing pool of knowledge related to oxo anion removal by surfactant-modified media.

11.2   Behavior of Surfactants in Aqueous Systems

It is imperative to understand the behavior of surfactants in aqueous systems before they are used in the development of inexpensive media for removal of contaminants. Surface-active agents or surfactants are characterized by their amphiphilic natures. When low amounts of surfactants are introduced in water, they dissociate to form a surfactant ion and a counter ion. Immediately after the dissociation the surfactant ions, which are composed of two moieties, a hydrophobic tail and a hydrophilic head, arrange themselves into a thermodynamically stable orientation by minimizing contact between the hydrophobic tail and the water.[36,37] This process consequently leads to formation of a monolayer in the air/water (or nonpolar solid/water) interface region with the hydrophobic tail extending towards the air, as illustrated for the cationic surfactant HDTMA in Figure 11.1. As the concentration of the surfactant increases and there are no longer available sites in the air/water or solid/water interface regions, the surfactants aggregate in solution to form micelles in an

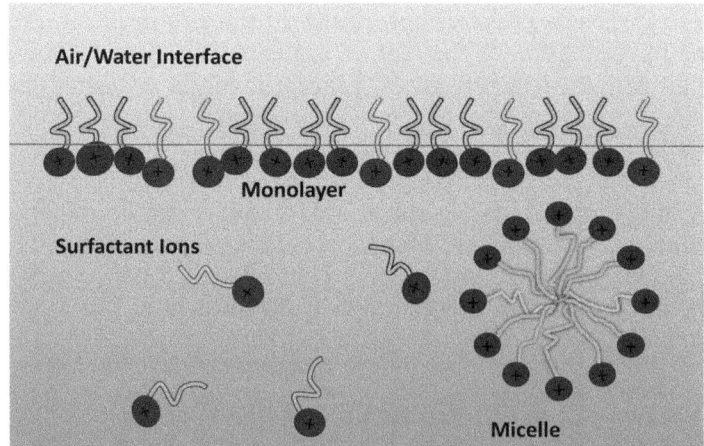

Figure 11.1   Cationic surfactant orientation in aqueous solution at the CMC.

attempt to minimize the contact between the water molecules and the hydrophobic tails. The concentration at which a complete monolayer has formed at the air/water surface (or nonpolar solid/water) interface region, and beyond which all additional surfactant added to the system forms micelles, is referred to as the critical micelle concentration (CMC).[36–38]

When high surface area solid media is introduced into an aqueous system, additional surface area becomes available for distribution of the surfactant. The surfactant orients on the surface such that it minimizes contact between the hydrophobic tail and the water. The orientation of the surfactant at the solid media surface depends upon the surfactant/solid media ratio and the surface properties (*e.g.* charge). This process could be exemplified by using HDTMA sorbing onto a negatively charged surface such as clinoptilolite or the oxidized surface of granular activated carbon.[3,5,16,39–43] Surfactants containing positive moieties and present in low concentrations would sorb onto a negative surface, often parallel to the solid media surface, as illustrated in Figure 11.2. With increase in surfactant concentration (*i.e.* increase in surfactant/solid media ratio), the hydrophobic interactions of the surfactant with the water molecules cause formation of a patchy bilayer and eventually a bilayer on the media surface as the most thermodynamically stable state (Figure 11.3).[5,7,17,20,44] In a system containing a high surface area solid, it is suggested that surfactant aggregation (admicelles and micelles) occurs upon complete formation of the bilayer, although formation of a patchy bilayer and monolayer on the media surface may occur in parallel.[16,45,46] The point at which all additional surfactant introduced to the system forms micelles is referred to as the dose critical micelle concentration (D_{CMC}). The D_{CMC} represents a point at which a complete formation of the bilayer could be considered under a given set of

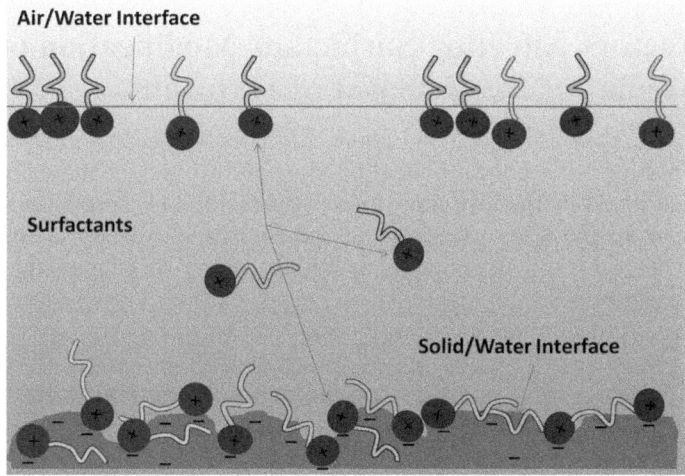

Figure 11.2 Sorption of surfactants onto to sorbent media at low surfactant concentrations.

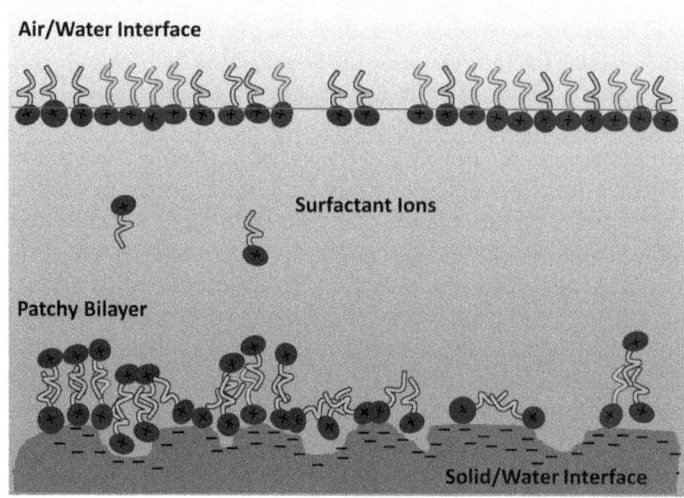

Figure 11.3 Patchy bilayer formation at concentrations below the dose critical micelle concentration.

conditions, although formation of a patchy bilayer and monolayer on the media surface may occur in parallel until the D_{CMC} is achieved.[5,16,40,47,48] Surfactant properties, types of bonds between the surfactant and the surface (*e.g.* ionic, London–van der Waals), and surface properties determine the sorption mechanisms, the D_{CMC}, and the level of bilayer formation.[31,41,45,49–52] Consequently, these factors affect the success of surfactant modification of solid media for removal of oxo anions. As such, they must be carefully considered during the development of surfactant modification research.

11.3   Factors Affecting Surfactant Modification of Solid Media for the Removal of Oxo Ions

There are a number of factors that could affect the surfactant modification of solid sorbent media for the removal of oxo ions. However, the majority of them can be summarized by the four most important factors: (1) specific surface area of the sorbent media; (2) surface charge/ion-exchange capacity of the sorbent media; (3) porosity and pore size distribution of the sorbent media; and (4) surfactant properties.

11.3.1   Influence of Specific Surface Area on Surfactant Modification of the Sorbent Media

Specific surface area, surface charge/ion-exchange capacity (IEC), and porosity (*i.e.* pore size distribution) represent the most important properties of the sorbent media that determine the aptitude for surfactant modification.

The magnitude of surfactant sorption onto solid sorbent media surfactants depends on the media's surface area that is available to interact with the surfactant.[53] A solid media with a high specific surface area generally has more sorption sites for surfactant bonding to occur than the one with a low specific surface area.[54] There is a direct relationship between the specific surface area and the D_{CMC} for a given surfactant and media with available binding sites. Considering that the specific surface area is a function of the media's particle size, any change in particle size affects this set relationship and causes redistribution of the surfactant on the surface of the media. This consequently affects the number of available oxo ion exchange sites, and consequently the extent of double layer formation. It is not uncommon for these relationships to be neglected when fabricating surfactant-modified media for treatment of oxo ions. An example of this negligence is represented in many studies where surface modifications, or ion-exchange processes, were conducted using batch equilibrium reactors under rapid shaking or agitation conditions, or where researchers used previously determined adsorption capacities calculated for a specific particle size and then used a different particle size.[18,27,55,56] The rapid agitation under batch conditions could cause attrition of the solid media and consequent change of the media's particle size distribution, yielding misleading results that suggest media performances similar to the performances in a packed bed reactor configuration, which is the typical setup for sorption or ion-exchange processes.

11.3.2 Influence of Surface Charge/Ion-Exchange Capacity on Surfactant Modification of the Sorbent Media

The media's surface charge under given pH conditions is directly related to the available sites where the surfactants could sorb and form a bilayer, as illustrated in Figure 11.3. Many natural sorbents, such as zeolites, and activated carbon media typically exhibit a negative surface charge in pH ranges that are usual for natural waters.[5,39,40] In zeolites, the surface charge is proportional to the total number of cation-exchange sites on the surface (*i.e.* the cationic-exchange capacity) that are available for bonding with surfactants containing positive moieties.[5,39,40,57–59] For example, the mineral content and Al/Si ratio influences the size of the lattice apertures, as well as the number of functional groups available for exchange in zeolites.[57–59] In activated carbon media the surface charge is proportional to the oxygen-containing functional groups that may be deprotonated or exhibit strong bipolar character.[60,61] The concentration of these groups on the surface of granular activated carbon (GAC) media depends on the base material from which the GAC is made. GAC can be created from a variety of base materials (lignite or bituminous coal, coconut, or rice husk) and the base material directly influences the functional groups and the pore size of the media.[61] In both cases, however, when the surfactant dose exceeds the cation exchange capacity (CEC) or saturates the available negatively charged surface groups, the negative surface charge begins to change from negative to positive due to the formation of a patchy bilayer on the media surface.[8,18,40,45] However, in activated carbon media the creation of a positive surface charge may commence immediately after introduction of the surfactants into the aqueous system through hydrophobic bonding between the surfactant's nonpolar moiety and the nonpolar sites on the activated carbon's surface.[62–64] The IEC (*e.g.* CEC) or the surface charge density are not always reported in studies, which makes it difficult to determine the dose at which the formation of bilayer begins or is completed.[3,17,18,21,22,24,27,34,46,47,55,56,65–80] Some studies utilize kinetic model testing to determine the maximum surfactant sorption capacity.[44,47] Kinetic studies have demonstrated that the maximum surfactant sorption capacity is approximately two times the IEC.[72] However, considering that often nonpolar sites present on the GAC surface could be engaged in the sorption of surfactants, the researchers discuss the modification process in terms of dose in excess of the CMC, instead the more adequate approach, which is to determine the D_{CMC}.[12,14,81–83]

11.3.3 Influence of Porosity and Pore Size Distribution on Surfactant Modification of the Sorbent Media

The high specific surface area of many sorbents suitable for surfactant modification is a direct consequence of the high porosity of these media.[84] The porosity, including the overall pore size distribution, represents another controlling factor that could significantly diminish the media's aptitude for surfactant modification. This hindering effect comes from the fact that much of the surface area, which contains available sorption sites for surfactants to reverse the polarity of the media's surface by creating a bilayer, is located in the microporous and mesoporous regions in many of these high surface area materials. As such, this surface area may not available for the surfactants to interact with the surface because some of the surfactants may be too large to fit within these pores (*e.g.* micropores); see Figure 11.4.[40,60] For example, fully

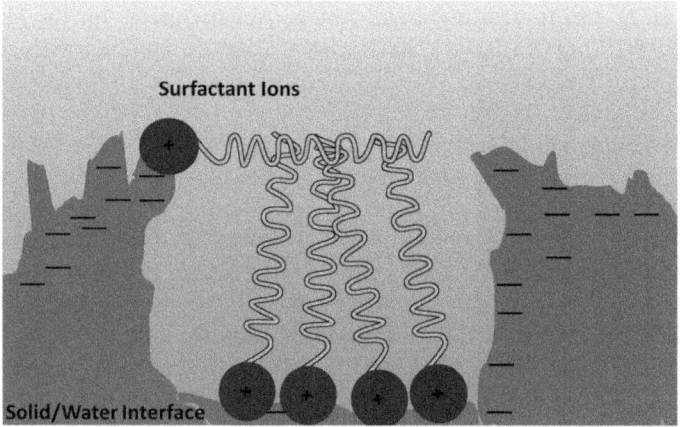

Figure 11.4 Large surfactant molecule interfering with a sorption site.

extended HDTMA has a length of ~ 3.5 nm, causing it to be unable to form a bilayer in the media's micropores and mesopores.[40,49] Furthermore, even if the bilayer is created in some of the small pores of the media, the diffusivity of oxo ions and their access to the available ion-exchange sites may be limited because of the reduced pore volume by the surfactant bilayer.[85] For example, a minimum of ~ 1 nm would be required to allow diffusion of a hydrated nitrate ion ($d_{hydrated} = 0.68$ nm) through the pores of the media that contain a bilayer.[3] So theoretically, one can approximate that pores with diameters of <8 nm may not be available for use in the surfactant modification and consequent ion-exchange to occur.[86] Nonetheless, a number of studies appear to neglect this factor, which could lead to unrealistic oxo anion removal performances.[3,4,6,7,10,11,14–16,19–24,27,29,33,35,39–41,44–49,53,56,60,63–69,71,73,78–83,85,87–120]

In some cases, the unrealistically high performances do not result from the ability of the modified media to remove oxo anions, but rather from the interferences that the surfactants may have with the analytical tools, or from the experimental errors that originate from partitioning of surfactants on the glassware/plasticware. The surfactant partitioned on the glassware/plasticware surface could contain the exchanged oxo anion, consequently yielding its lower bulk concentrations.

11.3.4 Influence of Surfactant Properties on Surfactant Modification of the Sorbent Media

The surfactant properties, such as its size, shape, and functional groups, could significantly affect the D_{CMC} and surfactant packing on the surface of a given sorbent media. The size of the surfactant, primarily dictated by the length of the hydrophobic moiety (*i.e.* length of the carbon chain) impacts the stability of the bilayer. Reports suggest that the stability of bilayers comprised of surfactants containing tails ranging between C-10 and C-16 is much greater than longer nonpolar moieties. Many studies considered this factor and utilized HDTMA as the surfactant of choice to modify media for the removal of oxo anionic contaminants such as arsenate, chromate, nitrate, and perchlorate from water.[1,3–6,8,14,15,18,20–22,24,25,28,29,31–34,40,77,83,116,121–124] There were studies that utilized surfactants that were not well suited, such as cetylpyridinium chloride, a 16-carbon chain with a large pyridine head,[7,12,13,25,26] or octadecyl-trimethylammonium chloride, a 18-carbon chain.[125]

It is not uncommon for long-chained quaternary ammonium surfactants to be prevented from entering the sorbent media's micropores, mesopores, and lower range macropores or access cation-exchange sites located deeper inside the media particles. For example, sorption of long-chain quaternary ammonium surfactants onto the surface of zeolite particles is limited exclusively to their larger macropores, as in Figure 11.4.[3,8,49,82] This arrangement often results in pore blockage and inability of the surfactant to access other macropores capable of providing sorption sites and room for bilayer formation. Pore blockage could also be developed by a surfactant's ionic moiety (*e.g.* quaternary amine) if this moiety exhibits a large enough size or inadequate configuration. For example, a surfactant with a cationic moiety larger than the average pore size will yield a lower surfactant sorption capacity and the inability to form a bilayer.[3,49,53,82,126] Furthermore, the configuration of the ionic moiety could affect the stability of the bilayer, which is affected by the electrostatic interactions that exist among the ionic moieties in a completely packed bilayer. These electrostatic interactions, and the configuration of the ionic moiety, ultimately depend on the functional group used in fabrication of the surfactant, which often represents the main factor determining the selection of a surfactant to fabricate media capable of removal of different types of ions (*e.g.* strong acid *vs.* weak acid oxo anions). Consequently, the selection of a surfactant often requires a tradeoff between selecting the best functional group for removal of a specific oxo anion and the extent of the media's surfactant modification.

11.4 Conclusions

When designing surfactant-modified media for removal of oxo ions, it is imperative to consider the factors affecting the specific surface area, surface charge/ion-exchange capacity, and porosity of given media, because the overall performance of the surfactant-modified media is strongly dependent on them. Sorbent media with high surface area that strongly exhibit macroporosity and have high ion-exchange capacity/strongly oxidized surface represent the ideal candidates for surfactant modification. Surfactants that offer minimum blockage of available media pores and sorption sites and maximum bilayer formation should be carefully selected when modifying given media for oxo anion removal. The D_{CMC} needs to be determined for every combination of media and surfactant before the surfactant modification process is conducted. It is imperative to minimize surface change of the media as a result of attrition during surfactant modification or oxo anion removal processes. Any change in the surface area has the potential to seriously impact the already optimized D_{CMC}. This could yield misleading results and prevent scaling-up the oxo anion removal process, which is the ultimate goal of all laboratory-based research.

References

1. K. G. Bhattacharyya and S. S. Gupta, *Adv. Colloid Interface Sci.*, 2008, **140**, 114.
2. D. Borah, S. Satokawa and S. K. Kato, *J. Colloid Interface Sci.*, 2008, **319**, 53.
3. R. S. Bowman, G. M. Haggerty, D. N. Huddleston and M. M. Flynn, in *Surfactant-Enhanced Subsurface Remediation; Emerging Technologies*, American Chemical Society, Washington, 1994, pp. 54–64.
4. R. S. Bowman, *Sorption of Anions, Cations, and Neutral Organics by SMZ*, New Mexico Tech, Department of Earth and Engineering Sciences, Socorro, NM, 2002.

5. R. S. Bowman, *Microporous Mesoporous Mater.*, 2003, **61**, 43.
6. V. Campos and P. M. Buchler, *Environ. Geol.*, 2007, **52**, 1187.
7. H. Choi, W. Jung, J. Cho, B. Ryu, J. Yang and K. Baek, *J. Hazard. Mater.*, 2009, **166**, 642.
8. P. Chutia, S. Kato, T. Kojima and S. Satokawa, *J. Hazard. Mater.*, 2009, **162**, 204.
9. M. P. Elizalde-Gonzalez, J. Matuusch and R. Wennrich, *J. Environ. Monit.*, 2001, **3**, 22.
10. S. Gammoudi, N. Frini-Srasra and E. Srasra, *Eng. Geol.*, 2012, **124**, 119.
11. H. Hong, W. Jiang, X. Zhang, L. Tie and Z. Li, *Appl. Clay Sci.*, 2008, **42**, 292.
12. H. Hong, H. Kim, K. Baek and J. Yang, *Desalination*, 2008, **223**, 221.
13. H. Hong, H. Kim, Y. Lee and J. Yang, *J. Hazard. Mater.*, 2009, **170**, 1242.
14. J. Iqbal, H. Kim, J. Yang, K. Baek and J. Yang, *Chemosphere*, 2007, **66**, 970.
15. S. Y. Lee and S. J. Kim, *Clays Clay Miner.*, 2002, **50**, 435.
16. Z. Li and R. S. Bowman, *Environ. Sci. Technol.*, 1997, **31**, 2407.
17. Z. Li, I. Anghel and R. S. Bowman, *J. Dispersion Sci. Technol.*, 1998, **17**, 843.
18. Z. Li, R. Beacher, Z. McManama and H. Hanlie, *Microporous Mesoporous Mater.*, 2007, **105**, 291.
19. Z. Li, D. Alesssi and L. Allen, *J. Environ. Qual.*, 2002, **31**, 1106.
20. Z. H. Li, *J. Environ. Qual.*, 1999, **28**, 1457.
21. M. G. Macedo-Miranda and M. T. Olguín, *J. Inclusion Phenom. Macrocyclic Chem.*, 2007, **59**, 131.
22. M. Majdan, S. Pikus, Z. Rzączyńska, M. Iwan, O. Maryuk, R. Kwiatkowski and H. Skrzypek, *J. Mol. Struct.*, 2006, **791**, 53.
23. M. Majdan, S. Pikus, A. Gajowiak, A. Gładysz-Płaska, H. Krzyżanowska, J. Żuk and M. Bujacka, *Appl. Surf. Sci.*, 2010, **256**, 5416.
24. W. Mozgawa, M. Król and T. Bajda, *J. Mol. Struct.*, 2011, **993**, 109.
25. R. Parette and F. S. Cannon, *Water Res.*, 2005, **39**, 4020.
26. J. Patterson, R. Parette, F. S. Cannon, C. Lutes and T. Henderson, *Environ. Eng. Sci.*, 2011, **28**, 249.
27. A. I. Perez Cordoves, M. Granada Valdes, J. C. Torres Fernandes, G. Pina Luis, J. A. Garcia-Calzon and M. E. DiazGarcia, *Microporous Mesoporous Mater.*, 2008, **109**, 38.
28. J. Schick, P. Caullet, J. Paillaud, J. Patarin and C. Mangold-Callarec, *Microporous Mesoporous Mater.*, 2011, **142**, 549.
29. J. Schick, P. Caullet, J. Paillaud, J. Patarin and C. Mangold-Callarec, *Microporous Mesoporous Mater.*, 2010, **132**, 395.
30. J. Su, H. Huang, X. Jin, X. Lu and Z. Chen, *J. Hazard. Mater.*, 2011, **185**, 63.
31. E. J. Sullivan, J. W. Carey and R. S. Bowman, *J. Colloid Interface Sci.*, 1998, **206**, 369.

32. E. J. Sullivan, R. S. Bowman and I. A. Legiec, *J. Environ. Qual.*, 2003, **32**, 2387.
33. U. Wingenfelder, G. Furrer and R. Schulin, *Microporous Mesoporous Mater.*, 2006, **95**, 265.
34. A. M. Yusof and N. A. Malek, *J. Hazard. Mater.*, 2009, **162**, 1019.
35. F. Zhang, W. Sun, X. Lu, H. Hu and J. Ni, *Environ. Eng. Manage. J.*, 2012, **10**, 1433.
36. R. Zana, in *Dynamics of Surfactant Self-Assemblies: Micelles, Micro-emulsions, Vesicles, and Lytropic Phases*, ed. R. Zana, Taylor & Francis, Boca Raton, FL, 2005, pp. 2–13.
37. D. Karsa, *Chemistry and Technology of Surfactants*, Blackwell, Oxford, 2006, pp. 2–25.
38. S. Heller, *Chem. Int.*, 2009, **31**, 7.
39. E. J. Sullivan, D. B. Hunter and R. S. Bowman, *Environ. Sci. Technol.*, 1998, **32**, 1948.
40. G. Haggerty and R. S. Bowman, *Environ. Sci. Technol.*, 1994, **28**, 452.
41. A. Gurses, S. Karaca, F. Aksakal and M. Acikyildiz, *Desalination*, 2010, **264**, 165.
42. Z. Li, *Langmuir*, 1999, **15**, 6438.
43. P. Sehgal, H. Doe and M. Bakshi, *J. Surfactants Deterg.*, 2002, **5**, 123.
44. S. Xu and S. Boyd, *Environ. Sci. Technol.*, 1995, **29**, 312.
45. C. K. Basar, A. Karagunduz, B. Keskinlelr and A. Cakici, *Appl. Surf. Sci.*, 2003, **218**, 169.
46. Z. Li, S. J. Roy, Y. Zou and R. S. Bowman, *Environ. Sci. Technol.*, 1998, **32**, 2628.
47. J. Wang, B. Han, H. Yan, Z. Li and R. K. Thomas, *Langmuir*, 1999, **15**, 8207.
48. A. V. Sineva, A. M. Parfenova and A. A. Fedorova, *Colloids Surf. A*, 2007, **306**, 68.
49. E. J. Sullivan, D. B. Hunter and R. S. Bowman, *Clays Clay Miner.*, 1997, **45**, 42.
50. S. M. Yakout and A. A. Nayl, *Carbon: Sci. Technol.*, 2009, **2**, 107.
51. M. Yalcin, A. Gurses, C. Dogar and M. Sozbilir, *Adsorption*, 2004, **10**, 339.
52. M. Majdan, O. Maryuk, A. Gładysz-Płaska, S. Pikus and R. Kwiatkowski, *J. Mol. Struct.*, 2008, **874**, 101.
53. M. Nadeem, A. Mahmood, S. A. Shahid, A. A. Shah, A. M. Khalid and G. McKay, *J. Hazard. Mater.*, 2006, **138**, 604.
54. A. M. Cooper, K. D. Hristovski, T. Moller, P. Westerhoff and P. Sylvester, *J. Hazard. Mater.*, 2010, **183**, 381.
55. A. Torabian, H. Kazemian, L. Seifi, G. Bidhendi, A. Azimi and S. Ghadiri, *Clean: Soil, Air, Water*, 2010, **38**, 77.
56. R. Cortés-Martínez, M. Solache-Ríos, V. Martínez-Miranda and R. Alfaro-Cuevas, *Water, Air, Soil Pollut.*, 2007, **183**, 85.
57. T. Eyde, *Min. Eng.*, 1997, **49**, 50.

58. M. Baacke and A. Kiss, in *Ion Exchangers*, ed. K. Dorfner and Walter de Gruyter, 1991, p. 473–492.
59. A. F. Masters and T. Maschmeyer, *Microporous Mesoporous Mater.*, 2011, **142**, 423.
60. A. Gurses, M. Yalcin, M. Sozbilir and C. Dogar, *Fuel Process Technol.*, 2003, **81**, 57.
61. *Datasheet Hydrodarco 3000, Granulated Activated Carbon*, Norit Americas, Marshall, TX, 2007.
62. M. Rosu, A. Marlina, A. Kaya and A. Schumpe, *Chem. Eng. Sci.*, 2007, **62**, 7336.
63. C. M. Gonzalez-Garcia, M. L. Gonzalez-Martin, R. Denoyel, A. M. Gallardo-Moreno, L. Labajos-Baroncano and J. M. Bruque, *J. Colloid Interface Sci.*, 2004, **278**, 257.
64. C. M. González-García, M. L. Gonzales-Martin, J. F. Gonzalez, E. Sabio, A. Ramiro and J. Ganan, *Powder Technol.*, 2004, **148**, 32.
65. S. Al-Asheh, *Sep. Purif. Technol.*, 2003, **33**, 1.
66. S. L. Bartelt-Hunt, S. E. Burns and J. A. Smith, *J. Colloid Interface Sci.*, 2003, **266**, 251.
67. R. Bowman, *Abstr. Papers Am. Chem. Soc.*, 1994, **207**, 178.
68. A. Celik, N. Yildiz and A. Calimli, *Rev. Chem. Eng.*, 2000, **16**, 301.
69. A. Celik, N. Yildiz and A. Calimli, *Rev. Chem. Eng.*, 1999, **15**, 349.
70. H. Cho, T. Lee, S. Hwang and J. Park, *Chemosphere*, 2005, **58**, 103.
71. R. Juang, S. Lin and K. Tsao, *J. Colloid Interface Sci.*, 2004, **269**, 46.
72. Z. Li, Y. Du and H. Hong, *Microporous Mesoporous Mater.*, 2008, **116**, 473.
73. Z. Li and L. Gallus, *Colloids Surf. A*, 2005, **264**, 61.
74. Z. Li and R. S. Bowman, *Environ. Sci. Technol.*, 1998, **32**, 2278.
75. Z. Li, C. Willms, S. Roy and R. S. Bowman, *Environ. Geosci.*, 2003, **10**, 37.
76. Z. Li, in *Proceedings of the 15th International Zeolite Conference*, ed. R. Xu, Z. Gao, J. Chen and W. Yan, Elsevier, Beijing, 2007, p. 2098–2103.
77. J. Mendoza-Barrón, A. Jacobo-Azuara, R. Leyva-Ramos, M. S. Berber-Mendoza, R. M. Guerrero-Coronado, L. Fuentes-Rubio and J. M. Martínez-Rosales, *Adsorption*, 2011, **17**, 489.
78. V. A. Oyanedel-Craver and J. A. Smith, *J. Hazard. Mater.*, 2006, **137**, 1102.
79. A. Z. Redding, S. E. Burns, R. T. Upson and E. F. Anderson, *J. Colloid Interface Sci.*, 2002, **250**, 261.
80. C. Volzone, J. O. Rinaldi and J. Ortiga, *J. Environ. Manage.*, 2006, **79**, 247.
81. M. M. Mohamed, *J. Colloid Interface Sci.*, 2004, **272**, 28.
82. M. Nadeem, M. Shabbir, M. A. Abdullah, A. A. Shah and G. McKay, *J. Hazard. Mater.*, 2009, **148**, 365.
83. J. Zhang and M. He, *Water Environ. Res.*, 2011, **83**, 15.
84. *Physicochemical Treatment Processes*, ed. L. K. Wang, Y. T. Hung and N. Shammas, Humana, Totowta, NJ, 2004, vol. 3.

85. M. Anbia and S. E. Moradi, *Desalin. Water Treat.*, 2010, **21**, 44.
86. L. Banasiak and A. I. Schäfer, *J. Membr. Sci.*, 2009, **334**, 101.
87. S. L. Bartelt-Hunt, J. A. Smith, S. E. Burns and A. J. Rabideau, *J. Geotech. Geoenviron. Eng.*, 2005, **131**, 848.
88. A. A. Basar, A. Karagunduz, A. Cakici and B. Kleskinler, *Water Res.*, 2004, **38**, 2117.
89. M. A. Fulazzaky, *Chem. Eng. J.*, 2011, **166**, 832.
90. L. M. Cotoruelo, M. D. Marques, J. Rodriguz-Mirasol, J. J. Rodriquez and T. Cordero, *J. Colloid Interface Sci.*, 2009, **332**, 39.
91. A. Jakubowska, *J. Colloid Interface Sci.*, 2010, **346**, 398.
92. Z. Li, H. Kirk Jones, P. Zhang and R. S. Bowman, *Chemosphere*, 2007, **68**, 1861.
93. R. Marsalek, J. Pospisil and B. Taraba, *Colloids Surf. A*, 2011, **383**, 80.
94. B. Sarkar, M. Megharaj, Y. Xi, G. S. Krishnamurti and R. Naidu, *J. Hazard. Mater.*, 2010, **184**, 448.
95. S. R. Taffarel and J. Rubio, *Miner. Eng.*, 2010, **23**, 771.
96. J. Xiao, Y. Zang, C. Wang, J. Zhang, C. Wang, Y. Bao and Z. Zhao, *Carbon*, 2005, **43**, 1032.
97. C. R. Altare, R. S. Bowman, L. E. Katz, K. A. Kinney and E. J. Sullivan, *Microporous Mesoporous Mater.*, 2007, **105**, 305.
98. M. D. Baeza-Alvarado and M. T. Olguin, *Microporous Mesoporous Mater.*, 2011, **139**, 81.
99. S. S. Banerjee, M. V. Joshi and R. V. Jayaram, *J. Colloid Interface Sci.*, 2006, **303**, 477.
100. I. A. Bazbaz, R. I. Yousef and S. M. Musleh, *Asian J. Chem.*, 2011, **23**, 2553.
101. R. S. Bowman, *Microporous Mesoporous Mater.*, 2007, **105**, 211.
102. Y. Chen, *J. Colloid Interface Sci.*, 1992, **153**, 244.
103. V. Javanovic, V. Dondur, L. J. Damjanovic, J. Zakrzewska and M. Tomasevic-Canovic, *Mater. Sci. Forum*, 2010, **518**, November, 16.
104. H. K. Karapanagioti, D. A. Sabatinit and R. S. Bowman, *Water Res.*, 2005, **39**, 699.
105. J. Lee, P. S. Rao, I. C. Poyer, R. M. Toole, M. D. Annable and K. Hatfield, *J. Contam. Hydrol.*, 2007, **92**, 208.
106. J. Lemic, M. Tomasevic-Canovic, M. Adamovic, D. Kovacevic and S. Milicevic, *Microporous Mesoporous Mater.*, 2007, **105**, 317.
107. Z. Liu, *Environ. Sci. Technol.*, 1991, **25**, 127.
108. Z. Liu, D. A. Edwards and R. G. Luthy, *Water Res.*, 1992, **26**, 1337.
109. Z. Li and H. Hong, *J. Hazard. Mater.*, 2009, **162**, 1487.
110. V. A. Oyanedal-Craver, M. Fuller and J. A. Smith, *J. Colloid Interface Sci.*, 2007, **309**, 485.
111. M. Rozic, D. I. Sipusic, L. Sekovani, S. Miljani, L. Curkovi and J. Hrenovi, *J. Colloid Interface Sci.*, 2008, **331**, 295.
112. S. Sharmasarkar, W. F. Jaynes and G. F. Vance, *Water, Air, Soil Pollut.*, 2000, **119**, 257.
113. G. Sheng, S. Xu and S. A. Boyd, *Water Res.*, 1996, **30**, 1483.

114. J. A. Simpson and R. S. Bowman, *J. Contam. Hydrol.*, 2009, **108**, 1.
115. S. Wang, W. Gong, X. Liu, B. Gao, Q. Yue and D. Zhang, *Chem. Res. Chin. Univ.*, 2006, **22**, 566.
116. W. Widajanti, U. Tresye and T. Y. Rika, *Makara Seri Sains*, 2011, **15**, 53.
117. J. Xu, N. Gao, Y. Deng, M. Sui and Y. Tang, *Desalination*, 2011, **275**, 87.
118. L. Yan, J. Wang, H. Yu, Q. Wei, B. Du and X. Shan, *Appl. Clay Sci.*, 2007, **37**, 226.
119. A. Yazdankhah, S. E. Moradi, S. Amirmahmoodi, M. Abbasian and S. E. Shoja, *Microporous Mesoporous Mater.*, 2010, **133**, 45.
120. W. Zhou, K. Zhu, H. Zhan, M. Jiang and H. Chen, *J. Hazard. Mater.*, 2003, **100**, 209.
121. C. K. Ahn, S. H. Woo and J. M. Park, *Carbon*, 2008, **46**, 1401.
122. Z. H. Li, T. Burt and R. S. Bowman, *Environ. Sci. Technol.*, 2000, **34**, 3756.
123. P. Zhang, D. Avudzega and R. Bowman, *J. Environ. Qual.*, 2007, **36**, 1069.
124. P. Zhang, X. Tao, Z. Li and R. S. Bowman, *Environ. Sci. Technol.*, 2002, **36**, 3597.
125. C. Huang, K. Chang, H. Ou, Y. Chiang and C. Wang, *Microporous Mesoporous Mater.*, 2011, **141**, 102.
126. A. Gladysz-Plaska, M. Majdan, S. Pikus and D. Sternik, *Chem. Eng. J.*, 2012, **179**, 140.

Subject Index

References to figures are given in *italic* type. References to tables are given in **bold** type.